Differential Equations

Textbooks in Mathematics
Series editors:
Al Boggess and Ken Rosen

CRYPTOGRAPHY: THEORY AND PRACTICE, FOURTH EDITION
Douglas R. Stinson and Maura B. Paterson

GRAPH THEORY AND ITS APPLICATIONS, THIRD EDITION
Jonathan L. Gross, Jay Yellen and Mark Anderson

COMPLEX VARIABLES: A PHYSICAL APPROACH WITH APPLICATIONS,
SECOND EDITION
Steven G. Krantz

GAME THEORY: A MODELING APPROACH
Richard Alan Gillman and David Housman

FORMAL METHODS IN COMPUTER SCIENCE
Jiacun Wang and William Tepfenhart

AN ELEMENTARY TRANSITION TO ABSTRACT MATHEMATICS
Gove Effinger and Gary L. Mullen

ORDINARY DIFFERENTIAL EQUATIONS: AN INTRODUCTION TO THE
FUNDAMENTALS, SECOND EDITION
Kenneth B. Howell

SPHERICAL GEOMETRY AND ITS APPLICATIONS
Marshall A. Whittlesey

COMPUTATIONAL PARTIAL DIFFERENTIAL PARTIAL EQUATIONS USING
MATLAB®, SECOND EDITION
Jichun Li and Yi-Tung Chen

AN INTRODUCTION TO MATHEMATICAL PROOFS
Nicholas A. Loehr

DIFFERENTIAL GEOMETRY OF MANIFOLDS, SECOND EDITION
Stephen T. Lovett

MATHEMATICAL MODELING WITH EXCEL
Brian Albright and William P. Fox

THE SHAPE OF SPACE, THIRD EDITION
Jeffrey R. Weeks

CHROMATIC GRAPH THEORY, SECOND EDITION
Gary Chartrand and Ping Zhang

PARTIAL DIFFERENTIAL EQUATIONS: ANALYTICAL METHODS AND
APPLICATIONS
Victor Henner, Tatyana Belozerova, and Alexander Nepomnyashchy

ADVANCED PROBLEM SOLVING USING MAPLE: APPLIED MATHEMATICS,
OPERATION RESEARCH, BUSINESS ANALYTICS, AND DECISION ANALYSIS
William P. Fox and William C. Bauldry

DIFFERENTIAL EQUATIONS: A MODERN APPROACH WITH WAVELETS
Steven G. Krantz

https://www.crcpress.com/Textbooks-in-Mathematics/book-series/CANDHTEXBOOMTH

Differential Equations
A Modern Approach with Wavelets

Steven G. Krantz

CRC Press
Taylor & Francis Group
Boca Raton London New York

CRC Press is an imprint of the
Taylor & Francis Group, an **Informa** business

CRC Press
Taylor & Francis Group
6000 Broken Sound Parkway NW, Suite 300
Boca Raton, FL 33487-2742

First issued in paperback 2022

© 2020 by Taylor & Francis Group, LLC
CRC Press is an imprint of Taylor & Francis Group, an Informa business

No claim to original U.S. Government works

ISBN 13: 978-1-03-247484-7 (pbk)
ISBN 13: 978-0-367-44409-9 (hbk)

DOI: 10.1201/9781003009504

**Visit the Taylor & Francis Web site at
http://www.taylorandfrancis.com**

**and the CRC Press Web site at
http://www.crcpress.com**

To Yves Meyer, with respect and admiration.

Contents

Preface for the Instructor

A course on ordinary differential equations is a standard part of the undergraduate mathematics curriculum throughout the world. Such a course often includes some introductory material on Fourier series and a few applications.

But the world is changing. Because of the theory of wavelets, Fourier analysis is ever more important and central. And applications are a driving force behind all of mathematics.

Thus it is appropriate to have a text that presents a more balanced picture. The text should have differential equations (both ordinary and partial), Fourier analysis, and applications in equal measure and with equal weight. This is such a text.

Certainly a sophomore-level course on differential equations can be taught from this book. Also an undergraduate-level course on Fourier analysis can be taught from this book. And the copious and substantial applications that we provide enrich both those points of view.

While this is a substantive book, I should stress that the text *does not* assume that the student knows the Lebesgue integral. The Riemann integral is used throughout. And we also do not assume that the student knows any functional analysis. Both Lebesgue measure theory and functional analysis are graduate-level topics, and inappropriate in the present context.

We likewise do not assume that the student has had a course in undergraduate real analysis—as from the books of Rudin or Krantz or the author's. We intend this book for a broad audience of mathematics and engineering and physics students.

To make the book timely and exciting, we include a substantial chapter on basic properties of wavelets, with applications to signal processing and image processing. This should give students and instructors alike a taste of what is happening in the subject today.

Since this is a textbook, we present copious examples. There are a great many figures—just because the subject of analysis, properly viewed, is quite visual. And there are substantive exercise sets. The text also contains on-the-fly exercises which should cause the student to pick up his/her pencil and do some calculations. Each chapter ends with a special collection of exercises (called Problems for Review and Discovery) that ties together the ideas in the chapter. There is a collection of Drill Exercises, a collection of Challenge Problems (problems which require some thought), and a collection of Problems for Discussion and Exploration (problems that are suitable for group work).

Of course there are solutions to selected exercises provided at the end of the book. We include also a Glossary and a Table of Notation.

It is a pleasure to thank the book's many insightful reviewers, who contributed numerous ideas and suggestions. Ken Rosen gave particularly cogent and detailed advice which I appreciate very much. I also thank my editor Bob Ross for his support and encouragement.

It is hoped that this text will find an enthusiastic audience, both among students and instructors. We look forward to hearing from the readership as the book goes into use.

Steven G. Krantz
St. Louis, Missouri

Preface for the Student

The purpose of this book is to teach you the interrelationships among differential equations, Fourier analysis, and wavelet theory. This is a lot of territory to cover, and the purpose of this brief Preface is to give you some guidance in the process.

Differential equations are the language of science. Most of the laws of nature are formulated in the language of differential equations. If you want to be an engineer, or a physicist, or a mathematician, or even a biologist, then you should learn to speak differential equations. This book will set you on that road.

Fourier analysis is one of our most powerful tools for analyzing functions. In Fourier analysis, we break a given function up into component parts—usually sines and cosines. And we can understand the structure of the function using this tool. A modern aspect of Fourier analysis is wavelet theory, which allows us to replace sines and cosines by more general core objects that are tailored to the problem at hand. Wavelet theory has revolutionized the practice of image processing, signal processing, and many other parts of modern technology.

Working with this book is a real intellectual adventure. It should help you to learn and to grow as a mathematical scientist. We wish you all the best in your journey.

Steven G. Krantz
St. Louis, Missouri

1

What Is a Differential Equation?

- The concept of a differential equation
- Characteristics of a solution
- Finding a solution
- Separable equations
- First-order linear equations
- Exact equations
- Orthogonal trajectories

1.1   Introductory Remarks

A *differential equation* is an equation relating some function f to one or more of its derivatives. An example is

$$\frac{d^2 f}{dx^2}(x) + 2x\frac{df}{dx}(x) + f^2(x) = \sin x. \tag{1.1.1}$$

Observe that this particular equation involves a function f together with its first and second derivatives. Any given differential equation may or may not involve f or any particular derivative of f. But, for an equation to be a *differential* equation, at least some derivative of f must appear. The objective in solving an equation like (1.1.1) is to *find the function f*. Thus we already perceive a fundamental new paradigm: When we solve an algebraic equation, we seek a number or perhaps a collection of numbers; but when we solve a differential equation we seek one or more *functions*.

As a simple example, consider the differential equation

$$y' = y.$$

It is easy to determine that any function of the form $y = Ce^x$ is a solution of this equation. For the derivative of the function is equal to itself. So we see that the solution set of this particular differential equation is an infinite *family* of functions parametrized by a parameter C. This phenomenon is quite typical of what we will see when we solve differential equations in the chapters that follow.

Many of the laws of nature—in physics, in engineering, in chemistry, in biology, and in astronomy—find their most natural expression in the language of differential equations. Put in other words, differential equations are the language of nature. Applications of differential equations also abound in mathematics itself, especially in geometry and harmonic analysis and modeling. Differential equations occur in economics and systems science and other fields of mathematical science.

It is not difficult to perceive why differential equations arise so readily in the sciences. If $y = f(x)$ is a given function, then the derivative df/dx can be interpreted as the rate of change of f with respect to x. In any process of nature, the variables involved are related to their rates of change by the basic scientific principles that govern the process—that is, by the laws of nature. When this relationship is expressed in mathematical notation, the result is usually a differential equation.

Certainly Newton's law of universal gravitation, Maxwell's field equations, the motions of the planets, and the refraction of light are important examples which can be expressed using differential equations. Much of our understanding of nature comes from our ability to solve differential equations. The purpose of this book is to introduce you to some of these techniques.

The following example will illustrate some of these ideas. According to Newton's second law of motion, the acceleration **a** of a body of mass m is proportional to the total force **F** acting on the body. The standard expression of this relationship is

$$\mathbf{F} = m \cdot \mathbf{a}. \tag{1.1.2}$$

Suppose in particular that we are analyzing a falling body. Express the height of the body from the surface of the Earth as $y(t)$ feet at time t. The only force acting on the body is that due to gravity. If g is the acceleration due to gravity (about -32 ft./sec.^{2} near the surface of the Earth) then the force exerted on the body has magnitude $m \cdot g$. And of course the acceleration is d^2y/dt^2. Thus Newton's law (1.1.2) becomes

$$m \cdot g = m \cdot \frac{d^2 y}{dt^2} \tag{1.1.3}$$

or

$$g = \frac{d^2 y}{dt^2} \,.$$

We may make the problem a little more interesting by supposing that air exerts a resisting force proportional to the velocity. If the constant of proportionality is k, then the total force acting on the body is $mg - k \cdot (dy/dt)$.

Then equation (1.1.3) becomes

$$m \cdot g - k \cdot \frac{dy}{dt} = m \cdot \frac{d^2 y}{dt^2}. \tag{1.1.4}$$

Equations (1.1.3) and (1.1.4) express the essential attributes of this physical system.

A few additional examples of differential equations are these:

$$(1 - x^2) \frac{d^2 y}{dx^2} - 2x \frac{dy}{dx} + p(p+1)y = 0; \tag{1.1.5}$$

$$x^2 \frac{d^2 y}{dx^2} + x \frac{dy}{dx} + (x^2 - p^2)y = 0; \tag{1.1.6}$$

$$\frac{d^2 y}{dx^2} + xy = 0; \tag{1.1.7}$$

$$(1 - x^2)y'' - xy' + p^2 y = 0; \tag{1.1.8}$$

$$y'' - 2xy' + 2py = 0; \tag{1.1.9}$$

$$\frac{dy}{dx} = k \cdot y; \tag{1.1.10}$$

$$\frac{d^3 y}{dx^3} + \left(\frac{dy}{dx}\right)^2 = y^3 + \sin x. \tag{1.1.11}$$

Equations (1.1.5)–(1.1.9) are called Legendre's equation, Bessel's equation, Airy's equation, Chebyshev's equation, and Hermite's equation, respectively. Each has a vast literature and a history reaching back hundreds of years. We shall touch on each of these equations later in the book. Equation (1.1.10) is the equation of exponential decay (or of biological growth).

Math Nugget

Adrien Marie Legendre (1752–1833) invented Legendre polynomials (the artifact for which he is best remembered) in the context of gravitational attraction of ellipsoids. Legendre was a fine French mathematician who suffered the misfortune of seeing most of his best work—in elliptic integrals, number theory, and the method of least squares— superseded by the achievements of younger and abler men. For instance, he devoted forty years to the study of elliptic integrals, and his two-volume treatise on the subject had scarcely appeared in print before the discoveries of Abel and Jacobi revolutionized the field. Legendre was remarkable for the generous spirit with which he repeatedly welcomed newer and better work that made his own obsolete.

Each of equations (1.1.5)–(1.1.9) is of second order, meaning that the highest derivative that appears is the second. Equation (1.1.10) is of first order. Equation (1.1.11) is of third order. Each equation is an *ordinary differential equation*, meaning that it involves a function of a single variable and the *ordinary derivatives* of that function.

A *partial differential equation* is one involving a function of two or more variables, and in which the derivatives are *partial derivatives*. These equations are more subtle, and more difficult, than ordinary differential equations. We shall say something about partial differential equations in Chapter 10.

Math Nugget

Friedrich Wilhelm Bessel (1784–1846) was a distinguished German astronomer and an intimate friend of Gauss. The two corresponded for many years. Bessel was the first man to determine accurately the distance of a fixed star (the star 61 Cygni). In 1844 he discovered the binary (or twin) star Sirius. The companion star to Sirius has the size of a planet but the mass of a star; its density is many thousands of times the density of water. It was the first dead star to be discovered, and occupies a special place in the modern theory of stellar evolution.

1.2   A Taste of Ordinary Differential Equations

In this section we look at two *very simple* examples to get a notion of what solutions to ordinary differential equations look like.

EXAMPLE 1.2.1 Let us solve the differential equation

$$y' = x.$$

Solution: This is certainly an equation involving a function and some of its derivatives. It is plain to see, just intuitively, that a solution is given by

$$y = \frac{x^2}{2}.$$

But that is not the only solution. In fact the general solution to this differential equation is

$$y = \frac{x^2}{2} + C \,,$$

for an arbitrary real constant C. ∎

So we see that the solution to this differential equation is an *infinite family of functions*. This is quite different from the situation when we were solving polynomials—when the solution set was a finite list of numbers.

The differential equation that we are studying here is what we call *first order*—simply meaning that the highest derivative that appears in the equation is one. So we expect there to be one free parameter in the solution, and indeed there is.

EXAMPLE 1.2.2 Let us solve the differential equation

$$y' = y \,.$$

Solution: Just by intuition, we see that $y = e^x$ is a solution of this ODE. But that is not the only solution. In fact

$$y = Ce^x$$

is the general solution. ∎

It is curious that the arbitrary constant in the previous example occurred additively, while the constant in this solution occurred multiplicatively. Why is that?

Here is another way that we might have discovered the solution to the differential equation in Example 1.2.2. Write the problem in Leibniz notation as

$$\frac{dy}{dx} = y \,.$$

Now manipulate the symbols to write this as

$$dy = ydx$$

or

$$\frac{dy}{y} = dx \,.$$

(At first it may seem odd to manipulate dy/dx as though it were a fraction. But we are simply using the shorthand $dy = (dy/dx)dx$.)

We can integrate both sides of the last equality to obtain

$$\ln y = x + C$$

or

$$y = e^C \cdot e^x = D \cdot e^x.$$

Thus we have rediscovered the solution to this ODE, and we have rather naturally discovered that the constant occurs multiplicatively.

1.3   The Nature of Solutions

An ordinary differential equation of order n is an equation involving an unknown function f together with its derivatives

$$\frac{df}{dx}, \frac{d^2 f}{dx^2}, \dots, \frac{d^n f}{dx^n}.$$

We might, in a more formal manner, express such an equation as

$$F\left(x, y, \frac{df}{dx}, \frac{d^2 f}{dx^2}, \dots, \frac{d^n f}{dx^n}\right) = 0.$$

How do we verify that a given function f is actually the solution of such an equation?

The answer to this question is best understood in the context of concrete examples.

EXAMPLE 1.3.1 Consider the differential equation

$$y'' - 5y' + 6y = 0.$$

Without saying how the solutions are actually *found*, verify that $y_1(x) = e^{2x}$ and $y_2(x) = e^{3x}$ are both solutions.

Solution: To verify this assertion, we note that

$$
\begin{aligned}
y_1'' - 5y_1' + 6y_1 &= 2 \cdot 2 \cdot e^{2x} - 5 \cdot 2 \cdot e^{2x} + 6 \cdot e^{2x} \\
&= [4 - 10 + 6] \cdot e^{2x} \\
&\equiv 0
\end{aligned}
$$

and

$$
\begin{aligned}
y_2'' - 5y_2' + 6y_2 &= 3 \cdot 3 \cdot e^{3x} - 5 \cdot 3 \cdot e^{3x} + 6 \cdot e^{3x} \\
&= [9 - 15 + 6] \cdot e^{3x} \\
&\equiv 0.
\end{aligned}
$$

∎

This process of verifying that a given *function* is a solution of the given differential equation is entirely new. The reader will want to practice and become accustomed to it. In the present instance, the reader may check that any function of the form

$$y(x) = c_1 e^{2x} + c_2 e^{3x} \tag{1.3.1}$$

(where c_1, c_2 are arbitrary constants) is also a solution of the differential equation.

An important obverse consideration is this: When you are going through the procedure to solve a differential equation, how do you know when you are finished? The answer is that the solution process is complete when all derivatives have been eliminated from the equation. For then you will have y expressed in terms of x, at least implicitly. Thus you will have found the sought-after function or functions.

For a large class of equations that we shall study in detail in the present book, we shall find a number of "independent" solutions equal to the order of the differential equation. Then we shall be able to form a so-called "general solution" by combining them as in (1.3.1). Picard's existence and uniqueness theorem tells us that our general solution is complete—there are no other solutions.

Sometimes the solution of a differential equation will be expressed as an *implicitly defined function*. An example is the equation

$$\frac{dy}{dx} = \frac{y^2}{1 - xy}, \tag{1.3.2}$$

which has solution

$$xy = \ln y + c. \tag{1.3.3}$$

Note here that the hallmark of what we call a *solution* is that it has no derivatives in it: it is a direct formula, relating y (the dependent variable) to x (the independent variable). To verify that (1.3.3) is indeed a solution of (1.3.2), let us differentiate:

$$\frac{d}{dx}[xy] = \frac{d}{dx}[\ln y + c]$$

hence

$$1 \cdot y + x \cdot \frac{dy}{dx} = \frac{dy/dx}{y}$$

or

$$\frac{dy}{dx}\left(\frac{1}{y} - x\right) = y.$$

In conclusion,

$$\frac{dy}{dx} = \frac{y^2}{1 - xy},$$

as desired.

   One unifying feature of the two examples that we have now seen of verifying solutions is this: When we solve an equation of order n, we expect n "independent solutions" (we shall have to say later just what this word "independent" means) and we expect n undetermined constants. In the first example, the equation was of order 2 and the undetermined constants were c_1 and c_2. In the second example, the equation was of order 1 and the undetermined constant was c.

Math Nugget

Sir George Biddell Airy (1801–1892) was Astronomer Royal of England for many years. He was a hard-working, systematic plodder whose sense of decorum almost deprived John Couch Adams of credit for discovering the planet Neptune. As a boy, Airy was notorious for his skill in designing peashooters. Although this may have been considered to be a notable start, and in spite of his later contributions to the theory of light (he was one of the first to identify the medical condition known as astigmatism), Airy seems to have developed into an excessively practical sort of scientist who was obsessed with elaborate numerical calculations. He had little use for abstract scientific ideas. Nonetheless, Airy functions still play a prominent role in differential equations, special function theory, and mathematical physics.

EXAMPLE 1.3.4 Verify that, for any choice of the constants A and B, the function
$$y = x^2 + Ae^x + Be^{-x}$$
is a solution of the differential equation
$$y'' - y = 2 - x^2 .$$

Solution: This solution set is typical of what we shall learn to find for a second-order linear equation. There are two free parameters in the solution (corresponding to the degree 2 of the equation). Now, if $y = x^2 + Ae^x + Be^{-x}$, then
$$y' = 2x + Ae^x - Be^{-x}$$
and
$$y'' = 2 + Ae^x + Be^{-x} .$$

Hence

$$y'' - y = \left[2 + Ae^x + Be^{-x}\right] - \left[x^2 + Ae^x + Be^{-x}\right] = 2 - x^2$$

as required. ∎

EXAMPLE 1.3.5 One of the useful things that we do in this subject is to use an "initial condition" to specify a particular solution. Often this initial condition will arise from some physical consideration.

For example, let us solve the (very simple) problem

$$y' = 2y, \quad y(0) = 1.$$

Solution: Of course the general solution of this differential equation is $y = C \cdot e^{2x}$. We seek the solution that has value 1 when $x = 0$. So we set

$$1 = y(0) = C \cdot e^{2 \cdot 0}.$$

This gives

$$1 = C.$$

So we find the particular solution

$$y = e^{2x}.$$ ∎

REMARK 1.3.6 One of the powerful and fascinating features of the study of differential equations is the geometric interpretation that we can often place on the solution of a problem. This connection is most apparent when we consider a first-order equation. Consider the equation

$$\frac{dy}{dx} = F(x, y). \tag{1.3.6.1}$$

We may think of equation (1.3.6.1) as assigning to each point (x, y) in the plane a slope dy/dx. For the purposes of drawing a picture, it is more convenient to think of the equation as assigning to the point (x, y) the *vector* $\langle 1, dy/dx \rangle$. See Figure 1.1. Figure 1.2 illustrates how the differential equation

$$\frac{dy}{dx} = x$$

assigns such a vector to each point in the plane. Figure 1.3 illustrates how the differential equation

$$\frac{dy}{dx} = -y$$

assigns such a vector to each point in the plane.

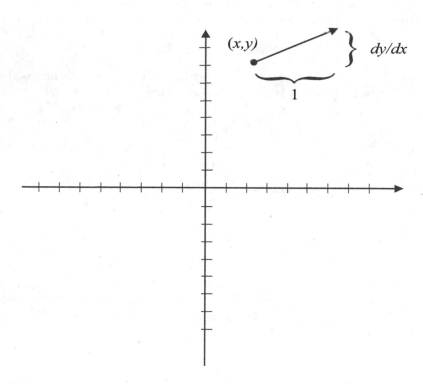

FIGURE 1.1
A vector field.

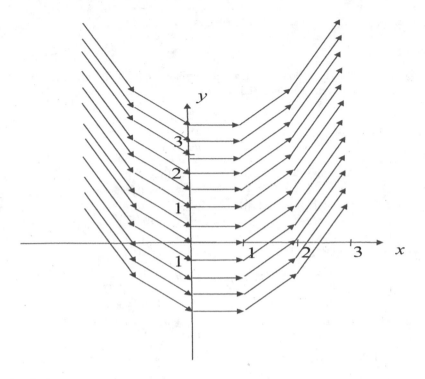

FIGURE 1.2
The vector field for $dy/dx = x$.

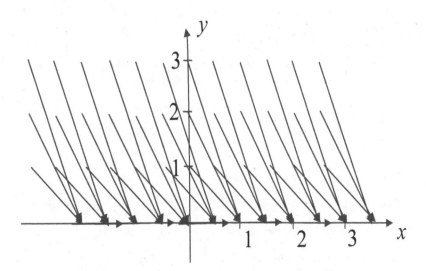

FIGURE 1.3
The vector field for $dy/dx = -y$.

Exercises

1. Verify that the following functions (explicit or implicit) are solutions of the corresponding differential equations.

   (a) $y = x^2 + c$ $y' = 2x$
   (b) $y = cx^2$ $xy' = 2y$
   (c) $y^2 - e^{2x} + c$ $yy' = e^{2x}$
   (d) $y = ce^{kx}$ $y' = ky$
   (e) $y = c_1 \sin 2x + c_2 \cos 2x$ $y'' + 4y = 0$
   (f) $y = c_1 e^{2x} + c_2 e^{-2x}$ $y'' - 4y = 0$
   (g) $y = c_1 \sinh 2x + c_2 \cosh 2x$ $y'' - 4y = 0$
   (h) $y = \arcsin xy$ $xy' + y = y'\sqrt{1 - x^2 y}$
   (i) $y = x \tan x$ $xy' = y + x^2 + y^2$
   (j) $x^2 = 2y^2 \ln y$ $y' = \dfrac{xy}{x^2 + y^2}$
   (k) $y^2 = x^2 - cx$ $2xyy' = x^2 + y^2$
   (l) $y = c^2 + c/x$ $y + xy' = x^4 (y')^2$
   (m) $y = ce^{y/x}$ $y' = y^2/(xy - x^2)$
   (n) $y + \sin y = x$ $(y \cos y - \sin y + x)y' = y$
   (o) $x + y = \arctan y$ $1 + y^2 + y^2 y' = 0$

2. Find the general solution of each of the following differential equations.

   (a) $y' = e^{3x} - x$ (f) $xy' = 1$
   (b) $y' = xe^{x^2}$ (g) $y' = \arcsin x$
   (c) $(1 + x)y' = x$ (h) $y' \sin x = 1$
   (d) $(1 + x^2)y' = x$ (i) $(1 + x^3)y' = x$
   (e) $(1 + x^2)y' = \arctan x$ (j) $(x^2 - 3x + 2)y' = x$

3. For each of the following differential equations, find the particular solution that satisfies the given initial condition.

   (a) $y' = xe^x$ $y = 3$ when $x = 1$
   (b) $y' = 2 \sin x \cos x$ $y = 1$ when $x = 0$
   (c) $y' = \ln x$ $y = 0$ when $x = e$
   (d) $(x^2 - 1)y' = 1$ $y = 0$ when $x = 2$
   (e) $x(x^2 - 4)y' = 1$ $y = 0$ when $x = 1$
   (f) $(x + 1)(x^2 + 1)y' = 2x^2 + x$ $y = 1$ when $x = 0$

4. Show that the function

$$y = e^{x^2} \int_0^x e^{-t^2} \, dt$$

   is a solution of the differential equation $y' = 2xy + 1$.

5. For the differential equation

$$y'' - 5y' + 4y = 0,$$

   carry out the detailed calculations required to verify these assertions:

(a) The functions $y = e^x$ and $y = e^{4x}$ are both solutions.

(b) The function $y = c_1 e^x + c_2 e^{4x}$ is a solution for any choice of constants c_1, c_2.

6. Verify that $x^2 y = \ln y + c$ is a solution of the differential equation $dy/dx = 2xy^2/(1 - x^2 y)$ for any choice of the constant c.

7. For which values of m will the function $y = y_m = e^{mx}$ be a solution of the differential equation

$$2y''' + y'' - 5y' + 2y = 0 ?$$

Find three such values m. Use the ideas in Exercise 5 to find a solution containing three arbitrary constants c_1, c_2, c_3.

1.4 Separable Equations

In this section we shall encounter our first general class of equations with the properties that

(i) we can immediately recognize members of this class of equations, and

(ii) we have a simple and direct method for (in principle) solving such equations.

This is the class of *separable equations*.

A first-order ordinary differential equation is *separable* if it is possible, by elementary algebraic manipulation, to arrange the equation so that all the dependent variables (usually the y variable) are on one side of the equation and all the independent variables (usually the x variable) are on the other side of the equation. Let us learn the method by way of some examples.

EXAMPLE 1.4.1 Solve the ordinary differential equation

$$y' = 2xy .$$

Solution: In the method of separation of variables—which is a method for *first-order* equations only—it is useful to write the derivative using Leibniz notation. Thus we have

$$\frac{dy}{dx} = 2xy .$$

We rearrange this equation as

$$\frac{dy}{y} = 2x \, dx .$$

(It should be noted here that we use the shorthand dy to stand for $\dfrac{dy}{dx}\,dx$ and we of course assume that $y \neq 0$.)

Now we can integrate both sides of the last displayed equation to obtain

$$\int \frac{dy}{y} = \int 2x\,dx\,.$$

We are fortunate in that both integrals are easily evaluated. We obtain

$$\ln|y| = x^2 + C\,.$$

(It is important here that we include the constant of integration.) Thus

$$y = e^{x^2 + C}\,.$$

We may rewrite this as

$$y = e^C \cdot e^{x^2} = D e^{x^2}\,. \tag{1.4.1.1}$$

■

Notice two important features of our final representation for the solution:

(i) We have re-expressed the constant e^c as the positive constant D. We will even allow D to be negative, so we no longer need to worry about the absolute values around y.

(ii) Our solution contains one free constant, as we may have anticipated since the differential equation is of order 1.

We invite the reader to verify that the solution in equation (1.4.1.1) actually satisfies the original differential equation.

REMARK 1.4.2 Of course it would be foolish to expect that all first-order differential equations will be separable. For example, the equation

$$\frac{dy}{dx} = x^2 + y^2$$

certainly is not separable. The property of being separable is rather special. But it is surprising that quite a few of the equations of mathematical physics turn out to be separable (as we shall see later in the book).

EXAMPLE 1.4.3 Solve the differential equation

$$xy' = (1 - 2x^2)\tan y\,.$$

Solution: We first write the equation in Leibniz notation. Thus

$$x \cdot \frac{dy}{dx} = (1 - 2x^2) \tan y.$$

Separating variables, we find that

$$\cot y \, dy = \left(\frac{1}{x} - 2x \right) dx.$$

Applying the integral to both sides gives

$$\int \cot y \, dy = \int \frac{1}{x} - 2x \, dx$$

or

$$\ln |\sin y| = \ln |x| - x^2 + C.$$

Again note that we were careful to include a constant of integration. We may express our solution as

$$\sin y = e^{\ln x - x^2 + C}$$

or

$$\sin y = D \cdot x \cdot e^{-x^2}.$$

The result is

$$y = \sin^{-1} \left(D \cdot x \cdot e^{-x^2} \right).$$

We invite the reader to verify that this is indeed a solution to the given differential equation. ∎

REMARK 1.4.4 Of course the technique of separable equations is one that is specifically designed for first-order equations. It makes no sense for second-order equations. Later in the book we shall learn techniques for reducing a second-order equation to a first-order; then it may happen that the separation-of-variables technique applies.

Exercises

1. Use the method of separation of variables to solve each of these ordinary differential equations.

   (a) $x^5 y' + y^5 = 0$
   (b) $y' = 4xy$
   (c) $y' + y \tan x = 0$
   (d) $(1 + x^2) \, dy + (1 + y^2) \, dx = 0$
   (e) $y \ln y \, dx - x \, dy = 0$

   (f) $xy' = (1 - 4x^2) \tan y$
   (g) $y' \sin y = x^2$
   (h) $y' - y \tan x = 0$
   (i) $xyy' = y - 1$
   (j) $xy^2 - y'x^2 = 0$

2. For each of the following differential equations, find the particular solution that satisfies the additional given property (called an *initial condition*).

    (a) $y'y = x + 1$ $y = 3$ when $x = 1$

    (b) $(dy/dx)x^2 = y$ $y = 1$ when $x = 0$

    (c) $\dfrac{y'}{1+x^2} = \dfrac{x}{y}$ $y = 3$ when $x = 1$

    (d) $y^2 y' = x + 2$ $y = 4$ when $x = 0$

    (e) $y' = x^2 y^2$ $y = 2$ when $x = -1$

    (f) $y'(1 + y) = 1 - x^2$ $y = -2$ when $x = -1$

3. For the differential equation

$$\frac{y''}{y'} = x^2,$$

make the substitution $y' = p$, $y'' = p'$ to reduce the order. Then solve the new equation by separation of variables. Now resubstitute and find the solution y of the original equation.

4. Use the method of Exercise 3 to solve the equation

$$y'' \cdot y' = x(1 + x)$$

subject to the initial conditions $y(0) = 1$, $y'(0) = 2$.

1.5   First-Order, Linear Equations

Another class of differential equations that is easily recognized and readily solved (at least in principle),[1] is that of first-order, linear equations.

    An equation is said to be *first-order linear* if it has the form

$$y' + a(x)y = b(x). \tag{1.5.1}$$

The "first-order" aspect is obvious: only first derivatives appear in the equation. The "linear" aspect depends on the fact that the left-hand side involves a differential operator that acts linearly on the space of differentiable functions. Roughly speaking, a differential equation is linear if y and its derivatives are not multiplied together, not raised to powers, and do not occur as the arguments of functions. For now, the reader should simply accept that an equation

---

[1] We throw in this caveat because it can happen, and frequently does happen, that we can write down integrals that represent solutions of our differential equation, but *we are unable to evaluate those integrals*. This is annoying, but there are numerical techniques that will address such an impasse.

of the form (1.5.1) is first-order linear, and that we will soon have a recipe for solving it.

As usual, we learn this new method by proceeding directly to the examples.

EXAMPLE 1.5.2 Consider the differential equation

$$y' + 2xy = x.$$

Find a complete solution.

Solution: First note that the equation definitely has the form (1.5.1) of a linear differential equation of first order. In particular, $a(x) = 2x$ and $b(x) = x$. So we may proceed.

We endeavor to multiply both sides of the equation by some function that will make each side readily integrable. It turns out that there is a trick that always works: You multiply both sides by $e^{\int a(x)\,dx}$.

Like many tricks, this one may seem unmotivated. But let us try it out and see how it works. Notice that $a(x) = 2x$ so that

$$\int a(x)\,dx = \int 2x\,dx = x^2.$$

(At this point we *could* include a constant of integration, but it is not necessary.) Thus $e^{\int a(x)\,dx} = e^{x^2}$. Multiplying both sides of our equation by this factor gives

$$e^{x^2} \cdot y' + e^{x^2} \cdot 2xy = e^{x^2} \cdot x$$

or

$$\left(e^{x^2} \cdot y\right)' = x \cdot e^{x^2}.$$

It is the last step that is a bit tricky. For a first-order linear equation, it is *guaranteed* that, if we multiply through by $e^{\int a(x)\,dx}$, then the left-hand side of the equation will end up being the derivative of $[e^{\int a(x)\,dx} \cdot y]$. Now of course we integrate both sides of the equation:

$$\int \left(e^{x^2} \cdot y\right)' dx = \int x \cdot e^{x^2}\,dx.$$

We can perform both the integrations: on the left-hand side we simply apply the fundamental theorem of calculus; on the right-hand side we do the integration as usual. The result is

$$e^{x^2} \cdot y = \frac{1}{2} \cdot e^{x^2} + C$$

or

$$y = \frac{1}{2} + Ce^{-x^2}.$$

Observe that, as we usually expect, the solution has one free constant (because the original differential equation was of order 1). We invite the reader to check in detail that this solution actually satisfies the original differential equation. ∎

REMARK 1.5.3 The last example illustrates a phenomenon that we shall encounter repeatedly in this book. That is the idea of a "general solution." Generally speaking, a differential equation will have an entire family of solutions. And, especially when the problem comes from physical considerations, we shall often have initial conditions that must be met by the solution. The family of solutions will depend on one or more parameters (in the last example there was one parameter C), and those parameters will often be determined by the initial conditions.

We shall see as the book develops that the amount of freedom built into the family of solutions—that is, the number of degrees of freedom provided by the parameters—meshes very nicely with the number of initial conditions that fit the problem (in the last example, one initial condition would be appropriate). Thus we shall generally be able to solve uniquely for numerical values of the parameters. Picard's Existence and Uniqueness Theorem gives a precise mathematical framework for the informal discussion in the present remark.

Summary of the Method

To solve a first-order linear equation

$$y' + a(x)y = b(x)\,,$$

multiply both sides of the equation by the "integrating factor" $e^{\int a(x)\,dx}$ and then integrate.

EXAMPLE 1.5.4 Solve the differential equation

$$x^2 y' + xy = x^3\,.$$

Solution: First observe that this equation is not in the standard form (equation (1.5.1)) for first-order linear. We render it so by multiplying through by a factor of $1/x^2$. Thus the equation becomes

$$y' + \frac{1}{x}y = x\,.$$

Now $a(x) = 1/x$, $\int a(x)\,dx = \ln|x|$, and $e^{\int a(x)\,dx} = |x|$. We multiply the differential equation through by this factor. In fact, in order to simplify the calculus, we shall restrict attention to $x > 0$. Thus we may eliminate the absolute value signs.

Thus

$$xy' + y = x^2\,.$$

Now, as is guaranteed by the theory, we may rewrite this equation as

$$\left(x \cdot y\right)' = x^2\,.$$

Applying the integral to both sides gives

$$\int \left(x \cdot y\right)'\,dx = \int x^2\,dx\,.$$

Now, as usual, we may use the fundamental theorem of calculus on the left; and we may simply integrate on the right. The result is

$$x \cdot y = \frac{x^3}{3} + C\,.$$

We finally find that our solution is

$$y = \frac{x^2}{3} + \frac{C}{x}\,.$$

You should plug this answer into the differential equation and check that it works. ∎

Exercises

1. Find the general solution of each of the following first-order, linear ordinary differential equations.

   (a) $y' - xy = 0$

   (b) $y' + xy = x$

   (c) $y' + y = \dfrac{1}{1 + e^{2x}}$

   (d) $y' + y = 2xe^{-x} + x^2$

   (e) $(2y - x^3)\,dx = x\,dy$

   (f) $y' + 2xy = 0$

   (g) $xy' - 3y = x^4$

   (h) $(1 + x^2)\,dy + 2xy\,dx = \cot x\,dx$

   (i) $y' + y \cot x = 2x \csc x$

   (j) $y - x + xy \cot x + xy' = 0$

2. For each of the following differential equations, find the particular solution that satisfies the given initial data.

(a)	$y' - xy = 0$	$y = 3$ when $x = 1$
(b)	$y' - 2xy = 6xe^{x^2}$	$y = 1$ when $x = 1$
(c)	$xy' + y = 3x^2$	$y = 2$ when $x = 2$
(d)	$y' - (1/x)y = x^2$	$y = 3$ when $x = 1$
(e)	$y' + 4y = e^{-x}$	$y = 0$ when $x = 0$
(f)	$x^2 y' + xy = 2x$	$y = 1$ when $x = 1$

3. The equation

$$\frac{dy}{dx} + P(x)y = Q(x)y^n$$

is known as Bernoulli's equation. It is linear when $n = 0$ or 1, otherwise not. In fact the equation can be reduced to a linear equation when $n > 1$ by the change of variables $z = y^{1-n}$. Use this method to solve each of the following equations.

(a) $xy' + y = x^4 y^3$ **(c)** $x\,dy + y\,dx = xy^2\,dx$
(b) $xy^2 y' + y^3 = x\cos x$ **(d)** $y' + xy = xy^4$

4. The usual Leibniz notation dy/dx implies that x is the independent variable and y is the dependent variable. In solving a differential equation, it is sometimes useful to reverse the roles of the two variables. Treat each of the following equations by reversing the roles of y and x:

(a) $(e^y - 2xy)y' = y^2$ **(c)** $xy' + 2 = x^3(y-1)y'$
(b) $y - xy' = y'y^2 e^y$ **(d)** $f(y)^2 \dfrac{dx}{dy} + 3f(y)f'(y)x = f'(y)$

5. We know from our solution technique that the general solution of a first-order linear equation is a family of curves of the form

$$y = c \cdot f(x) + g(x).$$

Show, conversely, that the differential equation of any such family is linear and first order.

6. Show that the differential equation $y' + Py = Qy \ln y$ can be solved by the change of variables $z = \ln y$. Apply this method to solve the equation

$$xy' = 2x^2 y + y \ln y.$$

7. One solution of the differential equation $y' \sin 2x = 2y + 2\cos x$ remains bounded as $x \to \pi/2$. Find this solution.

8. A tank contains 10 gallons of brine in which 2 pounds of salt are dissolved. New brine containing 1 pound of salt per gallon is pumped into the tank at the rate of 3 gallons per minute. The mixture is stirred and drained off at the rate of 4 gallons per minute. Find the amount $x - x(t)$ of salt in the tank at any time t.

9. A tank contains 40 gallons of pure water. Brine with 3 pounds of salt per gallon flows in at the rate of 2 gallons per minute. The thoroughly stirred mixture then flows out at the rate of 3 gallons per minute.

(a) Find the amount of salt in the tank when the brine in it has been reduced to 20 gallons.

(b) When is the amount of salt in the tank greatest?

1.6 Exact Equations

A great many first-order equations may be written in the form

$$M(x, y)\, dx + N(x, y)\, dy = 0.$$

(1.6.1)

This particular format is quite suggestive, for it brings to mind a family of curves. Namely, if it happens that there is a function $f(x, y)$ so that

$$\frac{\partial f}{\partial x} = M \qquad \text{and} \qquad \frac{\partial f}{\partial y} = N,$$

(1.6.2)

then we can rewrite the differential equation as

$$\frac{\partial f}{\partial x}\, dx + \frac{\partial f}{\partial y}\, dy = 0.$$

(1.6.3)

Of course the only way that such an equation can hold is if

$$\frac{\partial f}{\partial x} \equiv 0 \qquad \text{and} \qquad \frac{\partial f}{\partial y} \equiv 0.$$

And this entails that the function f be identically constant. In other words,

$$f(x, y) \equiv C.$$

This last equation describes a family of curves: for each fixed value of C, the equation expresses y implicitly as a function of x, and hence gives a curve. Refer to Figure 1.4 for an example. In later parts of this book we shall learn much from thinking of the set of solutions of a differential equation as a smoothly varying family of curves in the plane.

The method of solution just outlined is called the *method of exact equations*. It depends critically on being able to tell when an equation of the form (1.6.1) can be written in the form (1.6.3). This in turn begs the question of when (1.6.2) will hold.

Fortunately, we learned in calculus a complete answer to this question. Let us review the key points. First note that, if it is the case that

$$\frac{\partial f}{\partial x} = M \qquad \text{and} \qquad \frac{\partial f}{\partial y} = N,$$

(1.6.4)

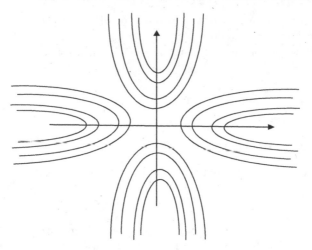

FIGURE 1.4
A family of curves defined by $x^2 - y^2 = c$.

then we see (by differentiation) that

$$\frac{\partial^2 f}{\partial y \partial x} = \frac{\partial M}{\partial y} \quad \text{and} \quad \frac{\partial^2 f}{\partial x \partial y} = \frac{\partial N}{\partial x}.$$

Since mixed partials of a smooth function may be taken in any order, we find
that a *necessary condition* for condition (1.6.4) to hold is that

$$\frac{\partial M}{\partial y} = \frac{\partial N}{\partial x}. \qquad (1.6.5)$$

We call (1.6.5) the *exactness condition*. This provides us with a useful test for
when the method of exact equations will apply.

It turns out that condition (1.6.5) is also sufficient—at least on a domain
with no holes. We refer the reader to any good calculus book (see, for instance,
[BLK]) for the details of this assertion. We shall use our worked examples to
illustrate the point.

EXAMPLE 1.6.6 Use the method of exact equations to solve

$$\frac{x}{2} \cdot \cot y \cdot \frac{dy}{dx} = -1.$$

Solution: First, we rearrange the equation as

$$2x \sin y \, dx + x^2 \cos y \, dy = 0.$$

Observe that the role of $M(x, y)$ is played by $2x \sin y$ and the role of
$N(x, y)$ is played by $x^2 \cos y$. Next we see that

$$\frac{\partial M}{\partial y} = 2x \cos y = \frac{\partial N}{\partial x}.$$

Thus our necessary condition (exactness) for the method of exact equations to work is satisfied. We shall soon see explicitly from our calculations that it is also sufficient.

We seek a function f such that $\partial f / \partial x = M(x, y) = 2x \sin y$ and $\partial f / \partial y = N(x, y) = x^2 \cos y$. Let us begin by concentrating on the first of these:

$$\frac{\partial f}{\partial x} = 2x \sin y \,,$$

hence

$$\int \frac{\partial f}{\partial x} \, dx = \int 2x \sin y \, dx \,.$$

The left-hand side of this equation may be evaluated with the fundamental theorem of calculus. Treating x and y as independent variables (which is part of this method), we can also compute the integral on the right. The result is

$$f(x, y) = x^2 \sin y + \phi(y) \,. \tag{1.6.6.1}$$

Now there is an important point that must be stressed. The reader should by now have expected a constant of integration to show up. But in fact our "constant of integration" is $\phi(y)$. This is because our integral was with respect to x, and therefore our constant of integration should be the most general possible expression *that does not depend on x*. That, of course, would be a function of y.

Now we differentiate both sides of (1.6.6.1) with respect to y to obtain

$$N(x, y) = \frac{\partial f}{\partial y} = x^2 \cos y + \phi'(y) \,.$$

But of course we already know that $N(x, y) = x^2 \cos y$. The upshot is that

$$\phi'(y) = 0$$

or

$$\phi(y) = D \,,$$

an ordinary constant.

Plugging this information into equation (1.6.6.1) now yields that

$$f(x, y) = x^2 \sin y + D \,.$$

We stress that *this is not the solution of the differential equation*. Before you proceed, please review the outline of the method of exact equations that preceded this example. Our job now is to set

$$f(x, y) = C \,.$$

So

$$x^2 \cdot \sin y + D = C$$

or

$$x^2 \cdot \sin y = \widetilde{C},$$

where $\widetilde{C} = C - D$.

This is in fact the solution of our differential equation, expressed implicitly. If we wish, we can solve for y in terms of x to obtain

$$y = \sin^{-1} \frac{\widetilde{C}}{x^2}.$$ ■

EXAMPLE 1.6.7 Use the method of exact equations to solve the differential equation

$$y^2 \, dx - x^2 \, dy = 0.$$

Solution: We first test the exactness condition:

$$\frac{\partial M}{\partial y} = 2y \neq -2x = \frac{\partial N}{\partial x}.$$

The exactness condition fails. As a result, this ordinary differential equation cannot be solved by the method of exact equations.   ■

It is a fact that, even when a differential equation fails the "exact equations test," it is always possible to multiply the equation through by an "integrating factor" so that it *will* pass the exact equations test. Unfortunately, it can be quite difficult to discover explicitly what that integrating factor might be. We shall learn more about the method of integrating factors later in the book.

EXAMPLE 1.6.8 Use the method of exact equations to solve

$$e^y \, dx + (xe^y + 2y) \, dy = 0.$$

Solution: First we check for exactness:

$$\frac{\partial M}{\partial y} = \frac{\partial}{\partial y}[e^y] = e^y = \frac{\partial}{\partial x}[xe^y + 2y] = \frac{\partial N}{\partial x}.$$

The exactness condition is verified, so we can proceed to solve for f:

$$\frac{\partial f}{\partial x} = M = e^y$$

hence

$$f(x, y) = x \cdot e^y + \phi(y).$$

But then

$$\frac{\partial}{\partial y} f(x, y) = \frac{\partial}{\partial y} (x \cdot e^y + \phi(y)) = x \cdot e^y + \phi'(y).$$

And this last expression must equal $N(x, y) = xe^y + 2y$. It follows that

$$\phi'(y) = 2y$$

or

$$\phi(y) = y^2 + D.$$

Altogether, then, we conclude that

$$f(x, y) = x \cdot e^y + y^2 + D.$$

We must not forget the final step. The solution of the differential equation is

$$f(x, y) = C$$

or

$$x \cdot e^y + y^2 = C - D \equiv \tilde{C}.$$

This time we must content ourselves with the solution expressed implicitly, since it is not feasible to solve for y in terms of x (at least not in an elementary, closed form). ∎

Exercises

Determine which of the following equations, in Exercises 1–19, is exact. Solve those that *are* exact.

1. $\left(x + \dfrac{2}{y} \right) dy + y\, dx = 0$

2. $(\sin x \tan y + 1)\, dx + \cos x \sec^2 y\, dy = 0$

3. $(y - x^3)\, dx + (x + y^3)\, dy = 0$

4. $(2y^2 - 4x + 5)\, dx = (4 - 2y + 4xy)\, dy$

5. $(y + y \cos xy)\, dx + (x + x \cos xy)\, dy = 0$

6. $\cos x \cos^2 y\, dx + 2 \sin x \sin y \cos y\, dy = 0$

7. $(\sin x \sin y - xe^y)\, dy = (e^y + \cos x \cos y)\, dx$

8. $-\dfrac{1}{y} \sin \dfrac{x}{y}\, dx + \dfrac{x}{y^2} \sin \dfrac{x}{y}\, dy = 0$

9. $(1 + y)\, dx + (1 - x)\, dy = 0$

10. $(2xy^3 + y \cos x)\, dx + (3x^2y^2 + \sin x)\, dy = 0$

11. $dx = \dfrac{y}{1 - x^2y^2}\, dx + \dfrac{x}{1 - x^2y^2}\, dy$

12. $(2xy^4 + \sin y)\, dx + (4x^2y^3 + x \cos y)\, dy = 0$

13. $\dfrac{y\, dx + x\, dy}{1 - x^2y^2} + x\, dx = 0$

14. $2x(1 + \sqrt{x^2 - y})\, dx = \sqrt{x^2 - y}\, dy$

15. $(x \ln y + xy)\, dx + (y \ln x + xy)\, dy = 0$

16. $(e^{y^2} - \csc y \csc^2 x)\, dx + (2xye^{y^2} - \csc y \cot y \cot x)\, dy = 0$

17. $(1 + y^2 \sin 2x)\, dx - 2y \cos^2 x\, dy = 0$

18. $\dfrac{x\, dx}{(x^2 + y^2)^{3/2}} + \dfrac{y\, dy}{(x^2 + y^2)^{3/2}} = 0$

19. $3x^2(1 + \ln y)\, dx + \left(\dfrac{x^3}{y} - 2y\right) dy = 0$

20.   Solve
$$\frac{y\, dx - x\, dy}{(x + y)^2} + dy = dx$$
as an exact equation by two different methods. Now reconcile the results.

21.   Solve
$$\frac{4y^2 - 2x^2}{4xy^2 - x^3}\, dx + \frac{8y^2 - x^2}{4y^3 - x^2 y}\, dy = 0\,.$$
as an exact equation. Later on (Section 1.8) we shall learn that we may also solve this equation as a homogeneous equation.

22.   For each of the following equations, find the value of n for which the equation is exact. Then solve the equation for that value of n.

   (a) $(xy^2 + nx^2y)\, dx + (x^3 + x^2y)\, dy = 0$

   (b) $(x + ye^{2xy})\, dx + nxe^{2xy}\, dy = 0$

1.7   Orthogonal Trajectories and Families of Curves

We have already noted that it is useful to think of the collection of solutions of a first-order differential equation as a family of curves. Refer, for instance, to the last example of the preceding section. We solved the differential equation

$$e^y\, dx + (xe^y + 2y)\, dy = 0$$

and found the solution set

$$x \cdot e^y + y^2 = \widetilde{C}\,. \tag{1.7.1}$$

For each value of $\widetilde{C}$, the equation describes a curve in the plane.

Conversely, if we are given a family of curves in the plane, then we can produce a differential equation from which the curves all come. Consider the example of the family

$$x^2 + y^2 = 2Cx\,. \tag{1.7.2}$$

The reader can readily see that this is the family of all circles tangent to the y-axis at the origin (Figure 1.5).

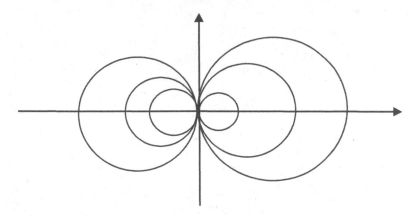

FIGURE 1.5
A family of circles.

We may differentiate the equation with respect to x, thinking of y as a function of x, to obtain

$$2x + 2y \cdot \frac{dy}{dx} = 2C .$$

Now the original equation (1.7.2) tells us that

$$x + \frac{y^2}{x} = 2C ,$$

and we may equate the two expressions for the quantity $2C$ (the point being to eliminate the constant C). The result is

$$2x + 2y \cdot \frac{dy}{dx} = x + \frac{y^2}{x}$$

or

$$\frac{dy}{dx} = \frac{y^2 - x^2}{2xy} . \tag{1.7.3}$$

In summary, we see that we can pass back and forth between a differential equation and its family of solution curves.

There is considerable interest, given a family $\mathcal{F}$ of curves, to find the corresponding family $\mathcal{G}$ of curves that are orthogonal (or perpendicular) to those of $\mathcal{F}$. For instance, if $\mathcal{F}$ represents the flow curves of an electric current, then $\mathcal{G}$ will be the equipotential curves for the flow. If we bear in mind that orthogonality of curves means orthogonality of their tangents, and that orthogonality of the tangent lines means simply that their slopes are negative reciprocals, then it becomes clear what we must do.

EXAMPLE 1.7.4 Find the orthogonal trajectories to the family of curves

$$x^2 + y^2 = C .$$

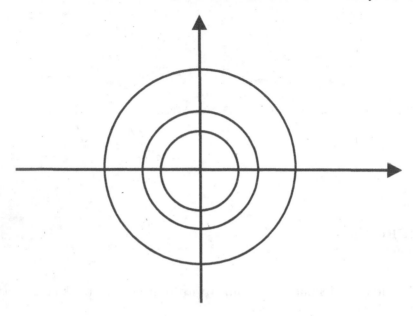

FIGURE 1.6
Circles centered at the origin.

Solution: First observe that we can differentiate the given equation to obtain

$$2x + 2y \cdot \frac{dy}{dx} = 0.$$

The constant C has disappeared, and we can take this to be the differential equation for the given family of curves (which in fact are all the circles centered at the origin—see Figure 1.6).

We rewrite the differential equation as

$$\frac{dy}{dx} = -\frac{x}{y}.$$

Now taking negative reciprocals, as indicated in the discussion right before this example, we obtain the new differential equation

$$\frac{dy}{dx} = \frac{y}{x}$$

for the family of orthogonal curves.

We may easily separate variables to obtain

$$\frac{1}{y}\, dy = \frac{1}{x}\, dx.$$

Applying the integral to both sides yields

$$\int \frac{1}{y}\, dy = \int \frac{1}{x}\, dx$$

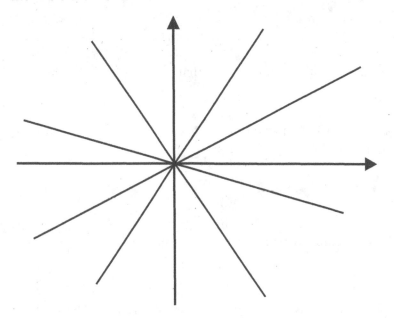

FIGURE 1.7
Lines through the origin.

or

$$\ln |y| = \ln |x| + C.$$

With some algebra, this simplifies to

$$|y| = D|x|$$

or

$$y = \pm Dx.$$

∎

The solution that we have found comes as no surprise: the orthogonal trajectories to the family of circles centered at the origin is the family of lines through the origin. See Figure 1.7.

EXAMPLE 1.7.5 Find the family of orthogonal trajectories to the curves

$$y = Cx^2.$$

Solution: We differentiate to find that

$$\frac{dy}{dx} = 2Cx$$

or

$$C = \frac{1}{2x} \cdot \frac{dy}{dx}.$$

But the original equation tells us that

$$C = \frac{y}{x^2}.$$

The point here is to eliminate C, and we can do so by equating the two expressions for C. Thus

$$\frac{1}{2x} \cdot \frac{dy}{dx} = \frac{y}{x^2}$$

or

$$\frac{dy}{dx} = \frac{2y}{x}.$$

Then the family of orthogonal curves will satisfy

$$\frac{dy}{dx} = -\frac{x}{2y}.$$

This equation is easily solved by separation of variables. The solution is

$$y = \pm\sqrt{C - x^2/2}.$$

We leave the details to the interested reader. ∎

Exercises

1. Sketch each of the following families of curves. In each case, find the family of orthogonal trajectories, and add those to your sketch.

   (a) $xy = c$ (d) $r = c(1 + \cos\theta)$
   (b) $y = cx^2$ (e) $y = ce^x$
   (c) $x + y = c$ (f) $x - y^2 = c$

2. What are the orthogonal trajectories of the family of curves $y = cx^4$? What are the orthogonal trajectories of the family of curves $y = cx^n$ for n a positive integer? Sketch both families of curves. How does the family of orthogonal trajectories change when n is increased?

3. Sketch the family $y^2 = 4c(x + c)$ of all parabolas with axis the x-axis and focus at the origin. Find the differential equation of this family. Show that this differential equation is unaltered if dy/dx is replaced by $-dx/dy$. What conclusion can be drawn from this fact?

4. In each of parts (a) through (f), find the family of curves that satisfy the given geometric condition (you should have six different answers for the six different parts of the problem):

(a) The part of the tangent cut off by the axes is bisected by the point of tangency.

(b) The projection on the x-axis of the part of the normal between (x, y) and the x-axis has length 1.

(c) The projection on the x-axis of the part of the tangent between (x, y) and the x-axis has length 1.

(d) The part of the tangent between (x, y) and the x-axis is bisected by the y-axis.

(e) The part of the normal between (x, y) and the y-axis is bisected by the x-axis.

(f) The point (x, y) is equidistant from the origin and the point of intersection of the normal with the x-axis.

5. A curve rises from the origin in the x-y plane into the first quadrant. The area under the curve from $(0, 0)$ to (x, y) is one third of the area of the rectangle with these points as opposite vertices. Find the equation of the curve.

6. Find the differential equation of each of the following one-parameter families of curves:

   (a) $y = x \sin(x + c)$

   (b) all circles through $(1, 0)$ and $(-1, 0)$

   (c) all circles with centers on the line $y = x$ and tangent to both axes

   (d) all lines tangent to the parabolas $x^2 = 4y$ [**Hint:** The slope of the tangent line at $(2a, a^2)$ is a.]

   (e) all lines tangent to the unit circle $x^2 + y^2 = 1$

7. Use your symbol manipulation software, such as `Maple` or `Mathematica`, to find the orthogonal trajectories to each of these families of curves:

   (a) $y = \sin x + cx^2$

   (b) $y = c \ln x + x, \quad x > 0$

   (c) $y = \dfrac{\cos x}{cx + \ln x}, \quad x > 0$

   (d) $y = \sin x + c \cos x$

1.8 Homogeneous Equations

The reader should be cautioned that the word "homogeneous" has two meanings in this subject (as mathematics is developed simultaneously by many people all over the world, and they do not always stop to cooperate on their choices of terminology).

One usage, which we shall see later, is that an ordinary differential equation is *homogeneous* when the right-hand side is zero; that is, there is no forcing term.

The other usage will be relevant to the present section. It bears on the "balance" of weight among the different variables. It turns out that a differential equation in which the x and y variables have a balanced presence is amenable to a useful change of variables. That is what we are about to learn.

First of all, a function $g(x, y)$ of two variables is said to be *homogeneous of degree* α, for α a real number, if

$$g(tx, ty) = t^\alpha g(x, y) \qquad \text{for all } t > 0 \,.$$

As examples, consider:

- Let $g(x, y) = x^2 + xy$. Then $g(tx, ty) = (tx)^2 + (tx) \cdot (ty) = t^2 x^2 + t^2 xy = t^2 \cdot g(x, y)$, so g is homogeneous of degree 2.

- Let $g(x, y) = \sin[x/y]$. Then $g(tx, ty) = \sin[(tx)/(ty)] = \sin[x/y] = t^0 \cdot g(x, y)$, so g is homogeneous of degree 0.

- Let $g(x, y) = \sqrt{x^2 + y^2}$. Then $g(tx, ty) = \sqrt{(tx)^2 + (ty)^2} = t\sqrt{x^2 + y^2} = t \cdot g(x, y)$, so g is homogeneous of degree 1.

If a function is not homogeneous in the sense just indicated, then we call it *inhomogeneous*.

In case a differential equation has the form

$$M(x, y)\, dx + N(x, y)\, dy = 0$$

and M, N have the *same degree of homogeneity*, then it is possible to perform the change of variable $z = y/x$ and make the equation separable (see Section 1.4). Of course we then have a well-understood method for solving the equation.

The next examples will illustrate the method.

EXAMPLE 1.8.1 Use the method of homogeneous equations to solve the equation

$$(x + y)\, dx - (x - y)\, dy = 0 \,.$$

Solution: This equation is *not* exact. However, observe that $M(x, y) = x + y$ and $N(x, y) = -(x - y)$ and each is homogeneous of degree 1. We thus rewrite the equation in the form

$$\frac{dy}{dx} = \frac{x + y}{x - y} \,.$$

Dividing numerator and denominator by x, we finally have

$$\frac{dy}{dx} = \frac{1 + \frac{y}{x}}{1 - \frac{y}{x}} \,. \tag{1.8.1.1}$$

The point of these manipulations is that the right-hand side is now plainly homogeneous of degree 0. We introduce the change of variable

$$z = \frac{y}{x} \tag{1.8.1.2}$$

hence
$$y = zx$$

and
$$\frac{dy}{dx} = z + x \cdot \frac{dz}{dx}.$$
(1.8.1.3)

Putting (1.8.1.2) and (1.8.1.3) into (1.8.1.1) gives
$$z + x\frac{dz}{dx} = \frac{1+z}{1-z}.$$

Of course this may be rewritten as
$$x\frac{dz}{dx} = \frac{1+z^2}{1-z}$$

or
$$\frac{1-z}{1+z^2}\,dz = \frac{dx}{x}.$$

Notice that we have separated the variables!! We apply the integral, and rewrite the left-hand side, to obtain
$$\int \frac{dz}{1+z^2} - \int \frac{z\,dz}{1+z^2} = \int \frac{dx}{x}.$$

The integrals are easily evaluated, and we find that
$$\arctan z - \frac{1}{2}\ln(1+z^2) = \ln x + C.$$

Now we return to our original notation by setting $z = y/x$. The result is
$$\arctan \frac{y}{x} - \frac{1}{2}\ln\left(1+\frac{y^2}{x^2}\right) = \ln x + C$$

or
$$\arctan \frac{y}{x} - \ln \sqrt{x^2+y^2} = C.$$

Thus we have expressed y implicitly as a function of x, all the derivatives are gone, and we have solved the differential equation. ∎

EXAMPLE 1.8.2 Solve the differential equation
$$xy' = 2x + 3y.$$

Solution: It is plain that the equation is first-order linear, and we encourage the reader to solve the equation by that method for practice and comparison purposes. Instead, developing the ideas of the present section, we shall use the method of homogeneous equations.

If we rewrite the equation as

$$-(2x + 3y)\,dx + x\,dy = 0\,,$$

then we see that each of $M = -(2x+3y)$ and $N = x$ is homogeneous of degree 1. Thus we render the equation as

$$\frac{dy}{dx} = \frac{2x + 3y}{x} = \frac{2 + 3(y/x)}{1} = 2 + 3\frac{y}{x}\,.$$

The right-hand side is homogeneous of degree 0, as we expect.

We set $z = y/x$ and $dy/dx = z + x[dz/dx]$. The result is

$$z + x \cdot \frac{dz}{dx} = 2 + 3\frac{y}{x} = 2 + 3z\,.$$

The equation separates, as we anticipate, into

$$\frac{dz}{2 + 2z} = \frac{dx}{x}\,.$$

This is easily integrated to yield

$$\frac{1}{2}\ln(1 + z) = \ln x + C$$

or

$$z = Dx^2 - 1\,.$$

Resubstituting $z = y/x$ gives

$$\frac{y}{x} = Dx^2 - 1$$

hence

$$y = Dx^3 - x\,. \qquad\qquad\qquad\qquad \blacksquare$$

Exercises

1. Verify that each of the following differential equations is homogeneous, and then solve it.

   (a) $(x^2 - 2y^2)\,dx + xy\,dy = 0$

   (b) $xy' - 3xy - 2y^2 = 0$

   (c) $x^2y' = 3(x^2 + y^2) \cdot \arctan\frac{y}{x} + xy$

   (d) $x\left(\sin\frac{y}{x}\right)\frac{dy}{dx} = y\sin\frac{y}{x} + x$

   (e) $xy' = y + 2xe^{-y/x}$

   (f) $(x - y)\,dx - (x + y)\,dy = 0$

   (g) $xy' = 2x - 6y$

   (h) $xy' = \sqrt{x^2 + y^2}$

   (i) $x^2y' = y^2 + 2xy$

   (j) $(x^3 + y^3)\,dx - xy^2\,dy = 0$

2. Use rectangular coordinates to find the orthogonal trajectories of the family of all circles tangent to the x-axis at the origin.

3. **(a)** If $ae \neq bd$ then show that h and k can be chosen so that the substitution $x = z - h$, $y = w - k$ reduces the equation

$$\frac{dy}{dx} = F\left(\frac{ax + by + c}{dx + ey + f}\right)$$

to a homogeneous equation.

   (b) If $ae = bd$ then show that there is a substitution that reduces the equation in **(a)** to one in which the variables are separable.

4. Solve each of the following differential equations.

   (a) $\dfrac{dy}{dx} = \dfrac{x + y + 4}{x - y - 6}$

   (b) $\dfrac{dy}{dx} = \dfrac{x + y + 4}{x + y - 6}$

   (c) $(2x - 2y)\,dx + (y - 1)\,dy = 0$

   (d) $\dfrac{dy}{dx} = \dfrac{x + y - 1}{x + 4y + 2}$

   (e) $(2x + 3y - 1)\,dx$
   $\qquad -4(x + 1)\,dy = 0$

5. By making the substitution $z = y/x^n$ (equivalently $y = zx^n$) and choosing a convenient value of n, show that the following differential equations can be transformed into equations with separable variables, and then solve them.

   (a) $\dfrac{dy}{dx} = \dfrac{1 - xy^2}{2x^2y}$

   (b) $\dfrac{dy}{dx} = \dfrac{2 + 3xy^2}{4x^2y}$

   (c) $\dfrac{dy}{dx} = \dfrac{y - xy^2}{x + x^2y}$

6. Show that a straight line through the origin intersects all integral curves of a homogeneous equation at the same angle.

7. Use your symbol manipulation software, such as Maple or Mathematica, to find solutions to each of the following homogeneous equations. (Note that these would be difficult to do by hand.)

   (a) $y' = \sin[y/x] - \cos[y/x]$

   (b) $e^{x/y}\,dx - \dfrac{y}{x}\,dy = 0$

   (c) $\dfrac{dy}{dx} = \dfrac{x^2 - xy}{y^2 \cos(x/y)}$

   (d) $y' = \dfrac{y}{x} \cdot \tan[y/x]$

1.9 Integrating Factors

We used a special type of integrating factor in Section 1.5 on first-order linear equations. At that time, we suggested that integrating factors may be applied in some generality to the solution of first-order differential equations. The trick is in *finding* the integrating factor.

In this section we shall discuss this matter in some detail, and indicate the uses and the limitations of the method of integrating factors.

First let us illustrate the concept of integrating factor by way of a concrete example. The differential equation

$$y\,dx + (x^2 y - x)\,dy = 0 \tag{1.9.1}$$

is plainly *not exact*, just because $\partial M/\partial y = 1$ while $\partial N/\partial x = 2xy - 1$, and these are unequal. However, if we multiply equation (1.9.1) through by a factor of $1/x^2$ then we obtain the equivalent equation

$$\frac{y}{x^2}\,dx + \left(y - \frac{1}{x}\right) = 0\,,$$

and this equation *is exact* (as the reader may easily verify by calculating $\partial M/\partial y$ and $\partial N/\partial x$). And of course we have a direct method (see Section 1.6) for solving such an exact equation.

We call the function $1/x^2$ in the last paragraph an *integrating factor*. It is obviously a matter of some interest to be able to find an integrating factor for any given first-order equation. So, given a differential equation

$$M(x,y)\,dx + N(x,y)\,dy = 0\,,$$

we wish to find a function $\mu(x,y)$ such that

$$\mu(x,y) \cdot M(x,y)\,dx + \mu(x,y) \cdot N(x,y)\,dy = 0$$

is exact. This entails

$$\frac{\partial(\mu \cdot M)}{\partial y} = \frac{\partial(\mu \cdot N)}{\partial x}\,.$$

Writing this condition out, we find that

$$\mu\frac{\partial M}{\partial y} + M\frac{\partial \mu}{\partial y} = \mu\frac{\partial N}{\partial x} + N\frac{\partial \mu}{\partial x}\,.$$

This last equation may be rewritten as

$$\frac{1}{\mu}\left(N\frac{\partial \mu}{\partial x} - M\frac{\partial \mu}{\partial y}\right) = \frac{\partial M}{\partial y} - \frac{\partial N}{\partial x}\,.$$

Now we use the method of wishful thinking: we suppose not only that an integrating factor μ exists, but in fact that one exists that only depends on the variable x (and not at all on y). Then the last equation reduces to

$$\frac{1}{\mu}\frac{d\mu}{dx} = \frac{\partial M/\partial y - \partial N/\partial x}{N}\,.$$

Notice that the left-hand side of this new equation is a function of x only. Hence so is the right-hand side. Call the right-hand side $g(x)$. Notice that g is something that we can always compute.

Thus

$$\frac{1}{\mu}\frac{d\mu}{dx} = g(x)$$

hence

$$\frac{d(\ln\mu)}{dx} = g(x)$$

or

$$\ln\mu = \int g(x)\,dx\,.$$

We conclude that, in case there is an integrating factor μ that depends on x only, then

$$\mu(x) = e^{\int g(x)\,dx}\,,$$

where

$$g(x) = \frac{\partial M/\partial y - \partial N/\partial x}{N}$$

can always be computed directly from the original differential equation.

Of course the best way to understand a new method like this is to look at some examples. This we now do.

EXAMPLE 1.9.2 Solve the differential equation

$$(xy - 1)\,dx + (x^2 - xy)\,dy = 0\,.$$

Solution: You may plainly check that this equation is not exact. It is also not separable. So we shall seek an integrating factor that depends only on x. Now

$$g(x) = \frac{\partial M/\partial y - \partial N/\partial x}{N} = \frac{[x] - [2x - y]}{x^2 - xy} = \frac{-x + y}{-x(-x + y)} = -\frac{1}{x}\,.$$

This g depends only on x, signaling that the methodology we just developed will actually work.

We set

$$\mu(x) = e^{\int g(x)\,dx} = e^{\int -1/x\,dx} = e^{-\ln x} = \frac{1}{x}\,.$$

This is our integrating factor. We multiply the original differential equation through by $1/x$ to obtain

$$\left(y - \frac{1}{x}\right)dx + (x - y)\,dy = 0\,.$$

The reader may check that *this* equation is certainly exact. We omit the details of solving this exact equation, since that technique was covered in Section 1.6. ∎

Of course the roles of y and x may be reversed in our reasoning for finding an integrating factor. In case the integrating factor μ depends only on y (and not at all on x) then we set

$$h(y) = -\frac{\partial M/\partial y - \partial N/\partial x}{M}$$

and define

$$\mu(y) = e^{\int h(y)\,dy}.$$

EXAMPLE 1.9.3 Solve the differential equation

$$y\,dx + (2x - ye^y)\,dy = 0.$$

Solution: First observe that the equation is not exact as it stands. Second,

$$\frac{\partial M/\partial y - \partial N/\partial x}{N} = \frac{1-2}{2x - ye^y} = \frac{-1}{2x - ye^y}$$

does *not* depend only on x. So instead we look at

$$-\frac{\partial M/\partial y - \partial N/\partial x}{M} = -\frac{1-2}{y} = \frac{1}{y},$$

and this expression depends only on y. So it will be our $h(y)$. We set

$$\mu(y) = e^{\int h(y)\,dy} = e^{\int 1/y\,dy} = y.$$

Multiplying the differential equation through by $\mu(y) = y$, we obtain the new equation

$$y^2\,dx + (2xy - y^2 e^y)\,dy = 0.$$

You may easily check that this new equation is exact, and then solve it by the method of Section 1.6. ∎

We conclude this section by noting that the differential equation

$$xy^3\,dx + yx^2\,dy = 0$$

has the properties that

- it is not exact;

- $\dfrac{\partial M/\partial y - \partial N/\partial x}{N}$ does not depend on x only;

- $-\dfrac{\partial M/\partial y - \partial N/\partial x}{M}$ does not depend on y only.

Thus the method of the present section is not a panacea. We shall not always be able to find an integrating factor. Still, the technique has its uses.

Exercises

1. Solve each of the following differential equations by finding an integrating factor.

   (a) $(3x^2 - y^2)\,dy - 2xy\,dx = 0$

   (b) $(xy - 1)\,dx + (x^2 - xy)\,dy = 0$

   (c) $x\,dy + y\,dx + 3x^3y^4\,dy = 0$

   (d) $e^x\,dx + (e^x \cot y + 2y \csc y)\,dy = 0$

   (e) $(x + 2)\sin y\,dx + x\cos y\,dy = 0$

   (f) $y\,dx + (x - 2x^2y^3)\,dy = 0$

   (g) $(x + 3y^2)\,dx + 2xy\,dy = 0$

   (h) $y\,dx + (2x - ye^y)\,dy = 0$

   (i) $(y \ln y - 2xy)\,dx + (x + y)\,dy = 0$

   (j) $(y^2 + xy + 1)\,dx + (x^2 + xy + 1)\,dy = 0$

   (k) $(x^3 + xy^3)\,dx + 3y^2\,dy = 0$

2. Show that if $(\partial M/\partial y - \partial N/\partial x)/(Ny - Mx)$ is a function $g(z)$ of the product $z = xy$, then

$$\mu = e^{\int g(z)\,dz}$$

   is an integrating factor for the differential equation

$$M(x, y)\,dx + N(x, y)\,dy = 0\,.$$

3. Under what circumstances will the differential equation

$$M(x, y)\,dx + N(x, y)\,dy = 0$$

   have an integrating factor that is a function of the sum $z = x + y$?

4. Solve the following differential equation by making the substitution $z = y/x^n$ (equivalently $y = xz^n$) and choosing a convenient value for n:

$$\frac{dy}{dx} = \frac{2y}{x} + \frac{x^3}{y} + x\tan\frac{y}{x^2}\,.$$

5. Use your symbol manipulation software, such as Maple or Mathematica, to write a routine for finding the integrating factor for a given differential equation.

1.10 Reduction of Order

It is a fact that virtually *any* ordinary differential equation can be transformed to a first-order *system* of equations. This is, in effect, just a notational trick, but it emphasizes the centrality of first-order equations and systems. In the present section, we shall learn how to reduce certain higher-order equations to first-order equations—ones which we can frequently solve.

We begin by concentrating on differential equations of order 2 (and thinking about how to reduce them to equations of order 1). In each differential equation in this section, x will be the independent variable and y the dependent variable. So a typical second-order equation will involve x, y, y', y''. The key to the success of each of the methods that we shall introduce in this section is that one of these four variables must be missing from the equation.

1.10.1   Dependent Variable Missing

In case the variable y is missing from our differential equation, we make the substitution $y' = p$. This entails $y'' = p'$. Thus the differential equation is reduced to first order.

EXAMPLE 1.10.1 Solve the differential equation

$$xy'' - y' = 3x^2$$

using reduction of order.

Solution: Notice that the dependent variable y is missing from the differential equation. We set $y' = p$ and $y'' = p'$, so that the equation becomes

$$xp' - p = 3x^2 \,.$$

Observe that this new equation is first-order linear. We think of x as the independent variable and p as the new dependent variable.

We write the equation in standard form as

$$p' - \frac{1}{x}p = 3x \,.$$

We may solve this equation by using the integrating factor $\mu(x) = e^{\int -1/x\,dx} = 1/x$. Thus

$$\frac{1}{x}p' - \frac{1}{x^2}p = 3$$

so

$$\left(\frac{1}{x}p\right)' = 3$$

or

$$\int \left(\frac{1}{x}p\right)' dx = \int 3\,dx \,.$$

Performing the integrations, we conclude that

$$\frac{1}{x}p = 3x + C \,,$$

hence

$$p(x) = 3x^2 + Cx.$$

Now we recall that $p = y'$, so we make that substitution. The result is

$$y' = 3x^2 + Cx,$$

hence

$$y = x^3 + \frac{C}{2}x^2 + D = x^3 + Ex^2 + D.$$

We invite the reader to confirm that this is the complete and general solution to the original differential equation. (Note that, since the original equation is second order, there are two undetermined constants in the solution.) ∎

EXAMPLE 1.10.2 Find the solution of the differential equation

$$[y']^2 = x^2 y''.$$

Solution: We note that y is missing, so we make the substitution $p = y'$, $p' = y''$. Thus the equation becomes

$$p^2 = x^2 p'$$

or

$$p^2 = x^2 \cdot \frac{dp}{dx}.$$

This equation is amenable to separation of variables.
The result is

$$\frac{dx}{x^2} = \frac{dp}{p^2},$$

which integrates to

$$-\frac{1}{x} = -\frac{1}{p} + E$$

or

$$p = \frac{x}{1 + Ex}$$

for some unknown constant E. We resubstitute $p = y'$ and write the equation as

$$\frac{dy}{dx} = \frac{x}{1 + Ex} = \frac{1/E + x}{1 + Ex} - \frac{1}{E} \cdot \frac{1}{1 + Ex} = \frac{1}{E} - \frac{1}{E} \cdot \frac{1}{1 + Ex}.$$

Now we integrate to obtain finally that

$$y(x) = \frac{x}{E} - \frac{1}{E^2} \ln(1 + Ex) + D$$

is the general solution of the original differential equation.
Note here that we have used our method of reduction of order to solve a nonlinear differential equation of second order. ∎

1.10.2   Independent Variable Missing

In case the variable x is missing from our differential equation, we make the substitution $y' = p$. This time the corresponding substitution for y'' will be a bit different. To wit,

$$y'' = \frac{dp}{dx} = \frac{dp}{dy}\frac{dy}{dx} = \frac{dp}{dy} \cdot p.$$

This change of variable will reduce our differential equation to first order. In the reduced equation, we treat p as the dependent variable and y as the independent variable.

EXAMPLE 1.10.3 Solve the differential equation

$$y'' + k^2 y = 0$$

(where it is understood that k is an unknown real constant).

Solution: We notice that the independent variable x is missing. So we make the substitution

$$y' = p, \quad y'' = p \cdot \frac{dp}{dy}.$$

The equation then becomes

$$p \cdot \frac{dp}{dy} + k^2 y = 0.$$

In this new equation we can separate variables:

$$p \, dp = -k^2 y \, dy$$

hence, integrating,

$$\frac{p^2}{2} = -k^2 \frac{y^2}{2} + C,$$

so that

$$p = \pm\sqrt{D - k^2 y^2} = \pm k\sqrt{E - y^2}.$$

Now we resubstitute $p = dy/dx$ to obtain

$$\frac{dy}{dx} = \pm k\sqrt{E - y^2}.$$

We can separate variables to obtain

$$\frac{dy}{\sqrt{E - y^2}} = \pm k \, dx$$

hence (integrating)

$$\sin^{-1}\frac{y}{\sqrt{E}} = \pm kx + F$$

or

$$\frac{y}{\sqrt{E}} = \sin(\pm kx + F)$$

thus

$$y = \sqrt{E}\sin(\pm kx + F).$$

Now we apply the sum formula for sine to rewrite the last expression as

$$y = \sqrt{E}\sin(\pm kx)\cos F + \sqrt{E}\cos(\pm kx)\sin F.$$

A moment's thought reveals that we may consolidate the constants and finally write our general solution of the differential equation as

$$y = A\sin(kx) + B\cos(kx).$$ ∎

We shall learn in the next chapter a different, and perhaps more expeditious, method of attacking examples of the last type. It should be noted quite plainly in the last example, and also in some of the earlier examples of the section, that the method of reduction of order basically transforms the problem of solving one second-order equation to a new problem of solving *two* first-order equations. Examine each of the examples we have presented and see whether you can say what the two new equations are.

In the next example, we shall solve a differential equation subject to an *initial condition*. This will be an important idea throughout the book. Solving a differential equation gives rise to a *family* of functions. Specifying an initial condition is a natural way to specialize down to a particular solution. In applications, these initial conditions will make good physical sense.

EXAMPLE 1.10.4 Use the method of reduction of order to solve the differential equation

$$y'' = y' \cdot e^y$$

with initial conditions $y(0) = 0$ and $y'(0) = 1$.

Solution: Noting that the dependent variable x is missing, we make the substitution

$$y' = p, \quad y'' = p \cdot \frac{dp}{dy}.$$

So the equation becomes

$$p \cdot \frac{dp}{dy} = p \cdot e^y.$$

We of course may separate variables, so the equation becomes

$$dp = e^y\, dy.$$

This is easily integrated to give

$$p = e^y + C.$$

Now we resubstitute $p = y'$ to find that

$$y' = e^y + C$$

or

$$\frac{dy}{dx} = e^y + C.$$

Because of the initial conditions $[dy/dx](0) = 1$ and $y(0) = 0$, we may conclude right away that

$$1 = e^0 + C$$

hence that $C = 0$. Thus our equation is

$$\frac{dy}{dx} = e^y$$

or

$$\frac{dy}{e^y} = dx.$$

This may be integrated to

$$-e^{-y} = x + D.$$

Of course we can rewrite the equation finally as

$$y = -\ln(-x + E).$$

Since $y(0) = 0$, we conclude that

$$y(x) = -\ln(-x + 1)$$

is the solution of our initial value problem.   ∎

Exercises

1. Solve the following differential equations using the method of reduction of order.

    (a) $yy'' + (y')^2 = 0$
    (b) $xy'' = y' + (y')^3$
    (c) $y'' - k^2 y = 0$
    (d) $x^2 y'' = y' + (y')^2$

    (e) $2yy'' = 1 + (y')^2$
    (f) $yy'' - (y')^2 = 0$
    (g) $xy'' + y' = 4x$

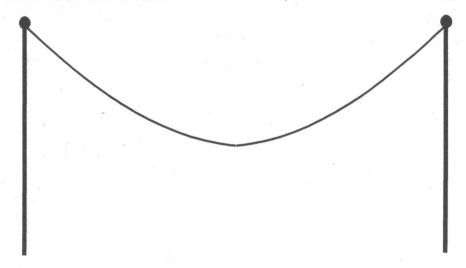

FIGURE 1.8
The hanging chain.

2. Find the specified particular solution of each of the following equations.
   (a) $(x^2 + 2y')y'' + 2xy' = 0$, $y = 1$ and $y' = 0$ when $x = 0$
   (b) $yy'' = y^2 y' + (y')^2$, $y = -1/2$ and $y' = 1$ when $x = 0$
   (c) $y'' = y' e^y$, $y = 0$ and $y' = 2$ when $x = 0$
3. Solve each of these differential equations using both methods of this section, and reconcile the results.

$$\text{(a)} \quad y'' = 1 + (y')^2 \qquad\qquad \text{(b)} \quad y'' + (y')^2 = 1$$

4. Inside the Earth, the force of gravity is proportional to the distance from the center. A hole is drilled through the Earth from pole to pole and a rock is dropped into the hole. This rock will fall all the way through the hole, pause at the other end, and return to its starting point. How long will the complete round trip take?

1.11   The Hanging Chain and Pursuit Curves

1.11.1   The Hanging Chain

Imagine a flexible steel chain, attached firmly at equal height at both ends, hanging under its own weight (see Figure 1.8). What shape will it describe as it hangs?

   This is a classical problem of mechanical engineering, and its analytical

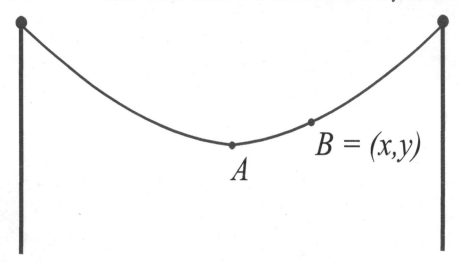

FIGURE 1.9
Analysis of the hanging chain.

solution involves calculus, elementary physics, and differential equations. We describe it here.

We analyze a portion of the chain between points A and B, as shown in Figure 1.9, where A is the lowest point of the chain and $B = (x, y)$ is a variable point.

We let

- T_1 be the horizontal tension at A;

- T_2 be the component of tension *tangent* to the chain at B;

- w be the weight of the chain per unit of length.

Here T_1, T_2, w are numbers. Figure 1.10 exhibits these quantities.

Notice that, if s is the length of the chain between two given points, then ws is the downward force of gravity on this portion of the chain; this is indicated in the figure. We use the symbol θ to denote the angle that the tangent to the chain at B makes with the horizontal.

By Newton's first law we may equate horizontal components of force to obtain

$$T_1 = T_2 \cos \theta. \qquad (1.11.1)$$

Likewise, we equate vertical components of force to obtain

$$ws = T_2 \sin \theta. \qquad (1.11.2)$$

Dividing the right side of (1.11.2) by the right side of (1.11.1) and the left side

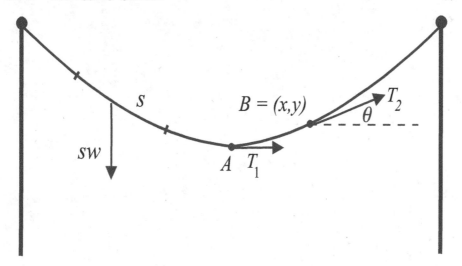

FIGURE 1.10
The quantities T_1 and T_2.

of (1.11.2) by the left side of (1.11.1) and equating gives

$$\frac{ws}{T_1} = \tan\theta\,.$$

Think of the hanging chain as the graph of a function: y is a function of x. Then y' at B equals $\tan\theta$ so we may rewrite the last equation as

$$y' = \frac{ws}{T_1}\,.$$

We can simplify this equation by a change of notation: set $q = y'$. Then we have

$$q(x) = \frac{w}{T_1}s(x)\,. \tag{1.11.3}$$

If Δx is an increment of x then $\Delta q = q(x+\Delta x)-q(x)$ is the corresponding increment of q and $\Delta s = s(x+\Delta x)-s(x)$ the increment in s. As Figure 1.11 indicates, Δs is well approximated by

$$\Delta s \approx \left((\Delta x)^2 + (y'\Delta x)^2\right)^{1/2} = \left(1+(y')^2\right)^{1/2}\Delta x = (1+q^2)^{1/2}\Delta x\,.$$

Thus, from (1.11.3), we have

$$\Delta q = \frac{w}{T_1}\Delta s \approx \frac{w}{T_1}(1+q^2)^{1/2}\Delta x\,.$$

Dividing by Δx and letting Δx tend to zero gives the equation

$$\frac{dq}{dx} = \frac{w}{T_1}(1+q^2)^{1/2}\,. \tag{1.11.4}$$

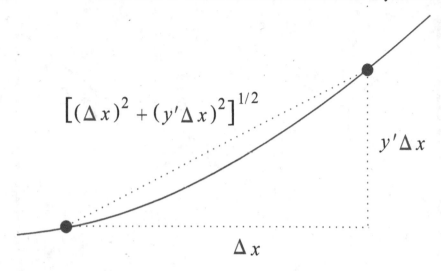

FIGURE 1.11
An increment of the chain.

This may be rewritten as

$$\int \frac{dq}{(1+q^2)^{1/2}} = \frac{w}{T_1} \int dx.$$

It is trivial to perform the integration on the right side of the equation, and a little extra effort enables us to integrate the left side (use the substitution $u = \tan \psi$, or else use inverse hyperbolic trigonometric functions). Thus we obtain

$$\sinh^{-1} q = \frac{w}{T_1}x + C.$$

We know that the chain has a horizontal tangent when $x = 0$ (this corresponds to the point A—Figure 1.10). Thus $q(0) = y'(0) = 0$. Substituting this into the last equation gives $C = 0$. Thus our solution is

$$\sinh^{-1} q(x) = \frac{w}{T_1}x$$

or

$$q(x) = \sinh\left(\frac{w}{T_1}x\right)$$

or

$$\frac{dy}{dx} = \sinh\left(\frac{w}{T_1}x\right).$$

Finally, we integrate this last equation to obtain

$$y(x) = \frac{T_1}{w}\cosh\left(\frac{w}{T_1}x\right) + D,$$

where D is a constant of integration. The constant D can be determined from the height h_0 of the point A from the x-axis:

$$h_0 = y(0) = \frac{T_1}{w} \cosh(0) + D$$

hence

$$D = h_0 - \frac{T_1}{w}.$$

Our hanging chain is thus completely described by the equation

$$y(x) = \frac{T_1}{w} \cosh\left(\frac{w}{T_1}x\right) + \left(h_0 - \frac{T_1}{w}\right).$$

This curve is called a *catenary*, from the Latin word for chain (*catena*). Catenaries arise in a number of other physical problems. The St. Louis arch is in the shape of a catenary.

Math Nugget

Many of the special curves of classical mathematics arise in problems of mechanics. The *tautochrone property* of the cycloid curve (that a bead sliding down the curve will reach the bottom in the same time, no matter where on the curve it begins) was discovered by the great Dutch scientist Christiaan Huygens (1629–1695). He published it in 1673 in his treatise on the theory of pendulum clocks, and it was well known to all European mathematicians at the end of the seventeenth century. When Johann Bernoulli published his discovery of the *brachistochrone* (that special curve connecting two points down which a bead will slide in the least possible time) in 1696, he expressed himself in the following exuberant language (of course, as was the custom of the time, he wrote in Latin): "With justice we admire Huygens because he first discovered that a heavy particle falls down along a common cycloid in the same time no matter from what point on the cycloid it begins its motion. But you will be petrified with astonishment when I say that precisely this cycloid, the *tautochrone* of Huygens, is our required brachistochrone."

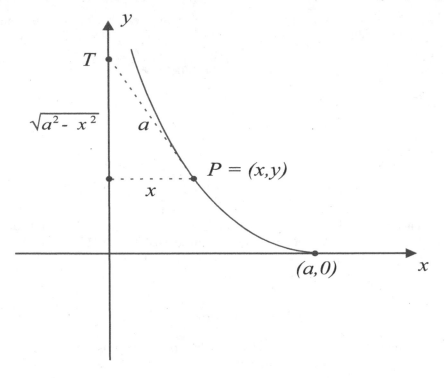

FIGURE 1.12
A tractrix.

1.11.2   Pursuit Curves

A submarine speeds across the ocean bottom in a particular path, and a destroyer at a remote location decides to engage in pursuit. What path does the destroyer follow? Problems of this type are of interest in a variety of applications. We examine a few examples. The first one is purely mathematical, and devoid of "real-world" trappings.

EXAMPLE 1.11.1 A point P is dragged along the x-y plane by a string PT of fixed length a. If T begins at the origin and moves along the positive y-axis, and if P starts at the point $(a, 0)$, then what is the path of P?

Solution: The curve described by the motion of P is called, in the classical literature, a *tractrix* (from the Latin *tractum*, meaning "drag"). Figure 1.12 exhibits the salient features of the problem.

Observe that we can calculate the slope of the pursuit curve at the point P in two ways: **(i)** as the derivative of y with respect to x and **(ii)** as the ratio of sides of the relevant triangle. This leads to the equation

$$\frac{dy}{dx} = -\frac{\sqrt{a^2 - x^2}}{x}.$$

This is a separable, first-order differential equation. We write

$$\int dy = - \int \frac{\sqrt{a^2 - x^2}}{x}\, dx\,.$$

Performing the integrations (the right-hand side requires the trigonometric substitution $x = a \sin \psi$), we find that

$$y = a \ln \left(\frac{a + \sqrt{a^2 - x^2}}{x} \right) - \sqrt{a^2 - x^2} + C$$

is the equation of the tractrix.[2]   ■

EXAMPLE 1.11.2 A rabbit begins at the origin and runs up the y-axis with speed a feet per second. At the same time, a dog runs at speed b from the point $(c, 0)$ in pursuit of the rabbit. What is the path of the dog?

Solution: At time t, measured from the instant both the rabbit and the dog start, the rabbit will be at the point $R = (0, at)$ and the dog at $D = (x, y)$. We wish to solve for y as a function of x. Refer to Figure 1.13.

The premise of a pursuit analysis is that the line through D and R is tangent to the path—that is, the dog will always run straight at the rabbit. This immediately gives the differential equation

$$\frac{dy}{dx} = \frac{y - at}{x}\,.$$

This equation is a bit unusual for us, since x and y are both unknown functions of t. First, we rewrite the equation as

$$xy' - y = -at\,.$$

(Here the $'$ on y stands for differentiation in x.)

We differentiate this equation with respect to x, which gives

$$xy'' = -a\frac{dt}{dx}\,.$$

Since s is arc length along the path of the dog, it follows that $ds/dt = b$. Hence

$$\frac{dt}{dx} = \frac{dt}{ds} \cdot \frac{ds}{dx} = -\frac{1}{b} \cdot \sqrt{1 + (y')^2}\,;$$

---

[2]This curve is of considerable interest in other parts of mathematics. If it is rotated about the y-axis then the result is a surface that gives a model for non-Euclidean geometry. The surface is called a *pseudosphere* in differential geometry. It is a surface of constant negative curvature (as opposed to a traditional sphere, which is a surface of constant positive curvature).

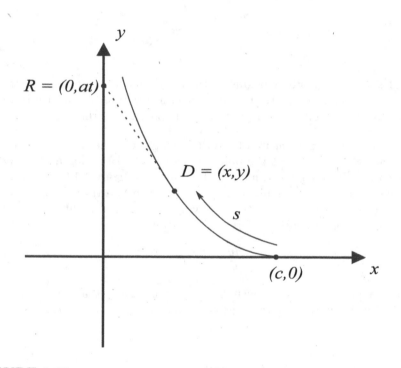

FIGURE 1.13
A pursuit curve.

here the minus sign appears because s decreases when x increases (see Figure 1.13). Of course we use the familiar expression for the derivative of arc length.

Combining the last two displayed equations gives

$$xy'' = \frac{a}{b}\sqrt{1 + (y')^2}\,.$$

For convenience, we set $k = a/b$, $y' = p$, and $y'' = dp/dx$ (the latter two substitutions being one of our standard reduction of order techniques). Thus we have

$$\frac{dp}{\sqrt{1 + p^2}} = k\frac{dx}{x}\,.$$

Now we may integrate, using the condition $p = 0$ when $x = c$. The result is

$$\ln\left(p + \sqrt{1 + p^2}\right) = \ln\left(\frac{x}{c}\right)^k.$$

When we solve for p, we find that

$$\frac{dy}{dx} = p = \frac{1}{2}\left\{\left(\frac{x}{c}\right)^k - \left(\frac{c}{x}\right)^k\right\}.$$

In order to continue the analysis, we need to know something about the relative sizes of a and b. Suppose, for example, that $a < b$ (so $k < 1$), meaning that the dog will certainly catch the rabbit. Then we can integrate the last equation to obtain

$$y(x) = \frac{1}{2}\left\{\frac{c}{k+1}\left(\frac{x}{c}\right)^{k+1} - \frac{c}{(1-k)}\left(\frac{c}{x}\right)^{k-1}\right\} + D.$$

Since $y = 0$ when $x = c$, we find that $D = ck$. Of course the dog catches the rabbit when $x = 0$. Since both exponents on x are positive, we can set $x = 0$ and solve for y to obtain $y = ck$ as the point at which the dog and the rabbit meet.   ∎

We invite the reader to consider what happens in this last example when $a = b$ and hence $k = 1$.

Exercises

1. Refer to our discussion of the shape of a hanging chain. Show that the tension T at an arbitrary point (x, y) on the chain is given by wy.

2. If the hanging chain supports a load of horizontal density $L(x)$, then what differential equation should be used in place of (1.11.4)?

3. What is the shape of a cable of negligible density (so that $w \equiv 0$) that supports a bridge of constant horizontal density given by $L(x) \equiv L_0$?

4. If the length of any small portion of an elastic cable of uniform density is proportional to the tension in it, then show that it assumes the shape of a parabola when hanging under its own weight.

5. A curtain is made by hanging thin rods from a cord of negligible density. If the rods are close together and equally spaced horizontally, and if the bottom of the curtain is trimmed so that it is horizontal, then what is the shape of the cord?

6. What curve lying above the x-axis has the property that the length of the arc joining any two points on it is proportional to the area under that arc?

7. Show that the tractrix discussed in Example 1.11.1 is orthogonal to the lower half of each circle with radius a and center on the positive y-axis.

8. (a) In Example 1.11.2, assume that $a < b$ (so that $k < 1$) and find y as a function of x. How far does the rabbit run before the dog catches him?

(b) Assume now that $a = b$, and find y as a function of x. How close does the dog come to the rabbit?

1.12   Electrical Circuits

We have alluded elsewhere in the book to the fact that our analyses of vibrating springs and other mechanical phenomena are analogous to the situation for electrical circuits. Now we shall examine this matter in some detail.

We consider the flow of electricity in the simple electrical circuit exhibited in Figure 1.14. The elements that we wish to note are these:

A. A source of electromotive force (emf) E—perhaps a battery or generator—which drives an electric charge and produces a current I. Depending on the nature of the source, E may be a constant or a function of time.

B. A resistor of resistance R, which opposes the current by producing a drop in emf of magnitude

$$E_R = RI.$$

This equation is called *Ohm's law*.

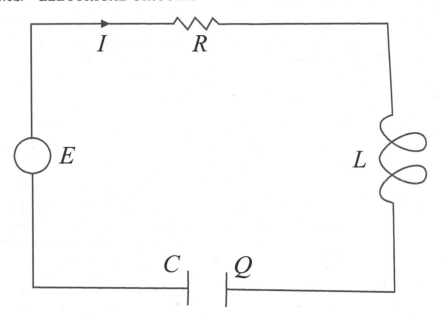

FIGURE 1.14
A simple electric circuit.

Math Nugget

Georg Simon Ohm (1787–1854) was a German physicist whose only significant contribution to science was his discovery of the law that now bears his name. When he announced it in 1827, it seemed too good to be true; sadly, it was not generally believed. Ohm was, as a consequence, deemed to be unreliable. He was subsequently so badly treated that he resigned his professorship at Cologne and lived for several years in obscurity and poverty. Ultimately, it was recognized that Ohm was right all along. So Ohm was vindicated. One of Ohm's students in Cologne was Peter Dirichlet, who later became one of the most distinguished German mathematicians of the nineteenth century.

C. An inductor of inductance L, which opposes any change in the current by producing a drop in emf of magnitude

$$E_L = L \cdot \frac{dI}{dt}.$$

D. A capacitor (or condenser) of capacitance C, which stores the charge Q. The charge accumulated by the capacitor resists the inflow of additional charge, and the drop in emf arising in this way is

$$E_C = \frac{1}{C} \cdot Q.$$

Furthermore, since the current is the rate of flow of charge, and hence the rate at which charge builds up on the capacitor, we have

$$I = \frac{dQ}{dt}.$$

Those unfamiliar with the theory of electricity may find it helpful to draw an analogy here between the current I and the rate of flow of water in a pipe. The electromotive force E plays the role of a pump producing pressure (voltage) that causes the water to flow. The resistance R is analogous to friction in the pipe—which opposes the flow by producing a drop in the pressure. The inductance L is a sort of inertia that opposes any change in flow by producing a drop in pressure if the flow is increasing and an increase in pressure if the flow is decreasing. To understand this last point, think of a cylindrical water storage tank that the liquid enters through a hole in the bottom. The deeper the water in the tank (Q), the harder it is to pump new water in; and the larger the base of the tank (C) for a given quantity of stored water, the shallower is the water in the tank and the easier to pump in new water.

These four circuit elements act together according to *Kirchhoff's Law*, which states that the algebraic sum of the electromotive forces around a closed circuit is zero. This physical principle yields

$$E - E_R - E_L - E_C = 0$$

or

$$E - RI - L\frac{dI}{dt} - \frac{1}{C}Q = 0,$$

which we rewrite in the form

$$L\frac{dI}{dt} + RI + \frac{1}{C}Q = E. \tag{1.12.1}$$

We may perform our analysis by regarding either the current I or the charge Q as the dependent variable (obviously time t will be the independent variable).

- In the first instance, we shall eliminate the variable Q from (1.12.1) by differentiating the equation with respect to t and replacing dQ/dt by I (since current is indeed the rate of change of charge). The result is

$$L\frac{d^2I}{dt^2} + R\frac{dI}{dt} + \frac{1}{C}I = \frac{dE}{dt}.$$

- In the second instance, we shall eliminate the I by replacing it by dQ/dt. The result is

$$L\frac{d^2Q}{dt^2} + R\frac{dQ}{dt} + \frac{1}{C}Q = E. \qquad (1.12.2)$$

Both these ordinary differential equations are second-order, linear with constant coefficients. We shall study these in detail in Section 2.1. For now, in order to use the techniques we have already learned, we assume that our system has no capacitor present. Then, integrating, the equation becomes

$$L\frac{dI}{dt} + RI = E. \qquad (1.12.3)$$

EXAMPLE 1.12.1 Solve equation (1.12.3) when an initial current I_0 is flowing and a constant emf E_0 is impressed on the circuit at time $t = 0$.

Solution: For $t \geq 0$ our equation is

$$L\frac{dI}{dt} + RI = E_0.$$

We can separate variables to obtain

$$\frac{dI}{E_0 - RI} = \frac{1}{L}\, dt.$$

We integrate and use the initial condition $I(0) = I_0$ to obtain

$$\ln(E_0 - RI) = -\frac{R}{L}t + \ln(E_0 - RI_0),$$

hence

$$I = \frac{E_0}{R} + \left(I_0 - \frac{E_0}{R}\right)e^{-Rt/L}.$$

We have learned that the current I consists of a *steady-state* component E_0/R and a *transient component* $(I_0 - E_0/R)e^{-Rt/L}$ that approaches zero as $t \to +\infty$. Consequently, Ohm's law $E_0 = RI$ is nearly true for t large. We also note that, if $I_0 = 0$, then

$$I = \frac{E_0}{R}(1 - e^{-Rt/L});$$

if instead $E_0 = 0$, then $I = I_0 e^{-Rt/L}$.

■

Exercises

1. In Example 1.12.1, with $I_0 = 0$ and $E_0 \neq 0$, show that the current in the circuit builds up to half its theoretical maximum in $(L \ln 2)/R$ seconds.
2. Solve equation (1.12.3) for the case in which the circuit has an initial current I_0 and the emf impressed at time $t = 0$ is given by
    (a) $E = E_0 e^{-kt}$
    (b) $E = E_0 \sin \omega t$
3. Consider a circuit described by equation (1.12.3) and show that
    (a) Ohm's law is satisfied whenever the current is at a maximum or minimum.
    (b) The emf is increasing when the current is at a minimum and decreasing when it is at a maximum.
4. If $L = 0$ in equation (1.12.2) and if $Q = 0$ when $t = 0$, then find the charge buildup $Q = Q(t)$ on the capacitor in each of the following cases:
    (a) E is a constant E_0
    (b) $E = E_0 e^{-t}$
    (c) $E = E_0 \cos \omega t$
5. Use equation (1.12.1) with $R = 0$ and $E = 0$ to find $Q = Q(t)$ and $I = I(t)$ for the discharge of a capacitor through an inductor of inductance L, with initial conditions $Q = Q_0$ and $I = 0$ when $t = 0$.

1.13   The Design of a Dialysis Machine

The purpose of the kidneys is to filter out waste from the blood. When the kidneys malfunction, the waste material can build up to dangerous levels and be poisonous to the system. Doctors will use a kidney dialysis machine (or *dialyzer*) to assist the kidneys in the cleansing process.

How does the dialyzer work? Blood flows from the patient's body into the machine. There is a cleansing fluid, called the dialyzate, that flows through the machine in the opposite direction to the blood. The blood and the dialyzate are separated by a semi-permeable membrane. See Figure 1.15.

The membrane in the dialyzer has minute pores which will not allow the passage of blood but *will* allow the passage of the waste matter (which has much smaller molecules). The design of the dialysis machine concerns the flow rate of the waste material through the membrane. That flow is determined by the differences in concentration (of the waste material) on either side of the membrane. Of course the flow is from high concentration to low concentration. Refer again to the figure.

Of course the physician and the patient care about the rate at which the

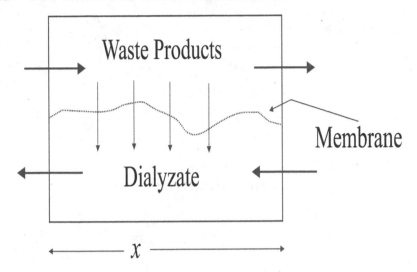

FIGURE 1.15
A dialysis machine.

waste material is removed. This rate will depend on **(i)** the flow rate of blood through the dialyzer, **(ii)** the flow rate of dialyzate through the dialyzer, **(iii)** the capacity of the dialyzer, and **(iv)** the permeability of the membrane.

It is convenient (and plausible) for our analysis here to take the capacity of the dialyzer and the permeability of the membrane to be fixed constants. Our analysis will focus on the dependence of the removal rate (of the waste) on the flow rate. We let x denote the horizontal position in the dialyzer (Figure 1.15). Our analysis centers on a small cross section of the dialyzer from position x to position to $x + \triangle x$. We refer to the cross section of the total flow pictured in Figure 1.16 as an "element" of the flow.

Clearly, from everything we have said so far, the most important variables for our analysis are the concentration of waste in the blood (call this quantity $p(x)$) and the concentration of waste in the dialyzate (call this quantity $q(x)$). There is in fact a standard physical law governing the passage of waste material through the membrane. This is *Fick's Law*. The enunciation is:

The amount of material passing through the membrane is proportional to the difference in concentration.

Let us examine Figure 1.16 in order to understand the movement of concentration of waste. The difference in concentration across $\alpha\epsilon$ (as one moves from the upper half of the figure to the lower half) is $p(x) - q(x)$; therefore the transfer of waste mass through a section of the membrane of width 1 and length $\triangle x$ from blood solution to dialyzate solution in unit time is approximately

$$k[p(x) - q(x)] \cdot \triangle x.$$

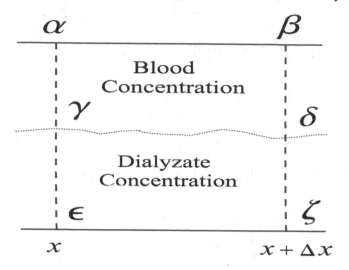

FIGURE 1.16
Cross section of the flow.

The constant of proportionality k is independent of x. Now we must consider the mass change in the "element" $\alpha\beta\zeta\epsilon$ in unit time. Now

$$\begin{matrix} \text{mass flow across } \alpha\gamma \\ \text{into element} \end{matrix} = \begin{matrix} \text{mass passing through} \\ \text{membrane } \gamma\delta \end{matrix} + \begin{matrix} \text{mass flow across } \beta\delta \\ \text{out of element.} \end{matrix}$$

Let F_B denote the constant rate of flow of blood through the dialyzer. Then we may express this relationship in mathematical language as

$$F_B \cdot p(x) = k[p(x) - q(x)]\,\triangle\,x + F_B \cdot p(x + \triangle x).$$

Rearranging this equation, we have

$$F_B \cdot \left(\frac{p(x + \triangle x) - p(x)}{\triangle x}\right) = -k[p(x) - q(x)].$$

Now it is natural to let $\triangle x \to 0$ to obtain the differential equation

$$F_B \frac{dp}{dx} = -k(p - q). \tag{1.12.4}$$

This last analysis, which led to equation (1.12.4), was based on an examination of the flow of the blood. We may perform a similar study of the flow of the dialyzate to obtain

$$-F_D \frac{dq}{dx} = k(p - q) \tag{1.12.5}$$

(note that the presence of the minus sign comes from the fact that the blood flows in the opposite direction as the dialyzate). Of course F_D is the flow rate of dialyzate through the machine.

Now we add equations (1.12.4) and (1.12.5) to obtain

$$\frac{dp}{dx} - \frac{dq}{dz} = -\frac{k}{F_B}(p-q) + \frac{k}{F_D}(p-q).$$

Notice that p and q occur in this equation in an antisymmetric manner. Thus it is advantageous to make the substitution $r = p - q$. We finally obtain

$$\frac{dr}{dx} = -\alpha r, \tag{1.12.6}$$

where $\alpha = k/F_B - k/F_D$.

This equation is easily solved with separation of variables. The result is

$$r(x) = Ae^{-\alpha x}, \tag{1.12.7}$$

where A is an arbitrary constant. Of course we wish to relate this solution to p and to q. Look again at equation (1.12.4). We see that

$$\frac{dp}{dx} = -\frac{k}{F_B}r = -\frac{k}{F_B}Ae^{-\alpha x}.$$

Integration yields

$$p = B + \frac{kA}{\alpha F_B}e^{-\alpha x}, \tag{1.12.8}$$

where B is an arbitrary constant. We now combine (1.12.7) and (1.12.8), recalling that $r = p - q = Ae^{-\alpha x}$, to obtain that

$$q = B + \frac{kA}{\alpha F_D}e^{-\alpha x}.$$

Finally, we must consider the initial conditions in the problem. We suppose that the blood has initial waste concentration p_0 and the dialyzate has initial waste concentration 0. Thus

$$p = p_0 \quad \text{at } x = 0$$
$$q = 0 \quad \text{at } x = L.$$

Here L is the length of the dialyzer machine. Some tedious algebra, applied in a by-now-familiar manner, finally tells us that

$$p(x) = p_0 \left(\frac{(e^{-\alpha L}/F_D) - (e^{-\alpha x}/F_B)}{(e^{-\alpha L}/F_D) - (1/F_B)}\right)$$

$$q(x) = \frac{p_0}{F_D}\left(\frac{e^{-\alpha L} - e^{-\alpha x}}{(e^{-\alpha L}/F_D) - (1/F_D)}\right).$$

These two equations represent a definitive analysis of the concentrations

of waste in the blood and in the dialyzate. To interpret these, we observe that
the amount of waste removed from the blood in unit time is

$$\int_0^L k[p(x) - q(x)]\, dx = -F_B \int_0^L \frac{dp}{dx}\, dx \qquad \text{by (1.12.4)}$$

$$= -F_B \int_{p_0}^{p(L)} dp$$

$$= -F_B[p_0 - p(L)].$$

Those who design dialyzers focus their attention on the "clearance" Cl,
which is defined to be

$$Cl = \frac{F_B}{p_0}[p_0 - p(L)].$$

Our equations for p and q yield, after some calculation, that

$$Cl = F_B \left(\frac{1 - e^{-\alpha L}}{1 - (F_B/F_D)e^{-\alpha L}} \right).$$

Here

$$\alpha L = \frac{kL}{F_B}(1 - F_B/F_D).$$

The actual design of a dialyzer would entail testing these theoretical results against experimental data, taking into account factors like the variation
of k with x, the depth of the channels, and variations in the membrane.

Problems for Review and Discovery

A. Drill Exercises

1. Find the general solution to each of the following differential equations.

(a) $xy' + y = x$

(b) $x^2 y' + y = x^2$

(c) $x^2 \dfrac{dy}{dx} = y$

(d) $\sec x \cdot \dfrac{dy}{dx} = \sec y$

(e) $\dfrac{dy}{dx} = \dfrac{x^2 + y^2}{x^2 - y^2}$

(f) $\dfrac{dy}{dx} = \dfrac{x + 2y}{2x - y}$

(g) $2xy\, dx + x^2\, dy = 0$

(h) $-\sin x \sin y\, dx + \cos x \cos y\, dy = 0$

2. Solve each of the following initial value problems.

(a) $xy' - y = 2x$, $y(0) = 1$

(b) $x^2 y' - 2y = 3x^2$, $y(1) = 2$

(c) $y^2 \dfrac{dy}{dx} = x$, $y(-1) = 3$

(d) $\csc x \cdot \dfrac{dy}{dx} = \csc y$, $y(\pi/2) = 1$

(e) $\dfrac{dy}{dx} = \dfrac{x+y}{x-y}$, $y(1) = 1$

(f) $\dfrac{dy}{dx} = \dfrac{x^2 + 2y^2}{x^2 - 2y^2}$, $y(0) = 1$

(g) $2x \cos y \, dx - x^2 \sin y \, dy = 0$, $y(1) = 1$

(h) $\dfrac{1}{y} dx - \dfrac{x}{y^2} dy = 0$, $y(0) = 2$

3. Find the orthogonal trajectories to the family of curves $y = c(x^2 + 1)$.

4. Use the method of reduction of order to solve each of the following differential equations.

(a) $y \cdot y'' - (y')^2 = 0$

(b) $xy'' = y' - 2(y')^3$

(c) $yy'' + y' = 0$

(d) $xy'' - 3y' = 5x$

B. Challenge Problems

1. A tank contains 50 gallons of brine in which 25 pounds of salt are dissolved. Beginning at time $t - 0$, water runs into this tank at the rate of 2 gallons per minute; the mixture flows out at the same rate through a second tank initially containing 50 gallons of pure water. When will the second tank contain the greatest amount of salt?

2. A natural extension of the first-order linear equation

$$y' = p(x) + q(x)y$$

is the *Riccati equation*

$$y' = p(x) + q(x)y + r(x)y^2.$$

In general, this equation cannot be explicitly solved by elementary methods. However, if a particular solution $y_1(x)$ is known, then the general solution has the form

$$y(x) = y_1(x) + z(x),$$

where $z(x)$ is the general solution of the associated Bernoulli equation

$$z' - (q + 2ry_1)z = rz^2.$$

Prove this assertion, and use this set of techniques to find the general solution of the equation

$$y' = \frac{y}{x} + x^3 y^2 - x^5. \qquad (*)$$

[**Hint:** The equation $(*)$ has $y_1(x) = x$ as a particular solution.]

3. The propagation of a single act in a large population (for instance, buy-
ing a Lexus rather than a Cadillac) often depends partly on external
circumstances (e.g., price, quality, and frequency-of-repair records) and
partly on a human tendency to imitate other people who have already
performed the same act. In this case the rate of increase of the propor-
tion $y(t)$ of people who have performed the act can be expressed by the
formula

$$\frac{dy}{dt} = (1 - y)[s(t) + Iy], \qquad\qquad (**)$$

where $s(t)$ measures the external stimulus and I is a constant called the
imitation coefficient.

   (a) Notice that $(**)$ is a Riccati equation (see the last exercise) and that
   $y \equiv 1$ is a particular solution. Use the result of the last exercise to
   find the Bernoulli equation satisfied by $z(t)$.

   (b) Find $y(t)$ for the case in which the external stimulus increases
   steadily with time, so that $s(t) = at$ for a positive constant a. Leave
   your answer in the form of an integral.

4. If Riccati's equation from Exercise 2 above has a known solution $y_1(x)$,
then show that the general solution has the form of the one-parameter
family of curves

$$y = \frac{cf(x) + g(x)}{cF(x) + G(x)} .$$

Show, conversely, that the differential equation of *any* one-parameter family of this form is a Riccati equation.

5. It begins to snow at a steady rate some time in the morning. A snow
plow begins plowing at a steady rate at noon. The plow clears twice as
much area in the first hour as it does in the second hour. When did it
start snowing?

C. Problems for Discussion and Exploration

1. A rabbit starts at the origin and runs up the right branch of the parabola
$y = x^2$ with speed a. At the same time a dog, running with speed b, starts
at the point $(c, 0)$ and pursues the rabbit. Write a differential equation
for the path of the dog.

2. Consider the initial value problem

$$\frac{dy}{dx} = \frac{\sin(xy)}{1 + x^2 + y^2} .$$

This equation cannot be solved by any of the methods presented in this
chapter. However, we can obtain some information about solutions by
using other methods.

   (a) On a large sheet of graph paper draw arrows indicating the direction
   of the curve at a large number of points. For example, at the point
   $(1, 1)$ the equation tells us that

   $$\frac{dy}{dx} = \frac{\sin 1}{3} .$$

Draw a little arrow with base at the point $(1,1)$ indicating that the curve is moving in the indicated direction.

Do the same at many other points. Connect these arrows with "flow lines." (There will be many distinct flow lines, corresponding to different initial conditions.) Thus you obtain a family of curves, representing the different solutions of the differential equation.

(b) What can you say about the nature of the flow lines that you obtained in part (a)? Are they curves that you can recognize? Are they polynomial curves? Exponential curves?

(c) What does your answer to part (b) tell you about this problem?

3. Suppose that the function $F(x, y)$ is continuously differentiable (i.e., continuous with continuous first derivatives). Show that the initial value problem

$$\frac{dy}{dx} = F(x, y), \quad \cdot y(0) = y_0$$

has at most one solution in a neighborhood of the origin.

2

Second-Order Linear Equations

- Second-order linear equations

- The nature of solutions of second-order linear equations

- General solutions

- Undetermined coefficients

- Variation of parameters

- Use of a known solution

- Vibrations and oscillations

- Electrical current

- Newton's law of gravitation

- Kepler's laws

- Higher-order equations

2.1 Second-Order Linear Equations with Constant Coefficients

Second-order linear equations are important because (considering Newton's second law) they arise frequently in engineering and physics. For instance, acceleration is given by the second derivative, and force is mass times acceleration.

In this section we learn about *second-order linear equations with constant coefficients*. The "linear" attribute means, just as it did in the first-order situation, that the unknown function and its derivatives are not multiplied together, are not raised to powers, and are not the arguments of other functions. So, for example,

$$y'' - 3y' + 6y = 0$$

is second-order linear while

$$\sin(y'') - y' + 5y = 0$$

and

$$y \cdot y'' + 4y' + 3y = 0$$

are *not*.

The "constant coefficient" attribute means that the coefficients in the equation are not functions—they are constants. Thus a second-order linear equation with constant coefficient will have the form

$$ay'' + by' + cy = d\,, \qquad\qquad (2.1.1)$$

where a, b, c, d are constants.

We in fact begin with the *homogeneous case*; this is the situation in which $d = 0$. We solve equation (2.1.1) by a process of organized guessing: any solution of (2.1.1) will be a function that cancels with its derivatives. Thus it is a function that is similar in form to its derivatives. Certainly exponentials fit this description. Thus we guess a solution of the form

$$y = e^{rx}\,.$$

Plugging this guess into (2.1.1) gives

$$a\left(e^{rx}\right)'' + b\left(e^{rx}\right)' + c\left(e^{rx}\right) = 0\,.$$

Calculating the derivatives, we find that

$$ar^2 e^{rx} + bre^{rx} + ce^{rx} = 0$$

or

$$[ar^2 + br + c] \cdot e^{rx} = 0\,.$$

Of course the exponential never vanishes. Thus this last equation can only be true (for all x) if

$$ar^2 + br + c = 0\,.$$

This is just a quadratic equation (called the *associated polynomial equation*),[1] and we may solve it using the quadratic formula. This process will lead to our solution set.

EXAMPLE 2.1.2 Solve the differential equation

$$y'' - 5y' + 4y = 0\,.$$

---

[1] Some texts will call this the *characteristic polynomial*, although that terminology has other meanings in mathematics.

Solution: Following the paradigm just outlined, we guess a solution of the form $y = e^{rx}$. This leads to the quadratic equation for r given by

$$r^2 - 5r + 4 = 0.$$

Of course this factors directly to

$$(r - 1)(r - 4) = 0,$$

so $r = 1, 4$.

Thus e^x and e^{4x} are solutions to the differential equation (you should check this assertion for yourself). A *general solution* is given by

$$y = A \cdot e^x + B \cdot e^{4x}, \tag{2.1.2.1}$$

where A and B are arbitrary constants. The reader may check that any function of the form (2.1.2.1) solves the original differential equation. Observe that our general solution has two undetermined constants, which is consistent with the fact that we are solving a second-order differential equation. ∎

REMARK 2.1.3 Again we see that the solving of a differential equation leads to a family of solutions. In the last example, that family is indexed by two parameters A and B. As we shall see below (especially Section 2.5), a typical physical problem will give rise to two initial conditions that determine those parameters. The Picard Existence and Uniqueness Theorem gives the mathematical underpinning for these ideas.

EXAMPLE 2.1.4 Solve the differential equation

$$2y'' + 6y' + 2y = 0.$$

Solution: The associated polynomial equation is

$$2r^2 + 6r + 2 = 0.$$

This equation does not factor in any obvious way, so we use the quadratic formula:

$$r = \frac{-6 \pm \sqrt{6^2 - 4 \cdot 2 \cdot 2}}{2 \cdot 2} = \frac{-6 \pm \sqrt{20}}{4} = \frac{-6 \pm 2\sqrt{5}}{4} = \frac{-3 \pm \sqrt{5}}{2}.$$

Thus the general solution to the differential equation is

$$y = A \cdot e^{\frac{-3+\sqrt{5}}{2} \cdot x} + B \cdot e^{\frac{-3-\sqrt{5}}{2} \cdot x}. \qquad \blacksquare$$

EXAMPLE 2.1.5 Solve the differential equation

$$y'' - 6y' + 9y = 0.$$

Solution: In this case the associated polynomial is

$$r^2 - 6r + 9 = 0.$$

This algebraic equation factors as $(r - 3) \cdot (r - 3) = 0$ hence has the single solution $r = 3$. But our differential equation is second order, and therefore we seek *two independent solutions*.

In the case that the associated polynomial has just one root, we find the other solution with an augmented guess: Our new guess is $y = x \cdot e^{3x}$. (See Section 2.4 for an explanation of where this guess comes from.) The reader may check for himself/herself that this new guess is also a solution. So the general solution of the differential equation is

$$y = A \cdot e^{3x} + B \cdot xe^{3x}. \qquad \blacksquare$$

As a prologue to our next example, we must review some ideas connected with complex exponentials. Recall that

$$e^x = 1 + x + \frac{x^2}{2!} + \frac{x^3}{3!} + \cdots = \sum_{j=0}^{\infty} \frac{x^j}{j!}.$$

This equation persists if we replace the real variable x by a complex variable z. Thus

$$e^z = 1 + z + \frac{z^2}{2!} + \frac{z^3}{3!} + \cdots = \sum_{j=0}^{\infty} \frac{z^j}{j!}.$$

Now write $z = x + iy$, and let us gather together the real and imaginary parts of this last equation:

$$
\begin{aligned}
e^z &= e^{x+iy} \\
&= e^x \cdot e^{iy} \\
&= e^x \cdot \left(1 + iy + \frac{(iy)^2}{2!} + \frac{(iy)^3}{3!} + \frac{(iy)^4}{4!} + \cdots \right) \\
&= e^x \cdot \left\{ \left(1 - \frac{y^2}{2!} + \frac{y^4}{4!} - + \cdots \right) + i \left(y - \frac{y^3}{3!} + \frac{y^5}{5!} - + \cdots \right) \right\} \\
&= e^x \left(\cos y + i \sin y \right).
\end{aligned}
$$

Taking $x = 0$ we obtain the famous identity

$$e^{iy} = \cos y + i \sin y.$$

This equation—much used in mathematics, engineering, and physics—is known as *Euler's formula*, in honor of Leonhard Euler (1707–1783). We shall also make considerable use of the more general formula

$$e^{x+iy} = e^x \Big(\cos y + i \sin y \Big) .$$

In using complex numbers, the reader should of course remember that the square root of a negative number is an *imaginary number*. For instance,

$$\sqrt{-4} = \pm 2i \qquad \text{and} \qquad \sqrt{-25} = \pm 5i .$$

EXAMPLE 2.1.6 Solve the differential equation

$$4y'' + 4y' + 2y = 0 .$$

Solution: The associated polynomial is

$$4r^2 + 4r + 2 = 0 .$$

We apply the quadratic formula to solve it:

$$r = \frac{-4 \pm \sqrt{4^2 - 4 \cdot 4 \cdot 2}}{2 \cdot 4} = \frac{-4 \pm \sqrt{-16}}{8} = \frac{-4 \pm 4\sqrt{-1}}{8} = \frac{-4 \pm 4i}{8} = \frac{-1 \pm i}{2} .$$

Thus the solutions to our differential equation are

$$y = e^{\frac{-1+i}{2} \cdot x} \qquad \text{and} \qquad y = e^{\frac{-1-i}{2} \cdot x} .$$

A general solution is given by

$$y = A \cdot e^{\frac{-1+i}{2} \cdot x} + B \cdot e^{\frac{-1-i}{2} \cdot x} .$$

Using Euler's formula, we may rewrite this general solution as

$$y = A \cdot e^{-x/2} e^{ix/2} + B \cdot e^{-x/2} e^{-ix/2}$$
$$= A \cdot e^{-x/2} [\cos x/2 + i \sin x/2] + B e^{-x/2} [\cos x/2 - i \sin x/2] . \qquad (2.1.6.1)$$

We shall now use some propitious choices of A and B to extract meaningful real-valued solutions. First choose $A = 1/2$, $B = 1/2$. Putting these values in equation (2.1.6.1) gives

$$y = e^{-x/2} \cos x/2 .$$

Now taking $A = -i/2$, $B = i/2$ gives the solution

$$y = e^{-x/2} \sin x/2 .$$

As a result of this little trick, we may rewrite our general solution as

$$y = E \cdot e^{-x/2} \cos x/2 + F \cdot e^{-x/2} \sin x/2 .$$

As usual, we invite the reader to plug this last solution into the differential equation to verify that it really works. ∎

We conclude this section with a last example of a homogeneous, second-order, linear ordinary differential equation with constant coefficients, and with complex roots, just to show how straightforward the methodology really is.

EXAMPLE 2.1.7 Solve the differential equation

$$y'' - 2y' + 5y = 0.$$

Solution: The associated polynomial is

$$r^2 - 2r + 5 = 0.$$

According to the quadratic formula, the solutions of this equation are

$$r = \frac{2 \pm \sqrt{(-2)^2 - 4 \cdot 1 \cdot 5}}{2} = \frac{2 \pm 4i}{2} = 1 \pm 2i.$$

Hence the roots of the associated polynomial are $r = 1 + 2i$ and $1 - 2i$.

According to what we have learned, two independent solutions to the differential equation are thus given by

$$y = e^x \cos 2x \qquad \text{and} \qquad y = e^x \sin 2x.$$

Therefore the general solution is given by

$$y = Ae^x \cos 2x + Be^x \sin 2x.$$

Please verify this solution for yourself. ∎

Exercises

1. Find the general solution of each of the following differential equations.

(a) $y'' + y' - 6y = 0$
(b) $y'' + 2y' + y = 0$
(c) $y'' + 8y = 0$
(d) $2y'' - 4y' + 8y = 0$
(e) $y'' - 4y' + 4y = 0$
(f) $y'' - 9y' + 20y = 0$
(g) $2y'' + 2y' + 3y = 0$
(h) $4y'' - 12y' + 9y = 0$
(i) $y'' + y' = 0$

(j) $y'' - 6y' + 25y = 0$
(k) $4y'' + 20y' + 25y = 0$
(l) $y'' + 2y' + 3y = 0$
(m) $y'' = 4y$
(n) $4y'' - 8y' + 7y = 0$
(o) $2y'' + y' - y = 0$
(p) $16y'' - 8y' + y = 0$
(q) $y'' + 4y' + 5y = 0$
(r) $y'' + 4y' - 5y = 0$

2. Find the solution of each of the following initial value problems:

(a) $y'' - 5y' + 6y = 0$, $y(1) = e^2$ and $y'(1) = 3e^2$
(b) $y'' - 6y' + 5y = 0$, $y(0) = 3$ and $y'(0) = 11$
(c) $y'' - 6y' + 9y = 0$, $y(0) = 0$ and $y'(0) = 5$
(d) $y'' + 4y' + 5y = 0$, $y(0) = 1$ and $y'(0) = 0$
(e) $y'' + 4y' + 2y = 0$, $y(0) = -1$ and $y'(0) = 2 + 3\sqrt{2}$
(f) $y'' + 8y' - 9y = 0$, $y(1) = 2$ and $y'(1) = 0$

3. Show that the general solution of the equation

$$y'' + Py' + Qy = 0$$

(where P and Q are constant) approaches 0 as $x \to +\infty$ if and only if P and Q are both positive.

4. Show that the derivative of any solution of

$$y'' + Py' + Qy = 0$$

(where P and Q are constant) is also a solution.

5. The equation

$$x^2 y'' + pxy' + qy = 0,$$

where p and q are constants, is known as *Euler's equidimensional equation*. Show that the change of variable $x = e^z$ transforms Euler's equation into a new equation with constant coefficients. Apply this technique to find the general solution of each of the following equations.

(a) $x^2 y'' + 3xy' + 10y = 0$ (f) $x^2 y'' + 2xy' - 6y = 0$
(b) $2x^2 y'' + 10xy' + 8y = 0$ (g) $x^2 y'' + 2xy' + 3y = 0$
(c) $x^2 y'' + 2xy' - 12y = 0$ (h) $x^2 y'' + xy' - 2y = 0$
(d) $4x^2 y'' - 3y = 0$ (i) $x^2 y'' + xy' - 16y = 0$
(e) $x^2 y'' - 3xy' + 4y = 0$

6. Find the differential equation of each of the following general solution sets.

(a) $Ae^x + Be^{-2x}$ (e) $Ae^{3x} + Be^{-x}$
(b) $A + Be^{2x}$ (f) $Ae^{-x} + Be^{-4x}$
(c) $Ae^{3x} + Be^{5x}$ (g) $Ae^{2x} + Be^{-2x}$
(d) $Ae^x \cos 3x + Be^x \sin 3x$ (h) $Ae^{-4x} \cos x + Be^{-4x} \sin x$

2.2   The Method of Undetermined Coefficients

"Undetermined coefficients" is a method of organized guessing. We have already seen guessing, in one form or another, serve us well in solving first-order linear equations and also in solving homogeneous second-order linear equations with constant coefficients. Now we shall expand the technique to cover *inhomogeneous* second-order linear equations.

We must begin by discussing what the solution to such an equation will look like. Consider an equation of the form

$$ay'' + by' + cy = f(x).\tag{2.2.1}$$

Suppose that we can find (by guessing or by some other means) a function $y = y_0(x)$ that satisfies this equation. We call y_0 a *particular solution* of the differential equation. Notice that it will *not* be the case that a constant multiple of y_0 will also solve the equation. In fact, if we consider $y = A \cdot y_0$ and plug this function into the equation, then we obtain

$$a[Ay_0]'' + b[Ay_0]' + c[Ay_0] = A[ay_0'' + by_0' + cy_0] = A \cdot f.$$

We see that, if $A \neq 1$, then $A \cdot y_0$ is *not* a solution. But we expect the solution of a second-order equation to have two free constants. Where will they come from?

The answer is that we must separately solve the associated *homogeneous equation*, which is

$$ay'' + by' + cy = 0.$$

If y_1 and y_2 are solutions of this equation then of course (as we learned in the last section) $A \cdot y_1 + B \cdot y_2$ will be a general solution of *this homogeneous equation*. But then the general solution of the original differential equation (2.2.1) will be

$$y = y_0 + A \cdot y_1 + B \cdot y_2.$$

We invite the reader to verify that, no matter what the choice of A and B, this y will be a solution of the original differential equation (2.2.1).

These ideas are best hammered home by the examination of some examples.

EXAMPLE 2.2.2 Find the general solution of the differential equation

$$y'' + y = \sin x.\tag{2.2.2.1}$$

Solution: We might guess that $y = \sin x$ or $y = \cos x$ is a particular solution of this equation. But in fact these are solutions of the homogeneous equation

$$y'' + y = 0$$

(as we may check by using the techniques of the last section, or just by direct verification). So if we want to find a particular solution of (2.2.2.1) then we must try a bit harder.

Inspired by our experience with the case of repeated roots for the second-order, homogeneous linear equation with constant coefficients (as in the last section), we instead will guess

$$y_0 = \alpha \cdot x \cos x + \beta \cdot x \sin x$$

for our particular solution. Notice that we allow arbitrary constants in front of the functions $x \cos x$ and $x \sin x$. These are the "undetermined coefficients" that we seek.

Now we simply plug the guess into the differential equation and see what happens. Thus

$$[\alpha \cdot x \cos x + \beta \cdot x \sin x]'' + [\alpha \cdot x \cos x + \beta \cdot x \sin x] = \sin x$$

or

$$\alpha\big(2(-\sin x)+x(-\cos x)\big)+\beta\big(2\cos x+x(-\sin x)\big)+\big[\alpha x \cos x+\beta x \sin x\big] = \sin x$$

or

$$(-2\alpha)\sin x + (2\beta)\cos x + (-\beta+\beta)x\sin x + (-\alpha+\alpha)x\cos x = \sin x\,.$$

We see that there is considerable cancellation, and we end up with

$$-2\alpha\sin x + 2\beta\cos x = \sin x\,.$$

The only way that this can be an identity in x is if $-2\alpha = 1$ and $2\beta = 0$ or $\alpha = -1/2$ and $\beta = 0$.

Thus our particular solution is

$$y_0 = -\frac{1}{2}x\cos x$$

and our general solution is

$$y = -\frac{1}{2}x\cos x + A\cos x + B\sin x\,. \qquad\blacksquare$$

REMARK 2.2.3 As usual, for a second-order equation we expect, and find, that there are two unknown parameters that parametrize the set of solutions (or the general solution). Notice that these *are not* the same as the "undetermined coefficients" that we used to find our particular solution.

EXAMPLE 2.2.4 Find the solution of

$$y'' - y' - 2y = 4x^2$$

that satisfies $y(0) = 0$ and $y'(0) = 1$.

Solution: The associated homogeneous equation is

$$y'' - y' - 2y = 0$$

and this has associated polynomial

$$r^2 - r - 2 = 0\,.$$

The roots are obviously $r = 2, -1$ and so the general solution of the homogeneous equation is $y = A \cdot e^{2x} + B \cdot e^{-x}$.

For a particular solution, our guess will be a polynomial. Guessing a second-degree polynomial makes good sense, since a guess of a higher-order polynomial is going to produce terms of high degree that we do not want. Thus we guess that $y_0(x) = \alpha x^2 + \beta x + \gamma$. Plugging this guess into the differential equation gives

$$[\alpha x^2 + \beta x + \gamma]'' - [\alpha x^2 + \beta x + \gamma]' - 2[\alpha x^2 + \beta x + \gamma] = 4x^2$$

or

$$[2\alpha] - [\alpha \cdot 2x + \beta] - [2\alpha x^2 + 2\beta x + 2\gamma] = 4x^2.$$

Grouping like terms together gives

$$-2\alpha x^2 + [-2\alpha - 2\beta]x + [2\alpha - \beta - 2\gamma] = 4x^2.$$

As a result, we find that

$$
\begin{aligned}
-2\alpha &= 4 \\
-2\alpha - 2\beta &= 0 \\
2\alpha - \beta - 2\gamma &= 0.
\end{aligned}
$$

This system is easily solved to yield $\alpha = -2$, $\beta = 2$, $\gamma = -3$. So our particular solution is $y_0(x) = -2x^2 + 2x - 3$. The general solution of the original differential equation is then

$$y(x) = (-2x^2 + 2x - 3) + A \cdot e^{2x} + Be^{-x}. \qquad (2.2.4.1)$$

Now we seek the solution that satisfies the initial conditions $y(0) = 0$ and $y'(0) = 1$. These translate to

$$0 = y(0) = (-2 \cdot 0^2 + 2 \cdot 0 - 3) + A \cdot e^0 + B \cdot e^0$$

and

$$1 = y'(0) = (-4 \cdot 0 + 2 - 0) + 2A \cdot e^0 - B \cdot e^0.$$

This gives the equations

$$
\begin{aligned}
0 &= -3 + A + B \\
1 &= 2 + 2A - B.
\end{aligned}
$$

Of course we can solve this system quickly to find that $A = 2/3, B = 7/3$.

In conclusion, the solution to our initial boundary value problem is

$$y(x) = (-2x^2 + 2x - 3) + \frac{2}{3} \cdot e^{2x} + \frac{7}{3} \cdot e^{-x}. \qquad \blacksquare$$

REMARK 2.2.5 Again notice that the undetermined coefficients α and β and γ that we used to guess the particular solution are *not* the same as the parameters that gave the two degrees of freedom in our general solution (2.2.4.1). Further notice that we *needed* those two degrees of freedom so that we could meet the two initial conditions.

Exercises

1. Find the general solution of each of the following equations.
   (a) $y'' + 3y' - 10y = 6e^{4x}$
   (b) $y'' + 4y = 3\sin x$
   (c) $y'' + 10y' + 25y = 14e^{-5x}$
   (d) $y'' - 2y' + 5y = 25x^2 + 12$
   (e) $y'' - y' - 6y = 20e^{-2x}$
   (f) $y'' - 3y' + 2y = 14\sin 2x - 18\cos 2x$
   (g) $y'' + y = 2\cos x$
   (h) $y'' - 2y' = 12x - 10$
   (i) $y'' - 2y' + y = 6e^x$
   (j) $y'' - 2y' + 2y = e^x \sin x$
   (k) $y'' + y' = 10x^4 + 2$

2. Find the solution of the differential equation that satisfies the given initial conditions.

   (a) $y'' - 3y' + y = x$, $y(0) = 1$, $y'(0) = 0$
   (b) $y'' + 4y' + 6y = \cos x$, $y(0) = 0$, $y'(0) = 2$
   (c) $y'' + y' + y = \sin x$, $y(1) = 1$, $y'(1) = 1$
   (d) $y'' - 3y' + 2y = 0$, $y(-1) = 0$, $y'(-1) = 1$
   (e) $y'' - y' + y = x^2$, $y(0) = 1$, $y'(0) = 0$
   (f) $y'' + 2y' + y = 1$, $y(1) = 1$, $y'(1) = 0$

3. If k and b are positive constants, then find the general solution of
$$y'' + k^2 y = \sin bx.$$

4. If y_1 and y_2 are solutions of
$$y'' + P(x)y' + Q(x)y = R_1(x)$$
and
$$y'' + P(x)y' + Q(x)y = R_2(x),$$
respectively, then show that $y = y_1 + y_2$ is a solution of
$$y'' + P(x)y' + Q(x)y = R_1(x) + R_2(x).$$

This is called the *principle of superposition*. Use this idea to find the general solution of

(a) $y'' + 4y = 4\cos 2x + 6\cos x + 8x^2$

(b) $y'' + 9y = 2\sin 3x + 4\sin x - 26e^{-2x}$

5. Use your symbol manipulation software, such as Maple or Mathematica, to write a routine for solving for the undetermined coefficients in the solution of an ordinary differential equation, once an initial guess is given.

2.3 The Method of Variation of Parameters

Variation of parameters is a method for producing a *particular solution* to an inhomogeneous equation by exploiting the (usually much simpler to find) solutions to the associated homogeneous equation.

Let us consider the differential equation

$$y'' + p(x)y' + q(x)y = r(x). \qquad (2.3.1)$$

Assume that, by some method or other, we have found the general solution of the associated homogeneous equation

$$y'' + p(x)y' + q(x)y = 0$$

to be

$$y = Ay_1(x) + By_2(x).$$

What we do now is to *guess* that a particular solution to the original equation (2.3.1) is

$$y_0(x) = v_1(x) \cdot y_1(x) + v_2(x) \cdot y_2(x) \qquad (2.3.2)$$

for some choice of functions v_1, v_2.

Now let us analyze this guess. We calculate that

$$y_0' = [v_1'y_1 + v_1 y_1'] + [v_2'y_2 + v_2 y_2'] = [v_1'y_1 + v_2'y_2] + [v_1 y_1' + v_2 y_2']. \qquad (2.3.3)$$

We also need to calculate the second derivative of y_0. But we do not want the extra complication of having second derivatives of v_1 and v_2. So we shall mandate that the first expression in brackets on the far right-hand side of (2.3.3) is identically zero. Thus we have

$$v_1'y_1 + v_2'y_2 \equiv 0. \qquad (2.3.4)$$

Hence

$$y_0' = v_1 y_1' + v_2 y_2' \qquad (2.3.5)$$

and we can now calculate that

$$y_0'' = [v_1'y_1' + v_1y_1''] + [v_2'y_2' + v_2y_2'']. \tag{2.3.6}$$

Now let us substitute (2.3.2), (2.3.5), and (2.3.6) into the differential equation. The result is

$$\left([v_1'y_1' + v_1y_1''] + [v_2'y_2' + v_2y_2''] \right) + p(x) \cdot \left(v_1y_1' + v_2y_2' \right) + q(x) \cdot \left(v_1y_1 + v_2y_2 \right) = r(x).$$

After some algebraic manipulation, this becomes

$$v_1 \left(y_1'' + py_1' + qy_1 \right) + v_2 \left(y_2'' + py_2' + qy_2 \right) + v_1'y_1' + v_2'y_2' = r.$$

Since y_1, y_2 are solutions of the homogeneous equation, the expressions in parentheses vanish. The result is

$$v_1'y_1' + v_2'y_2' = r. \tag{2.3.7}$$

At long last we have two equations to solve in order to determine what v_1 and v_2 must be. Namely, we focus on equations (2.3.4) and (2.3.7) to obtain

$$v_1'y_1 + v_2'y_2 = 0,$$

$$v_1'y_1' + v_2'y_2' = r.$$

In practice, these can be solved for v_1', v_2', and then integration tells us what v_1, v_2 must be.

As usual, the best way to understand a new technique is by way of some examples.

EXAMPLE 2.3.8 Find the general solution of

$$y'' + y = \csc x.$$

Solution: Of course the general solution to the associated homogeneous equation is familiar. It is

$$y(x) = A \sin x + B \cos x.$$

We of course think of $y_1(x) = \sin x$ and $y_2(x) = \cos x$. In order to find a particular solution, we need to solve the equations

$$
\begin{aligned}
v_1' \sin x + v_2' \cos x &= 0 \\
v_1'(\cos x) + v_2'(-\sin x) &= \csc x.
\end{aligned}
$$

This is a simple algebra problem, and we find that

$$v_1'(x) = \cot x \qquad \text{and} \qquad v_2'(x) = -1.$$

As a result,

$$v_1(x) = \ln|\sin x| \qquad \text{and} \qquad v_2(x) = -x \,.$$

(As the reader will see, we do not need any constants of integration.)

The final result is then that a particular solution of our differential equation is

$$y_0(x) = v_1(x)y_1(x) + v_2(x)y_2(x) = (\ln|\sin x|) \cdot \sin x + (-x) \cdot \cos x \,.$$

We invite the reader to check that this solution actually works. The general solution of the original differential equation is thus

$$y(x) = \left\{ (\ln|\sin x|) \cdot \sin x + (-x) \cdot \cos x \right\} + A \sin x + B \cos x \,. \qquad \blacksquare$$

REMARK 2.3.9 The method of variation of parameters has the advantage—over the method of undetermined coefficients—of not involving any guessing. It is a direct method that always leads to a solution. However, the integrals that we may need to perform to carry the technique to completion may, for some problems, be rather difficult.

EXAMPLE 2.3.10 Solve the differential equation

$$y'' - y' - 2y = 4x^2$$

using the method of variation of parameters.

Solution: The reader will note that, in the last section (Example 2.2.4), we solved this same equation using the method of undetermined coefficients (or organized guessing). Now we shall solve it a second time by our new method.

As we saw before, the homogeneous equation has the general solution

$$y = Ae^{2x} + Be^{-x} \,.$$

Thus $y_1(x) = e^{2x}$ and $y_2(x) = e^{-x}$.

Hence we solve the system

$$v_1' e^{2x} + v_2' e^{-x} = 0 \,,$$

$$v_1'[2e^{2x}] + v_2'[-e^{-x}] = 4x^2 \,.$$

The result is

$$v_1'(x) = \frac{4}{3} x^2 e^{-2x} \qquad \text{and} \qquad v_2'(x) = -\frac{4}{3} x^2 e^x \,.$$

We may use integration by parts to then determine that

$$v_1(x) = -\frac{2x^2}{3} e^{-2x} - \frac{2x}{3} e^{-2x} - \frac{1}{3} e^{-2x}$$

and

$$v_2(x) = -\frac{4x^2}{3}e^x + \frac{8x}{3}e^x - \frac{8}{3}e^x.$$

We finally see that a particular solution to our differential equation is

$$
\begin{aligned}
y_0(x) &= v_1(x) \cdot y_1(x) + v_2(x)y_2(x) \\
&= \left(-\frac{2x^2}{3}e^{-2x} - \frac{2x}{3}e^{-2x} - \frac{1}{3}e^{-2x}\right) \cdot e^{2x} \\
&\quad + \left(-\frac{4x^2}{3}e^x + \frac{8x}{3}e^x - \frac{8}{3}e^x\right] \cdot e^{-x} \\
&= \left(-\frac{2x^2}{3} - \frac{2x}{3} - \frac{1}{3}\right) + \left(-\frac{4x^2}{3} + \frac{8x}{3} - \frac{8}{3}\right) \\
&= -2x^2 + 2x - 3.
\end{aligned}
$$

In conclusion, the general solution of the original differential equation is

$$y(x) = \left\{-2x^2 + 2x - 3\right\} + Ae^{2x} + Be^{-x}.$$

As you can see, this is the same answer that we obtained in Section 2.2, Example 2.2.4, by the method of undetermined coefficients. ∎

REMARK 2.3.11 Of course the method of variation of parameters is a technique for finding a *particular solution* of a differential equation. The general solution of the associated homogeneous equation must be found by a different technique.

Exercises

1. Find a particular solution of each of the following differential equations.

   (a) $y'' + 4y = \tan 2x$ (d) $y'' + 2y' + 5y = e^{-x}\sec 2x$
   (b) $y'' + 2y' + y = e^{-x}\ln x$ (e) $2y'' + 3y' + y = e^{-3x}$
   (c) $y'' - 2y' - 3y = 64xe^{-x}$ (f) $y'' - 3y' + 2y = (1 + e^{-x})^{-1}$

2. Find a particular solution of each of the following differential equations.

   (a) $y'' + y = \sec x$ (e) $y'' + y = \tan x$
   (b) $y'' + y = \cot^2 x$ (f) $y'' + y = \sec x \tan x$
   (c) $y'' + y = \cot 2x$ (g) $y'' + y = \sec x \csc x$
   (d) $y'' + y = x \cos x$

3. Find a particular solution of

$$y'' - 2y' + y = 2x,$$

first by inspection and then by variation of parameters.

4. Find a particular solution of

$$y'' - y' - 6y = e^{-x},$$

first by undetermined coefficients and then by variation of parameters.

5. Find the general solution of each of the following equations.

(a) $(x^2 - 1)y'' - 2xy' + 2y = (x^2 - 1)^2$

(b) $(x^2 + x)y'' + (2 - x^2)y' - (2 + x)y = x(x + 1)^2$

(c) $(1 - x)y'' + xy' - y = (1 - x)^2$

(d) $xy'' - (1 + x)y' + y = x^2 e^{2x}$

(e) $x^2 y'' - 2xy' + 2y = xe^{-x}$

6. Use your symbol manipulation software, such as Maple or Mathematica, to find a particular solution to an ordinary differential equation, once the solutions to the homogeneous equation are given.

2.4   The Use of a Known Solution to Find Another

Consider a general, second-order, linear, homogeneous equation of the form

$$y'' + p(x)y' + q(x)y = 0. \qquad (2.4.1)$$

It often happens—and we have seen this in our earlier work—that one can either guess or elicit one solution to the equation. But finding the second independent solution is more difficult. In this section we introduce a method for finding that second solution.

In fact we exploit a notational trick that served us well in Section 2.3 on variation of parameters. Namely, we shall assume that we have found the one solution y_1 and we shall suppose that the second solution we seek is $y_2 = v \cdot y_1$ for some undetermined function v. Our job, then, is to find v.

Assuming, then, that y_1 is a solution of (2.4.1), we shall substitute $y_2 = v \cdot y_1$ into (2.4.1) and see what this tells us about calculating v. We see that

$$[v \cdot y_1]'' + p(x) \cdot [v \cdot y_1]' + q(x) \cdot [v \cdot y_1] = 0$$

or

$$[v'' \cdot y_1 + 2v' \cdot y_1' + v \cdot y_1''] + p(x) \cdot [v' \cdot y_1 + v \cdot y_1'] + q(x) \cdot [v \cdot y_1] = 0.$$

We rearrange this identity to find that

$$v \cdot [y_1'' + p(x) \cdot y_1' + q(x)y_1] + [v'' \cdot y_1] + [v' \cdot (2y_1' + p(x) \cdot y_1)] = 0.$$

Now we are *assuming* that y_1 is a solution of the differential equation (2.4.1), so the first expression in brackets must vanish. As a result,

$$[v'' \cdot y_1] + [v' \cdot (2y_1' + p \cdot y_1)] = 0.$$

In the spirit of separation of variables, we may rewrite this equation as

$$\frac{v''}{v'} = -2\frac{y_1'}{y_1} - p.$$

Integrating once, we find that

$$\ln v' = -2\ln y_1 - \int p(x)\, dx$$

or

$$v' = \frac{1}{y_1^2} e^{-\int p(x)\, dx}.$$

Applying the integral one last time yields

$$v = \int \frac{1}{y_1^2} e^{-\int p(x)\, dx}\, dx.$$

What does this mean? It tells us that a *second solution* to our differential equation, given that y_1 is one solution, is given by

$$y_2(x) = v(x) \cdot y_1(x) = \left[\int \frac{1}{y_1^2} e^{-\int p(x)\, dx}\, dx \right] \cdot y_1(x). \qquad (2.4.2)$$

In order to really understand what this means, let us apply the method to some particular differential equations.

EXAMPLE 2.4.3 Find the general solution of the differential equation

$$y'' - 6y' + 9y = 0.$$

Solution: When we first encountered this type of equation in Section 2.1, we learned to study the associated polynomial

$$r^2 - 6r + 9 = 0.$$

Unfortunately, the polynomial has only the repeated root $r = 3$, so we find just the one solution $y_1(x) = e^{3x}$. Where do we find another?

In Section 2.1, we found the second solution by guessing. Now we have a more systematic way of finding that second solution, and we use it to test out the new methodology. Observe that $p(x) = -6$ and $q(x) = 9$. According to formula (2.4.1), we can find a second solution $y_2 = v \cdot y_1$ with

$$
\begin{aligned}
v &= \int \frac{1}{y_1^2} e^{-\int p(x)\, dx}\, dx \\
&= \int \frac{1}{[e^{3x}]^2} e^{-\int -6\, dx}\, dx \\
&= \int e^{-6x} \cdot e^{6x}\, dx \\
&= \int 1\, dx = x\,.
\end{aligned}
$$

Thus the second solution to our differential equation is $y_2 = v \cdot y_1 = x \cdot e^{3x}$ and the general solution is therefore

$$
y = A \cdot e^{3x} + B \cdot x e^{3x}\,.
$$

This reaffirms what we learned in Section 2.1 by a different and more elementary technique.   ■

  Next we turn to an example of a nonconstant coefficient equation.

EXAMPLE 2.4.4 Find the general solution of the differential equation

$$
x^2 y'' + x y' - y = 0\,.
$$

Solution: Differentiating a monomial once lowers the degree by 1 and differentiating it twice lowers the degree by 2. So it is natural to guess that this differential equation has a power of x as a solution. And $y_1(x) = x$ works.
  We use formula (2.4.2) to find a second solution of the form $y_2 = v \cdot y_1$. First we rewrite the equation in the standard form as

$$
y'' + \frac{1}{x} y' - \frac{1}{x^2} y = 0
$$

and we note then that $p(x) = 1/x$ and $q(x) = -1/x^2$. Thus

$$
\begin{aligned}
v(x) &= \int \frac{1}{y_1^2} e^{-\int p(x)\, dx}\, dx \\
&= \int \frac{1}{x^2} e^{-\int 1/x\, dx}\, dx \\
&= \int \frac{1}{x^2} e^{-\ln x}\, dx \\
&= \int \frac{1}{x^2} \cdot \frac{1}{x}\, dx \\
&= \int \frac{1}{x^3}\, dx \\
&= -\frac{1}{2x^2}\,.
\end{aligned}
$$

In conclusion, $y_2 = v \cdot y_1 = [-1/(2x^2)] \cdot x = -1/(2x)$ and the general solution is

$$y(x) = A \cdot x + B \cdot \left(-\frac{1}{2x}\right).$$ ∎

Exercises

1. Use the method of this section to find y_2 and the general solution of each of the following equations from the given solution y_1.

   (a) $y'' + y = 0$, $\quad y_1(x) = \sin x$

   (b) $y'' - y = 0$, $\quad y_1(x) = e^x$

2. The equation $xy'' + 3y' = 0$ has the obvious solution $y_1 \equiv 1$. Find y_2 and find the general solution.

3. Verify that $y_1 = x^2$ is one solution of $x^2 y'' + xy' - 4y = 0$, and then find y_2 and the general solution.

4. The equation

   $$(1 - x^2)y'' - 2xy' + 2y = 0 \qquad (*)$$

   is a special case, corresponding to $p = 1$, of the Legendre equation

   $$(1 - x^2)y'' - 2xy' + p(p+1)y = 0.$$

   Equation ($*$) has $y_1 = x$ as a solution. Find the general solution.

5. The equation

   $$x^2 y'' + xy' + \left(x^2 - \frac{1}{4}\right)y = 0 \qquad (\star)$$

   is a special case, corresponding to $p = 1/2$, of the Bessel equation

   $$x^2 y'' + xy' + (x^2 - p^2)y = 0.$$

   Verify that $y_1(x) = x^{-1/2}\sin x$ is a solution of ($\star$) for $x > 0$ and find the general solution.

6. For each of the following equations, $y_1(x) = x$ is one solution. In each case, find the general solution.

   (a) $y'' - \dfrac{x}{x-1}y' + \dfrac{1}{x-1}y = 0$

   (b) $x^2 y'' + 2xy' - 2y = 0$

   (c) $x^2 y'' - x(x+2)y' + (x+2)y = 0$

7. Find the general solution of the differential equation

   $$y'' - x^2 y' + xy = 0.$$

8. Verify that one solution of the equation

$$xy'' - (2x+1)y' + (x+1)y = 0$$

is $y_1(x) = e^x$. Find the general solution.

9. If y_1 is a nonzero solution of the differential equation

$$y'' + P(x)y' + Q(x)y = 0$$

and if $y_2 = v \cdot y_1$, with

$$v(x) = \int \frac{1}{y_1^2} \cdot e^{-\int p\,dx}\,dx\,,$$

then show that y_1 and y_2 are linearly independent (that is, y_1 is not a multiple of y_2).

2.5 Vibrations and Oscillations

When a physical system in stable equilibrium is disturbed, then it is subject to forces that tend to restore the equilibrium. The result can lead to oscillations or vibrations. It is described by an ordinary differential equation of the form

$$\frac{d^2x}{dt^2} + p(t) \cdot \frac{dx}{dt} + q(t)x = r(t)\,.$$

In this section we shall learn how and why such an equation models the physical system we have described, and we shall see how its solution sheds light on the physics of the situation.

2.5.1 Undamped Simple Harmonic Motion

Our basic example will be a cart of mass M attached to a nearby wall by means of a spring. See Figure 2.1.

The spring exerts no force when the cart is at its rest position $x = 0$ (notice that, contrary to custom, we are locating the origin to the right of the wall). According to Hooke's law, if the cart is displaced a distance x, then the spring exerts a proportional force $F_s = -kx$, where k is a positive constant known as Hooke's constant. Observe that, if $x > 0$, then the cart is moved to the right and the spring pulls to the left; so the force is negative. Obversely, if $x < 0$ then the cart is moved to the left and the spring resists with a force to the right; so the force is positive.

Newton's second law of motion says that the mass of the cart times its acceleration equals the force acting on the cart. Thus

$$M \cdot \frac{d^2x}{dt^2} = F_s = -k \cdot x\,.$$

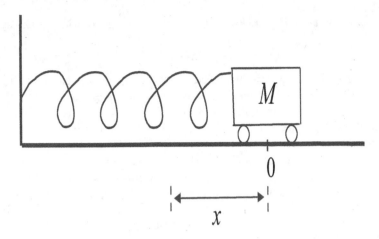

FIGURE 2.1
Hooke's law.

As a result,

$$\frac{d^2x}{dt^2} + \frac{k}{M}x = 0\,.$$

It is useful to let $a = \sqrt{k/M}$ (both k and M are positive) and thus to write the equation as

$$\frac{d^2x}{dt^2} + a^2x = 0\,.$$

Of course this is a familiar differential equation for us, and we can write its general solution immediately:

$$x(t) = A\sin at + B\cos at\,.$$

Now suppose that the cart is pulled to the right to an initial position of $x = x_0 > 0$ and then is simply released (with initial velocity 0). Then we have the initial conditions

$$x(0) = x_0 \qquad \text{and} \qquad \frac{dx}{dt}(0) = 0\,.$$

Thus

$$
\begin{aligned}
x_0 &= A\sin(a \cdot 0) + B\cos(a \cdot 0) \\
0 &= Aa\cos(a \cdot 0) - Ba\sin(a \cdot 0)
\end{aligned}
$$

or

$$
\begin{aligned}
x_0 &= B \\
0 &= A \cdot a\,.
\end{aligned}
$$

We conclude that $B = x_0$, $A = 0$, and we find the solution of the system to be

$$x(t) = x_0 \cos at.$$

In other words, if the cart is displaced a distance x_0 and released, then the result is a simple harmonic motion (described by the cosine function) with *amplitude* x_0 (i.e., the cart glides back and forth, x_0 units to the left of the origin and then x_0 units to the right), and with period $T = 2\pi/a$ (which means that the motion repeats itself every $2\pi/a$ units of time).

The *frequency* f of the motion is the number of cycles per unit of time, hence $f \cdot T = 1$, or $f = 1/T = a/(2\pi)$. It is useful to substitute back in the actual value of a so that we can analyze the physics of the system. Thus

$$\text{amplitude} = x_0$$

$$\text{period} = T = \frac{2\pi\sqrt{M}}{\sqrt{k}}$$

$$\text{frequency} = f = \frac{\sqrt{k}}{2\pi\sqrt{M}}.$$

We see that, if the stiffness k of the spring is increased, then the period becomes smaller and the frequency increases. Likewise, if the mass M of the cart is increased then the period increases and the frequency decreases.

2.5.2 Damped Vibrations

It probably has occurred to the reader that the physical model in the last subsection is not realistic. Typically, a cart that is attached to a spring and released, just as we have described, will enter a harmonic motion that *dies out over time*. In other words, resistance and friction will cause the system to be damped. Let us add that information to the system.

Physical considerations make it plausible to postulate that the resistance is proportional to the velocity of the moving cart. Thus

$$F_d = -c\frac{dx}{dt},$$

where F_d denotes damping force and $c > 0$ is a positive constant that measures the resistance of the medium (air or water or oil, etc.). Notice, therefore, that when the cart is traveling to the right then $dx/dt > 0$ and therefore the force of resistance is negative (i.e., in the other direction). Likewise, when the cart is traveling to the left then $dx/dt < 0$ and the force of resistance is positive.

Since the total of all the forces acting on the cart equals the mass times the acceleration, we now have

$$M \cdot \frac{d^2x}{dt^2} = F_s + F_d.$$

In other words,

$$\frac{d^2x}{dt^2} + \frac{c}{M} \cdot \frac{dx}{dt} + \frac{k}{M} \cdot x = 0 \, .$$

Because of convenience and tradition, we again take $a = \sqrt{k/M}$ and we set $b = c/(2M)$. Thus the differential equation takes the form

$$\frac{d^2x}{dt^2} + 2b \cdot \frac{dx}{dt} + a^2 \cdot x = 0 \, .$$

This is a second-order, linear, homogeneous ordinary differential equation with constant coefficients. The associated polynomial is

$$r^2 + 2br + a^2 = 0 \, ,$$

and it has roots

$$r_1, r_2 = \frac{-2b \pm \sqrt{4b^2 - 4a^2}}{2} = -b \pm \sqrt{b^2 - a^2} \, .$$

Now we must consider three cases.

CASE A. $c^2 - 4kM > 0$:   In other words, $b^2 - a^2 > 0$. We are assuming that the frictional force (which depends on c) is significantly larger than the stiffness of the spring (which depends on k). Thus we would expect the system to damp heavily. In any event, the calculation of r_1, r_2 involves the square root of a *positive real number* which is smaller than b, so that the values of $-b \pm \sqrt{b^2 - a^2}$ are definitely negative. Thus r_1, r_2 are distinct real (and negative) roots of the associated polynomial equation.

Thus the general solution of our system in this case is

$$x = Ae^{r_1 t} + Be^{r_2 t} \, ,$$

where (we repeat) r_1, r_2 are negative real numbers. We apply the initial conditions $x(0) = x_0$, $dx/dt(0) = 0$, just as in the last section (details are left to the reader). The result is the particular solution

$$x(t) = \frac{x_0}{r_1 - r_2} \left(r_1 e^{r_2 t} - r_2 e^{r_1 t} \right) . \tag{2.5.1}$$

Notice that, in this heavily damped system, no oscillation occurs (i.e., there are no sines or cosines in the expression for $x(t)$). The system simply dies out. Figure 2.2 exhibits the graph of the function in (2.5.1).

CASE B. $c^2 - 4kM = 0$:   In other words, $b^2 - a^2 = 0$. This is the critical case, where the resistance balances the force of the spring. We see that $b = a$ (both are known to be positive) and $r_1 = r_2 = -b = -a$. We know, then, that the general solution to our differential equation is

$$x(t) = Ae^{-at} + Bte^{-at} \, .$$

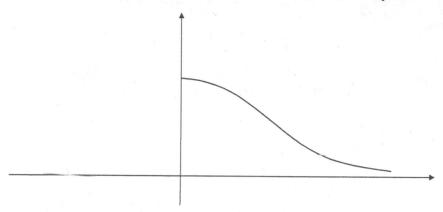

FIGURE 2.2
The motion dies out.

When the standard initial conditions are imposed, we find the particular solution

$$x(t) = x_0 \cdot e^{-at}(1 + at).$$

We see that this differs from the situation in **CASE A** by the factor $(1 + at)$. That factor of course attenuates the damping, but there is *still no oscillatory motion*. We call this the *critical case*. The graph of our new $x(t)$ is quite similar to the graph already shown in Figure 2.2.

   If there is any small decrease in the viscosity, however slight, then the system will begin to vibrate (as one would expect). That is the next, and last, case that we examine.

CASE C. $c^2 - 4kM < 0$:   This says that $b^2 - a^2 < 0$. Now $0 < b < a$ and the calculation of r_1, r_2 entails taking the square root of a negative number. Thus r_1, r_2 are the conjugate complex numbers $-b \pm i\sqrt{a^2 - b^2}$. We set $\alpha = \sqrt{a^2 - b^2} > 0$.

   Now the general solution of our system, as we well know, is

$$x(t) = e^{-bt}\left(A \sin \alpha t + B \cos \alpha t \right).$$

If we evaluate A, B according to our usual initial conditions, then we find the particular solution

$$
\begin{aligned}
x(t) &= \frac{x_0}{\alpha} e^{-bt}\left(b \sin \alpha t + \alpha \cos \alpha t \right) \\
&= \frac{x_0 \sqrt{\alpha^2 + b^2}}{\alpha} e^{-bt} \left(\frac{b}{\sqrt{\alpha^2 + b^2}} \sin \alpha t + \frac{\alpha}{\sqrt{\alpha^2 + b^2}} \cos \alpha t \right).
\end{aligned}
$$

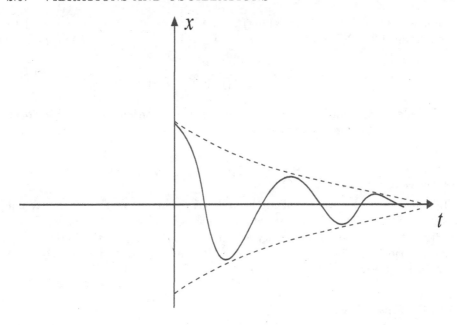

FIGURE 2.3
A damped vibration.

It is traditional and convenient to set $\theta = \arctan(b/\alpha)$. It follows that

$$\frac{b}{\sqrt{\alpha^2 + b^2}} = \sin\theta$$

and

$$\frac{\alpha}{\sqrt{\alpha^2 + b^2}} = \cos\theta.$$

With this notation, we can express the last equation in the form

$$x(t) = \frac{x_0\sqrt{\alpha^2 + b^2}}{\alpha} e^{-bt} \left(\sin\theta\sin\alpha t + \cos\theta\cos\alpha t\right)$$

$$= \frac{x_0\sqrt{\alpha^2 + b^2}}{\alpha} e^{-bt} \cos(\alpha t - \theta). \tag{2.5.2}$$

As you can see, there is oscillation because of the presence of the cosine function. The amplitude (the expression that appears in front of cosine) clearly falls off—rather rapidly—with t because of the presence of the exponential. The graph of this function is exhibited in Figure 2.3.

Of course this function is *not* periodic—it is dying off, and not repeating itself. What *is true*, however, is that the graph crosses the t-axis (the equilibrium position $x = 0$) at regular intervals. If we consider this interval T (which is not a "period," strictly speaking) as the time required for one complete cycle, then $\alpha T = 2\pi$ so

$$T = \frac{2\pi}{\alpha} = \frac{2\pi}{\sqrt{k/M - c^2/(4M^2)}}. \qquad (2.5.3)$$

We define the number f, which plays the role of "frequency" with respect to the indicated time interval, to be

$$f = \frac{1}{T} = \frac{1}{2\pi}\sqrt{\frac{k}{M} - \frac{c^2}{4M^2}}.$$

This number is commonly called the *natural frequency* of the system. When the viscosity vanishes, then our solution clearly reduces to the one we found earlier when there was no viscosity present. We also see that the frequency of the vibration is reduced by the presence of damping; increasing the viscosity further reduces the frequency.

2.5.3 Forced Vibrations

The vibrations that we have considered so far are called *free vibrations* because all the forces acting on the system are internal to the system itself. We now consider the situation in which there is an *external force* $F_e = f(t)$ acting on the system. This force could be an external magnetic field (acting on the steel cart) or vibration of the wall, or perhaps a stiff wind blowing. Again setting mass times acceleration equal to the resultant of all the forces acting on the system, we have

$$M \cdot \frac{d^2 x}{dt^2} = F_s + F_d + F_e,$$

where $F_e = f(t)$ is the external force.

Taking into account the definitions of the various forces, we may write the differential equation as

$$M\frac{d^2 x}{dt^2} + c\frac{dx}{dt} + kx = f(t).$$

So we see that the equation describing the physical system is second-order linear, and that the external force gives rise to an inhomogeneous term on the right. An interesting special case occurs when $f(t) = F_0 \cdot \cos \omega t$, in other words when that external force is periodic. Thus our equation becomes

$$M\frac{d^2 x}{dt^2} + c\frac{dx}{dt} + kx = F_0 \cdot \cos \omega t. \qquad (2.5.4)$$

If we can find a particular solution of this equation, then we can combine it with the information about the solution of the associated homogeneous equation in the last subsection and then come up with the general solution of the differential equation. We shall use the method of undetermined coefficients. Considering the form of the right-hand side, our guess will be

$$x(t) = A\sin\omega t + B\cos\omega t.$$

Substituting this guess into the differential equation gives

$$M\frac{d^2}{dt^2}[A\sin\omega t + B\cos\omega t] + c\frac{d}{dt}[A\sin\omega t + B\cos\omega t]$$
$$+k[A\sin\omega t + B\cos\omega t] = F_0 \cdot \cos\omega t.$$

With a little calculus and a little algebra we are led to the algebraic equations

$$\omega cA + (k - \omega^2 M)B = F_0$$
$$(k - \omega^2 M)A - \omega cB = 0.$$

We solve for A and B to obtain

$$A = \frac{\omega cF_0}{(k - \omega^2 M)^2 + \omega^2 c^2} \quad \text{and} \quad B = \frac{(k - \omega^2 M)F_0}{(k - \omega^2 M)^2 + \omega^2 c^2}.$$

Thus we have found the particular solution

$$x_0(t) = \frac{F_0}{(k - \omega^2 M)^2 + \omega^2 c^2}\left(\omega c\sin\omega t + (k - \omega^2 M)\cos\omega t\right).$$

Calculating as above, we may write this in a more useful form with the notation $\phi = \arctan[\omega c/(k - \omega^2 M)]$. Thus

$$x_0(t) = \frac{F_0}{\sqrt{(k - \omega^2 M)^2 + \omega^2 c^2}} \cdot \cos(\omega t - \phi). \tag{2.5.5}$$

If we assume that we are dealing with the underdamped system, which is **CASE C** of the last subsection, we find that the general solution of our differential equation with a periodic external forcing term is

$$x(t) = e^{-bt}\left(A\cos\alpha t + B\sin\alpha t\right)$$
$$+\frac{F_0}{\sqrt{(k - \omega^2 M)^2 + \omega^2 c^2}} \cdot \cos(\omega t - \phi).$$

We see that, as long as some damping is present in the system (that is, b is nonzero and positive), then the first term in the definition of $x(t)$ is clearly transient (i.e., it dies as $t \to \infty$ because of the exponential term). Thus, as time goes on, the motion assumes the character of the second term in $x(t)$, which is the *steady-state* term. So we can say that, for large t, the physical nature of the general solution to our system is more or less like that of the particular solution $x_0(t)$ that we found. The frequency of this forced vibration equals the impressed frequency (originating with the external forcing term) $\omega/2\pi$. The amplitude is the coefficient

$$\frac{F_0}{\sqrt{(k - \omega^2 M)^2 + \omega^2 c^2}}. \tag{2.5.6}$$

This expression for the amplitude depends on all the relevant physical constants, and it is enlightening to analyze it a bit. Observe, for instance, that if the viscosity c is very small and if ω is close to $\sqrt{k/M}$ (so that $k - \omega^2 M$ is very small) then the motion is lightly damped and the external (impressed) frequency $\omega/2\pi$ is close to the natural frequency

$$\frac{1}{2\pi}\sqrt{\frac{k}{M} - \frac{c^2}{4M^2}}.$$

Then the amplitude is very large (because we are dividing by a number close to 0). This phenomenon is known as *resonance*. There are classical examples of resonance.[2] For instance, several years ago there was a celebration of the anniversary of the Golden Gate Bridge (built in 1937), and many thousands of people marched in unison across the bridge. The frequency of their footfalls was so close to the natural frequency of the bridge (thought of as a suspended string under tension) that the bridge nearly fell apart. A famous incident at the Tacoma Narrows Bridge has been attributed to resonance, although more recent studies suggest a more complicated combination of effects (see the movie of this disaster at `http://www.ketchum.org/bridgecollapse.html`).

2.5.4　A Few Remarks about Electricity

It is known that if a periodic electromotive force, $E = E_0$, acts in a simple circuit containing a resistor, an inductor, and a capacitor, then the charge Q on the capacitor is governed by the differential equation

$$L\frac{d^2Q}{dt^2} + R\frac{dQ}{dt} + \frac{1}{C}Q = E_0\cos\omega t.$$

This equation is of course quite similar to equation (2.5.4) for the oscillating cart with external force. In particular, the following correspondences (or analogies) are suggested:

$$\text{mass } M \longleftrightarrow \text{ inductance } L;$$
$$\text{viscosity } c \longleftrightarrow \text{ resistance } R;$$
$$\text{stiffness of spring } k \longleftrightarrow \text{ reciprocal of capacitance } \frac{1}{C};$$
$$\text{displacement } x \longleftrightarrow \text{ charge } Q \text{ on capacitor}.$$

---

[2]One of the basic ideas behind filter design is resonance.

The analogy between the mechanical and electrical systems renders identical the mathematical analysis of the two systems, and enables us to carry over at once all mathematical conclusions from the first to the second. In the given electric circuit we therefore have a critical resistance below which the free behavior of the circuit will be vibratory with a certain natural frequency, a forced steady-state vibration of the charge Q, and resonance phenomena that appear when the circumstances are favorable.

Math Nugget

Charles Proteus Steinmetz (1865–1923) was a mathematician, inventor, and electrical engineer. He pioneered the use of complex numbers in the study of electrical circuits. After he left Germany (on account of his socialist political activities) and emigrated to America, he was employed by the General Electric Company. He soon solved some of GE's biggest problems—to design a method to mass-produce electric motors, and to find a way to transmit electricity more than 3 miles. With these contributions alone Steinmetz had a massive impact on mankind.

Steinmetz was a dwarf, crippled by a congenital deformity. He lived in pain, but was well liked for his humanity and his sense of humor, and certainly admired for his scientific prowess. The following Steinmetz story comes from the Letters section of *Life* Magazine (May 14, 1965):

Sirs: In your article on Steinmetz (April 23) you mentioned a consultation with Henry Ford. My father, Burt Scott, who was an employee of Henry Ford for many years, related to me the story behind the meeting. Technical troubles developed with a huge new generator at Ford's River Rouge plant. His electrical engineers were unable to locate the difficulty so Ford solicited the aid of Steinmetz. When "the little giant" arrived at the plant, he rejected all assistance, asking only for a notebook, pencil and cot. For two straight days and nights he listened to the generator and made countless computations. Then he asked for a ladder, a measuring tape, and a piece of chalk. He laboriously ascended the ladder, made careful measurements, and put a chalk mark on the side of the generator. He descended and told his skeptical audience to remove a plate from the side of the generator [at the marked spot] and take out 16 windings from the field coil at that location. The corrections were made and the generator then functioned perfectly. Subsequently Ford received a bill for $10,000 signed by Steinmetz for G.E. Ford returned the bill acknowledging the good job done by Steinmetz but respectfully requesting an itemized statement. Steinmetz replied as follows: Making chalk mark on generator $1. Knowing where to make mark $9,999. Total due $10,000.

Exercises

1. Consider the forced vibration in the underdamped case, and find the impressed frequency for which the amplitude attains a maximum. Will such an impressed frequency necessarily exist? This value of the impressed frequency, when it exists, is called the *resonance frequency*. Show that the resonance frequency is always less than the natural frequency.

2. Consider the underdamped free vibration described by formula (2.5.2). Show that x assumes maximum values for $t = 0, T, 2T, \ldots$, where T is the "period," as given in formula (2.5.3). If x_1 and x_2 are any two successive maxima, then show that $x_1/x_2 = e^{bT}$. The logarithm of this quantity, or bT, is known as the *logarithmic decrement* of the vibration.

3. A spherical buoy of radius r floats half-submerged in water. If it is depressed slightly, then a restoring force equal to the weight of the displaced

water presses it upward; and if it is then released, it will bob up and down. Find the period of oscillation if the friction of the water is negligible.

4. A cylindrical buoy 2 feet in diameter floats with its axis vertical in fresh water of density 62.4 lb./ft.^{3} When depressed slightly and released, its period of oscillation is observed to be 1.9 seconds. What is the weight of the buoy?

5. Suppose that a straight tunnel is drilled through the Earth between two points on its surface. The tunnel passes through the center of the earth. If tracks are laid, then—neglecting friction—a train placed in the tunnel at one end will roll through the Earth under its own weight, stop at the other end, and return. Show that the time required for a single, complete round trip is the same for all such tunnels (no matter what the beginning and ending points), and estimate its value. If the tunnel is $2L$ miles long, then what is the greatest speed attained by the train on its journey?

6. The cart in Figure 2.1 weighs 128 pounds and is attached to the wall by a spring with spring constant $k = 64$ lb./ft. The cart is pulled 6 inches in the direction away from the wall and released with no initial velocity. Simultaneously, a periodic external force $F_e = f(t) = 32 \sin 4t$ lb is applied to the cart. Assuming that there is no air resistance, find the position $x = x(t)$ of the cart at time t. Note particularly that $|x(t)|$ assumes arbitrarily large values as $t \to +\infty$. This phenomenon is known as *pure resonance* and is caused by the fact that the forcing function has the same period as the free vibrations of the unforced system.

7. Use your symbol manipulation software, such as Maple or Mathematica, to solve the ordinary differential equation with the given damping term and forcing term. In each instance you should assume that both the damping and the forcing terms occur on the right-hand side of the differential equation and that $t > 0$.

   (a) damping $= -e^t \, dx/dt$, $f = \sin t + \cos 2t$
   (b) damping $= -\ln t \, dx/dt$, $f = e^t$
   (c) damping $= -[e^t] \cdot \ln t \, dx/dt$, $f = \cos 2t$
   (d) damping $= -t^3 \, dx/dt$, $f = e^{-t}$

2.6 Newton's Law of Gravitation and Kepler's Laws

Newton's law of universal gravitation is one of the great ideas of modern physics. It underlies so many important physical phenomena that it is part of the bedrock of science. In this section we show how Kepler's laws of planetary motion can be derived from Newton's gravitation law. It might be noted that Johannes Kepler himself (1571–1630) used thousands of astronomical observations (made by his teacher Tycho Brahe (1546–1601)) in order to formulate his laws. Kepler was a follower of Copernicus, who postulated that the planets orbited about the sun, but Brahe held the more traditional Ptolemaic view that the Earth was the center of the orbits. Brahe did not want to let Kepler use his data, for he feared that Kepler would use the data to promote the Copernican theory. As luck would have it, Brahe died from a burst bladder after a night of excessive beer drinking at a social function. So Kepler was able to get the valuable numbers from Tycho Brahe's family.

KEPLER'S LAWS OF PLANETARY MOTION

I. The orbit of each planet is an ellipse with the sun at one focus (Figure 2.4).

II. The segment from the center of the sun to the center of an orbiting planet sweeps out area at a constant rate (Figure 2.5).

III. The square of the period of revolution of a planet is proportional to the cube of the length of the major axis of its elliptical orbit, with the same constant of proportionality for any planet. (Figure 2.6).

Interestingly, Copernicus believed that the orbits were circles (rather than ellipses, as we now know them to be). Newton determined how to derive the laws of motion analytically, and he was able to *prove* that the orbits must be ellipses (although it should be noted that the ellipses are very nearly circular—their eccentricity is very close to 0). Furthermore, the eccentricity of an elliptical orbit has an important physical interpretation. The present section explores all these ideas.

It turns out that the eccentricities of the ellipses that arise in the orbits of the planets are very small, so that the orbits are nearly circles, but they are definitely *not* circles. That is the importance of Kepler's first law.

The second law tells us that, when the planet is at its apogee (furthest from the sun), then it is traveling relatively slowly, whereas at its perigee (nearest point to the sun), it is traveling relatively rapidly—Figure 2.7. In fact the second law is valid for any central force, and Newton knew this important fact.

The third law allows us to calculate the length of a year on any given planet from knowledge of the shape of its orbit.

In this section we shall learn how to derive Kepler's three laws from Newton's inverse square law of gravitational attraction. To keep matters as simple

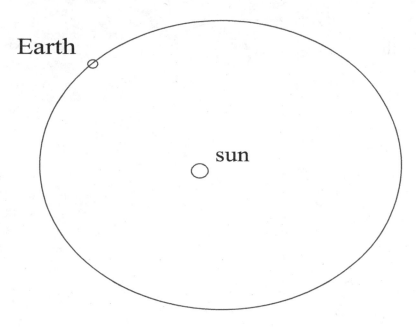

FIGURE 2.4
The elliptical orbit of the Earth about the sun.

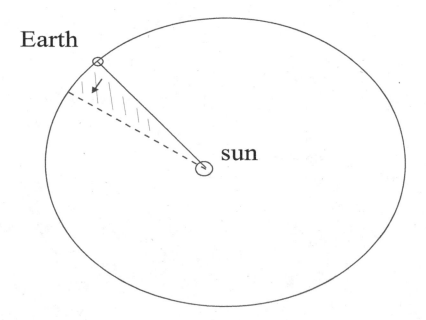

FIGURE 2.5
Area is swept out at a constant rate.

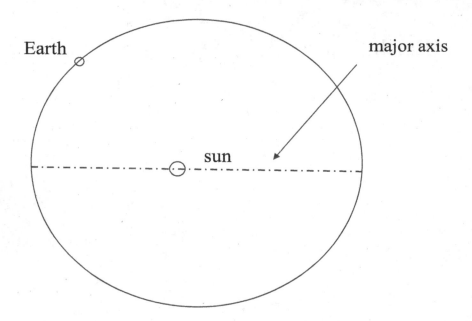

FIGURE 2.6
The square of the period is proportional to the cube of the major axis.

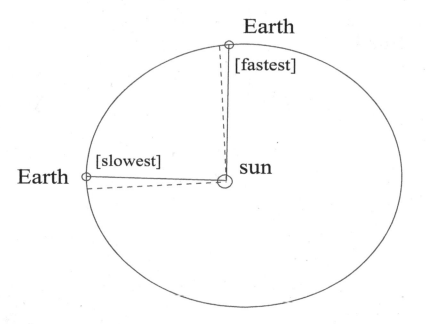

FIGURE 2.7
Apogee motion vs. perigee motion.

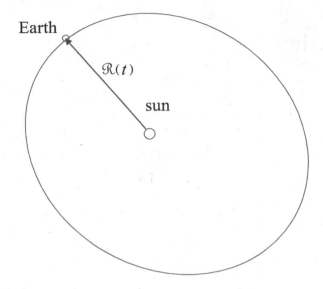

FIGURE 2.8
Polar coordinate system for the motion of the Earth about the sun.

as possible, we shall assume that our solar system contains a fixed sun and just one planet (the Earth, for instance). The problem of analyzing the gravitation influence of three or more planets on each other is incredibly complicated and is still not thoroughly understood.

The argument that we present is due to S. Kochen, and is used with his permission.

2.6.1 Kepler's Second Law

It is convenient to derive the second law first. We use a polar coordinate system with the origin at the center of the sun. We analyze a single planet which orbits the sun, and we denote the position of that planet at time t by the vector $\mathcal{R}(t)$. The only physical facts that we shall use in this portion of the argument are Newton's second law and the self-evident assertion that the gravitational force exerted by the sun on a planet is a vector parallel to $\mathcal{R}(t)$. See Figure 2.8.

If $\mathbf{F}$ is force, m is the mass of the planet (Earth), and $\mathbf{a}$ is its acceleration then Newton's second law says that

$$\mathbf{F} = m\mathbf{a} = m\mathcal{R}''(t).$$

We conclude that $\mathcal{R}(t)$ is parallel to $\mathcal{R}''(t)$ for every value of t.

Now

$$\frac{d}{dt}\left(\mathcal{R}(t) \times \mathcal{R}'(t)\right) = \left[\mathcal{R}'(t) \times \mathcal{R}'(t)\right] + \left[\mathcal{R}(t) \times \mathcal{R}''(t)\right].$$

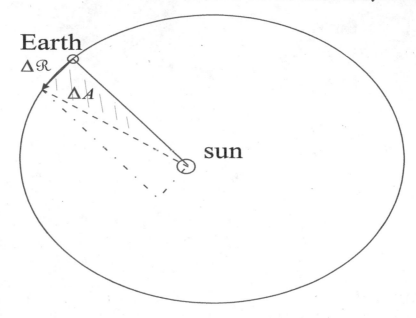

FIGURE 2.9
The increment of area.

Note that the first of these terms on the right is zero because the cross product of any vector with itself is zero. The second is zero because $\mathcal{R}(t)$ is parallel to $\mathcal{R}''(t)$ for every t. We conclude that

$$\frac{d}{dt}\left(\mathcal{R}(t) \times \mathcal{R}'(t)\right) = 0$$

hence

$$\mathcal{R}(t) \times \mathcal{R}'(t) = \mathbf{C}, \qquad (2.6.1)$$

where $\mathbf{C}$ is a constant vector. Notice that this already guarantees that $\mathcal{R}(t)$ and $\mathcal{R}'(t)$ always lie in the same plane, hence that the orbit takes place in a plane.

Now let Δt be an increment of time, $\Delta\mathcal{R}$ the corresponding increment of position, and ΔA the increment of area swept out. Look at Figure 2.9.

We see that ΔA is approximately equal to half the area of the parallelogram determined by the vectors $\mathcal{R}$ and $\Delta\mathcal{R}$. The area of this parallelogram is $\|\mathcal{R} \times \Delta\mathcal{R}\|$. Thus

$$\frac{\Delta A}{\Delta t} \approx \frac{1}{2}\frac{\|\mathcal{R} \times \Delta\mathcal{R}\|}{\Delta t} = \frac{1}{2}\left\|\mathcal{R} \times \frac{\Delta\mathcal{R}}{\Delta t}\right\|.$$

Letting $\Delta t \to 0$ gives

$$\frac{dA}{dt} = \frac{1}{2}\left\|\mathcal{R} \times \frac{d\mathcal{R}}{dt}\right\| = \frac{1}{2}\|\mathbf{C}\| = \text{constant}.$$

We conclude that area $A(t)$ is swept out at a constant rate. That is Kepler's second law.

2.6.2 Kepler's First Law

Now we write $\mathcal{R}(t) = r(t)\mathbf{u}(t)$, where $\mathbf{u}$ is a unit vector pointing in the same direction as $\mathcal{R}$ and r is a positive, scalar-valued function representing the length of $\mathcal{R}$. We use Newton's inverse square law for the attraction of two bodies. If one body (the sun) has mass M and the other (the planet) has mass m then Newton says that the force exerted by gravity on the planet is

$$-\frac{GmM}{r^2}\mathbf{u}.$$

Here G is a universal gravitational constant. Refer to Figure 2.10. Because this force is also equal to $m\mathcal{R}''$ (by Newton's second law), we conclude that

$$m\mathcal{R}'' = -\frac{GmM}{r^2}\mathbf{u}.$$

or

$$\mathcal{R}'' = -\frac{GM}{r^2}\mathbf{u}.$$

Also

$$\mathcal{R}'(t) = \frac{d}{dt}(r\mathbf{u}) = r'\mathbf{u} + r\mathbf{u}'$$

and

$$0 = \frac{d}{dt}1 = \frac{d}{dt}(\mathbf{u}\cdot\mathbf{u}) = 2\mathbf{u}\cdot\mathbf{u}'.$$

Therefore

$$\mathbf{u} \perp \mathbf{u}'. \tag{2.6.2}$$

Now, using (2.6.2), and the derivation of $\mathbf{C}$ from our discussion of Kepler's second law, we calculate

$$
\begin{aligned}
\mathcal{R}'' \times \mathbf{C} &= \mathcal{R}'' \times (\mathcal{R} \times \mathcal{R}'(t)) \\
&= -\frac{GM}{r^2}\mathbf{u} \times (r\mathbf{u} \times (r'\mathbf{u} + r\mathbf{u}')) \\
&= -\frac{GM}{r^2}\mathbf{u} \times (r\mathbf{u} \times r\mathbf{u}') \\
&= -GM(\mathbf{u} \times (\mathbf{u} \times \mathbf{u}')).
\end{aligned}
$$

We can determine the vector $\mathbf{u} \times (\mathbf{u} \times \mathbf{u}')$. For, with formula (2.6.2), we see that $\mathbf{u}$ and $\mathbf{u}'$ are perpendicular and that $\mathbf{u} \times \mathbf{u}'$ is perpendicular to both of these. Because $\mathbf{u} \times (\mathbf{u} \times \mathbf{u}')$ is perpendicular to the first and last of these three, it must therefore be parallel to $\mathbf{u}'$. It also has the same length as $\mathbf{u}'$

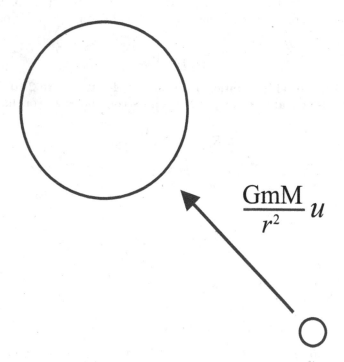

FIGURE 2.10
Newton's universal law of gravitation.

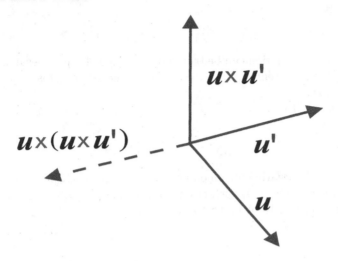

FIGURE 2.11
Calculations with $\mathbf{u}'$.

and, by the right-hand rule, points in the opposite direction. Look at Figure 2.11. We conclude that $\mathbf{u} \times (\mathbf{u} \times \mathbf{u}') = -\mathbf{u}'$, hence that

$$\mathcal{R}'' \times \mathbf{C} = GM\mathbf{u}'.$$

If we antidifferentiate this last equality we obtain

$$\mathcal{R}'(t) \times \mathbf{C} = GM(\mathbf{u} + \mathbf{K}),$$

where $\mathbf{K}$ is a constant vector of integration.

Thus we have

$$\mathcal{R} \cdot (\mathcal{R}'(t) \times \mathbf{C}) = r\mathbf{u}(t) \cdot GM(\mathbf{u}(t) + \mathbf{K}) = GMr(1 + \mathbf{u}(t) \cdot \mathbf{K}),$$

because $\mathbf{u}(t)$ is a unit vector. If $\theta(t)$ is the angle between $\mathbf{u}(t)$ and $\mathbf{K}$ then we may rewrite our equality as

$$\mathcal{R} \cdot (\mathcal{R}' \times \mathbf{C}) = GMr(1 + \|\mathbf{K}\| \cos\theta).$$

By a standard triple product formula,

$$\mathcal{R} \cdot (\mathcal{R}'(t) \times \mathbf{C}) = (\mathcal{R} \times \mathcal{R}'(t)) \cdot \mathbf{C},$$

which in turn equals

$$\mathbf{C} \cdot \mathbf{C} = \|\mathbf{C}\|^2.$$

(Here we have used the fact, which we derived in the proof of Kepler's second law, that $\mathcal{R} \times \mathcal{R}' = \mathbf{C}$.)

Thus
$$\|\mathbf{C}\|^2 = GMr(1 + \|\mathbf{K}\| \cos\theta).$$

(Notice that this equation can be true only if $\|\mathbf{K}\| \leq 1$—just because we want the expression in parentheses to always be nonnegative. This fact will come up again below.)

We conclude that
$$r = \frac{\|\mathbf{C}\|^2}{GM} \cdot \left(\frac{1}{1 + \|\mathbf{K}\| \cos\theta}\right).$$

This is the polar equation for an ellipse of eccentricity $\|\mathbf{K}\|$. (Exercises 4 and 5 will say a bit more about such polar equations.)

We have verified Kepler's first law.

2.6.3   Kepler's Third Law

Look at Figure 2.12. The length $2a$ of the major axis of our elliptical orbit is equal to the maximum value of r plus the minimum value of r. From the equation for the ellipse we see that these occur, respectively, when $\cos\theta$ is $+1$ and when $\cos\theta$ is -1. Thus

$$2a = \frac{\|\mathbf{C}\|^2}{GM} \cdot \frac{1}{1 - \|\mathbf{K}\|} + \frac{\|\mathbf{C}\|^2}{GM} \cdot \frac{1}{1 + \|\mathbf{K}\|} = \frac{2\|\mathbf{C}\|^2}{GM(1 - \|\mathbf{K}\|^2)}.$$

We conclude that
$$\|\mathbf{C}\| = \left(aGM(1 - \|\mathbf{K}\|^2)\right)^{1/2}. \tag{2.6.3}$$

Now recall from our proof of the second law that
$$\frac{dA}{dt} = \frac{1}{2}\|\mathbf{C}\|.$$

Then, by antidifferentiating, we find that
$$A(t) = \frac{1}{2}\|\mathbf{C}\|t.$$

(There is no constant term since $A(0) = 0$.) Let $\mathcal{A}$ denote the total area inside the elliptical orbit and T the time it takes to sweep out one orbit. Then
$$\mathcal{A} = A(T) = \frac{1}{2}\|\mathbf{C}\|T.$$

Solving for T we obtain
$$T = \frac{2\mathcal{A}}{\|\mathbf{C}\|}.$$

But the area inside an ellipse with major axis $2a$ and minor axis $2b$ is
$$\mathcal{A} = \pi ab = \pi a^2(1 - e^2)^{1/2},$$

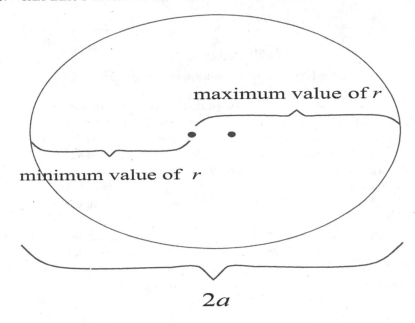

FIGURE 2.12
Analysis of the major axis.

where e is the eccentricity of the ellipse. This equals $\pi a^2(1 - \|\mathbf{K}\|^2)^{1/2}$ by Kepler's first law. Therefore

$$T = \frac{2\pi a^2(1 - \|\mathbf{K}\|^2)^{1/2}}{\|\mathbf{C}\|}.$$

Finally, we may substitute (2.6.3) into this last equation to obtain

$$T = \frac{2\pi a^{3/2}}{(GM)^{1/2}}$$

or

$$\frac{T^2}{a^3} = \frac{4\pi^2}{GM}.$$

This is Kepler's third law.

Math Nugget

Johannes Kepler and the Motion of the Planets

Johannes Kepler (1571–1630) is said to have been an ener-
getic and affectionate man. He married his first wife in part
for her money, and soon realized the error of his ways. When
she died, he decided to apply scientific methods in the selec-
tion of a second wife: he carefully analyzed and compared
the virtues and defects of several ladies before selecting his
second partner in matrimony. That marriage too was an
unhappy one.

Kepler's scientific career also had its ups and downs.
His attempt at collaboration with his hero Tycho Brahe fell
victim to the incompatibility of their strong personalities. In
his position as Royal Astronomer in Prague, a post which
he inherited from Tycho Brahe, he was often unpaid.

It appears that Kepler's personal frustration, his terrific
energy, and his scientific genius found a focus in questions
about planetary motion. Kepler formulated his three laws
by studying many years' worth of data about the motion of
the planets that had been gathered by Tycho Brahe. It is
amazing that he could stare at hundreds of pages of numerical data and come up with the three elegant laws that we
have discussed here.

Kepler could have simplified his task considerably by using
the tables of logarithms that John Napier (1550-1617) and
his assistants were developing at the time. But Kepler could
not understand Napier's mathematical justifications for his
tables, so he refused to use them.

Later, Newton conceived the idea that Kepler's laws could be *derived*, using calculus, from his inverse square law of gravitational attraction. In fact it seems clear that this problem is one of the main reasons that Newton developed the calculus. Newton's idea was a fantastic insight: that physical laws could be derived from a set of physical axioms was a new technique in the philosophy of science. On top of this, considerable technical proficiency was required to actually carry out the program. Newton's derivation of Kepler's laws, presented here in a modernized and streamlined form, is a model for the way that mathematical physics is done today.

EXAMPLE 2.6.1 The planet Uranus describes an elliptical orbit about the sun. It is known that the semimajor axis of this orbit has length 2870×10^6 kilometers. The gravitational constant is $G = 6.637 \times 10^{-8} \text{ cm}^3/(g \cdot \text{sec}^2)$. Finally, the mass of the sun is 2×10^{33} grams. Determine the period of the orbit of Uranus.

Solution: Refer to the explicit formulation of Kepler's third law that we proved above. We have

$$\frac{T^2}{a^3} = \frac{4\pi^2}{GM}.$$

We must be careful to use consistent units. The gravitational constant G is given in terms of grams, centimeters, and seconds. The mass of the sun is in grams. We convert the semimajor axis to centimeters: $a = 2870 \times 10^{11}$ cm. $= 2.87 \times 10^{14}$ cm. Then we calculate that

$$
\begin{aligned}
T &= \left(\frac{4\pi^2}{GM} \cdot a^3\right)^{1/2} \\
&= \left(\frac{4\pi^2}{(6.637 \times 10^{-8})(2 \times 10^{33})} \cdot (2.87 \times 10^{14})^3\right)^{1/2} \\
&\approx [70.308 \times 10^{17}]^{1/2} \text{sec}. \\
&= 26.516 \times 10^8 \text{ sec}.
\end{aligned}
$$

Notice how the units mesh perfectly so that our answer is in seconds. There are 3.16×10^7 seconds in an Earth year. We divide by this number to find that the time of one orbit of Uranus is

$$T \approx 83.9 \text{ Earth years}. \qquad \blacksquare$$

Exercises

1. It is common to take the "mean distance" of a planet from the sun to be the length of the semimajor axis of the elliptical orbit. That is because this number is the average of the least distance to the sun and the greatest distance to the sun. Now you can answer these questions:

    (a) Mercury's "year" is 88 Earth days. What is Mercury's mean distance from the sun?

    (b) The mean distance of the planet Saturn from the sun is 9.54 astronomical units.[3] What is Saturn's period of revolution about the sun?

2. Show that the speed v of a planet at any point of its orbit is given by

$$v^2 = k \left(\frac{2}{r} - \frac{1}{a} \right).$$

3. Suppose that the Earth explodes into fragments which fly off at the same speed in different directions into orbits of their own. Use Kepler's third law and the result of Exercise 2 to show that all fragments that do not fall into the sun or escape from the solar system will eventually reunite later at the same point where they began to diverge (i.e., where the explosion took place).

4. Kepler's first law may be written as

$$r = \frac{h^2/k}{1 + e\cos\theta}.$$

    Prove this assertion. Kepler's second law may be written as

$$r^2 \frac{d\theta}{dt} = h.$$

    Prove this assertion too. Let $\mathbf{F}$ be the central attractive force that the sun exerts on the planet and F its magnitude. Now verify these statements:

    (a) $F_\theta = 0$

    (b) $\dfrac{dr}{dt} = \dfrac{ke}{h} \sin\theta$

    (c) $\dfrac{d^2r}{dt^2} = \dfrac{ke\cos\theta}{r^2}$

    (d) $F_r = -\dfrac{mk}{r^2} = -G\dfrac{Mm}{r^2}$

    Use these facts to prove that a planet of mass m is attracted toward the origin (the center of the sun) with a force whose magnitude is inversely proportional to the square of r. (Newton's discovery of this fact caused him to formulate his law of universal gravitation and to investigate its consequences.)

---

[3]Here one astronomical unit is the Earth's mean distance from the sun, which is $93,000,000$ miles or $150,000,000$ kilometers.

5. It is common to take $h = \|\mathbf{C}\|$ and $k = GM$. Kepler's third law may then be formulated as

$$T^2 = \frac{4\pi^2 a^2 b^2}{h^2} = \left(\frac{4\pi^2}{k}\right) a^3 \qquad (*)$$

(remember that $b^2 = a^2(1 - e^2) = a^2(1 - \|\mathbf{K}\|^2)$). Prove this formula.

In working with Kepler's third law, it is customary to measure T in Earth years and a in astronomical units (see Exercise 1, the footnote, for the definition of this term). With these convenient units of measure, $(*)$ takes the simpler form $T^2 = a^3$. What is the period of revolution T of a planet whose mean distance from the sun is

(a) twice that of the Earth?

(b) three times that of the Earth?

(c) 25 times that of the Earth?

6. Use your symbol manipulation software, such as `Maple` or `Mathematica`, to calculate the orbit of a planet having mass m about a "sun" of mass M, assuming that the planet is given an initial velocity of v_0.

2.7 Higher-Order Coupled Harmonic Oscillators

We treat here some aspects of higher-order equations that bear a similarity to what we have learned about second-order examples. We shall concentrate primarily on linear equations with constant coefficients. As usual, we illustrate the ideas with a few key examples.

We consider an equation of the form

$$y^{(n)} + a_{n-1}y^{(n-1)} + \cdots + a_1 y^{(1)} + a_0 y = f. \qquad (2.7.1)$$

Here a superscript $^{(j)}$ denotes a jth derivative and f is some continuous function. This is a linear, constant-coefficient, ordinary differential equation of order n.

Following what we learned about second-order equations, we expect the general solution of (2.7.1) to have the form

$$y = y_0 + y_g,$$

where y_0 is a particular solution of (2.7.1) and y_g is the general solution of the associated homogeneous equation

$$y^{(n)} + a_{n-1}y^{(n-1)} + \cdots + a_1 y^{(1)} + a_0 y = 0. \qquad (2.7.2)$$

Furthermore, we expect that y_g will have the form

$$y_g = A_1 y_1 + A_2 y_2 + \cdots + A_{n-1} y_{n-1} + A_n y_n,$$

where the y_j are "independent" solutions of (2.7.2) and the A_j are arbitrary constants.

We begin by studying the homogeneous equation (2.7.2) and seeking the general solution y_g. Again following the paradigm that we developed for second-order equations, we guess a solution of the form $y = e^{rx}$. Substituting this guess into (2.7.2), we find that

$$e^{rx} \cdot \left(r^n + a_{n-1}r^{n-1} + \cdots + a_1 r + a_0 \right) = 0 \, .$$

Thus we are led to solving the *associated polynomial*

$$r^n + a_{n-1}r^{n-1} + \cdots + a_1 r + a_0 = 0 \, .$$

The fundamental theorem of algebra tells us that every polynomial of degree n has a total of n complex roots $r_1, r_2, \ldots, r_n$ (there may be repetitions in this list). Thus the polynomial factors as

$$(r - r_1) \cdot (r - r_2) \cdots (r - r_{n-1}) \cdot (r - r_n) \, .$$

In practice there may be some difficulty in *actually finding* the complete set of roots of a given polynomial. For instance, it is known that for polynomials of degree 5 and greater there is no elementary formula for the roots. Let us pass over this sticky point for the moment, and continue to comment on the theoretical setup.

I. Distinct Real Roots: For a given associated polynomial, if the polynomial roots $r_1, r_2, \ldots, r_n$ are distinct and real, then we can be sure that

$$e^{r_1 x}, e^{r_2 x}, \ldots, e^{r_n x}$$

are n distinct solutions to the differential equation (2.7.2). It then follows, just as in the order-2 case, that

$$y_g = A_1 e^{r_1 x} + A_2 e^{r_2 x} + \cdots + A_n e^{r_n x}$$

is the general solution to (2.7.2) that we seek.

II. Repeated Real Roots: If the roots are real, but two of them are equal (say that $r_1 = r_2$), then of course $e^{r_1 x}$ and $e^{r_2 x}$ are *not* distinct solutions of the differential equation. Just as in the case of order-2 equations, what we do in this case is manufacture two distinct solutions of the form $e^{r_1 x}$ and $x \cdot e^{r_1 x}$.

More generally, if several of the roots are equal, say $r_1 = r_2 = \cdots = r_k$, then we manufacture distinct solutions of the form $e^{r_1 x}, x \cdot e^{r_1 x}, x^2 \cdot e^{r_1 x}, \ldots, x^{k-1} \cdot e^{r_1 x}$.

III. Complex Roots: We have been assuming that the coefficients of the original differential equation ((2.7.1) or (2.7.2)) are all real. This being the case, any complex roots of the associated polynomial will occur in conjugate pairs $a + ib$ and $a - ib$. Then we have distinct solutions $e^{(a+ib)x}$ and $e^{(a-ib)x}$. Then we can use Euler's formula and a little algebra, just as we did in the order-2 case, to produce distinct real solutions $e^{ax} \cos bx$ and $e^{ax} \sin bx$.

In the case that a complex root is repeated to order k, then we take

$$e^{ax} \cos bx, xe^{ax} \cos bx, \ldots, x^{k-1} e^{ax} \cos bx$$

and

$$e^{ax} \sin bx, xe^{ax} \sin bx, \ldots, x^{k-1} e^{ax} \sin bx$$

as solutions of the ordinary differential equation.

EXAMPLE 2.7.3 Find the general solution of the differential equation

$$y^{(4)} - 5y^{(2)} + 4y = 0 .$$

Solution: The associated polynomial is

$$r^4 - 5y^2 + 4 = 0 .$$

Of course we may factor this as $(r^2 - 4)(r^2 - 1) = 0$ and then as

$$(r - 2)(r + 2)(r - 1)(r + 1) = 0 .$$

We find, therefore, that the general solution of our differential equation is

$$y(x) = A_1 e^{2x} + A_2 e^{-2x} + A_3 e^{x} + A_4 e^{-x} . \qquad \blacksquare$$

EXAMPLE 2.7.4 Find the general solution of the differential equation

$$y^{(4)} - 8y^{(2)} + 16y = 0 .$$

Solution: The associated polynomial is

$$r^4 - 8r^2 + 16 = 0 .$$

This factors readily as $(r^2 - 4)(r^2 - 4) = 0$, and then as

$$(r - 2)^2 (r + 2)^2 = 0 .$$

Thus the root 2 is repeated and the root -2 is repeated. According to our discussion in part **II**, the general solution of the differential equation is then

$$y(x) = A_1 e^{2x} + A_2 xe^{2x} + A_3 e^{-2x} + A_4 xe^{-2x} . \qquad \blacksquare$$

EXAMPLE 2.7.5 Find the general solution of the differential equation

$$y^{(4)} - 2y^{(3)} + 2y^{(2)} - 2y^{(1)} + y = 0.$$

Solution: The associated polynomial is

$$r^4 - 2r^3 + 2r^2 - 2r + 1 = 0.$$

We notice, just by inspection, that $r_1 = 1$ is a solution of this polynomial equation. Thus $r - 1$ divides the polynomial. In fact

$$r^4 - 2r^3 + 2r^2 - 2r + 1 = (r - 1) \cdot (r^3 - r^2 + r - 1).$$

But we again see that $r_2 = 1$ is a root of the new third-degree polynomial. Dividing out $r - 1$ again, we obtain a quadratic polynomial that we can solve directly.

The end result is

$$r^4 - 2r^3 + 2r^2 - 2r + 1 = (r - 1)^2 \cdot (r^2 + 1) = 0$$

or

$$(r - 1)^2 (r - i)(r + i) = 0.$$

The roots are 1 (repeated), i, and $-i$. As a result, we find that the general solution of the differential equation is

$$y(x) = A_1 e^x + A_2 x e^x + A_3 \cos x + A_4 \sin x. \qquad \blacksquare$$

EXAMPLE 2.7.6 Find the general solution of the equation

$$y^{(4)} - 5y^{(2)} + 4y = \sin x. \qquad (2.7.6.1)$$

Solution: In fact we found the general solution of the associated homogeneous equation in Example 2.7.1. To find a particular solution of (2.7.6.1), we use undetermined coefficients and guess a solution of the form $y = \alpha \cos x + \beta \sin x$. A little calculation reveals then that $y_p(x) = (1/10) \sin x$ is the particular solution that we seek. As a result,

$$y(x) = \frac{1}{10} \sin x + A_1 e^{2x} + A_2 e^{-2x} + A_3 e^x + A_4 e^{-x}$$

is the general solution of (2.7.6.1). $\qquad \blacksquare$

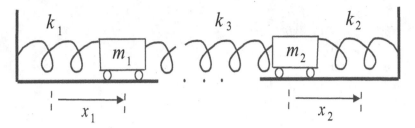

FIGURE 2.13
Coupled harmonic oscillators.

EXAMPLE 2.7.7 (Coupled Harmonic Oscillators) Linear equations of order greater than 2 arise in physics by the elimination of variables from simultaneous systems of second-order equations. We give here an example that arises from coupled harmonic oscillators. Accordingly, let two carts of masses m_1, m_2 be attached to left and right walls as in Figure 2.13 with springs having spring constants k_1, k_2. If there is no damping and the carts are unattached, then of course when the carts are perturbed we have two separate harmonic oscillators.

But if we connect the carts, with a spring having spring constant k_3, then we obtain *coupled harmonic oscillators*. In fact Newton's second law of motion can now be used to show that the motions of the coupled carts will satisfy these differential equations:

$$m_1 \frac{d^2 x_1}{dt^2} = -k_1 x_1 + k_3(x_2 - x_1).$$
$$m_2 \frac{d^2 x_2}{dt^2} = -k_2 x_2 - k_3(x_2 - x_1).$$

We can solve the first equation for x_2,

$$x_2 = \frac{1}{k_3} \left(x_1[k_1 + k_3] + m_1 \frac{d^2 x_1}{dt^2} \right),$$

and then substitute into the second equation. The result is a fourth order equation for x_1. ∎

Exercises

In each of Exercises 1–15, find the general solution of the given differential equation.

1. $y''' - 3y'' + 2y' = 0$

2. $y''' - 3y'' + 4y' - 2y = 0$

3. $y''' - y = 0$

4. $y''' + y = 0$

5. $y''' + 3y'' + 3y' + y = 0$

6. $y^{(4)} + 4y''' + 6y'' + 4y' + y = 0$

7. $y^{(4)} - y = 0$

8. $y^{(4)} + 5y'' + 4y = 0$

9. $y^{(4)} - 2a^2 y'' + a^4 y = 0$

10. $y^{(4)} + 2a^2 y'' + a^4 y = 0$

11. $y^{(4)} + 2y''' + 2y'' + 2y' + y = 0$

12. $y^{(4)} + 2y''' - 2y'' - 6y' + 5y = 0$

13. $y''' - 6y'' + 11y' - 6y = 0$

14. $y^{(4)} + y''' - 3y'' - 5y' - 2y = 0$

15. $y^{(5)} - 6y^{(4)} - 8y''' + 48y'' + 16y' - 96y = 0$

16. Find the general solution of $y^{(4)} = 0$. Now find the general solution of $y^{(4)} = \sin x + 24$.

17. Find the general solution of $y''' - 3y'' + 2y' = 10 + 42e^{3x}$.

18. Find the solution of $y''' - y' = 1$ that satisfies the initial conditions $y(0) = y'(0) = y''(0) = 4$.

19. Show that the change of independent variable given by $x = e^z$ transforms the third-order Euler equidimensional equation

$$x^3 y''' + a_2 x^2 y'' + a_1 x y' + a_0 y = 0$$

into a third-order linear equation with constant coefficients. Solve the following equations by this method.

(a) $x^3 y''' + 3x^2 y'' = 0$

(b) $x^3 y''' + x^2 y'' - 2xy' + 2y = 0$

(c) $x^3 y''' + 2x^2 y'' + xy' - y = 0$

20. In determining the drag on a small sphere moving at a constant speed through a viscous fluid, it is necessary to solve the differential equation

$$x^3 y^{(4)} + 8x^2 y''' + 8xy'' - 8y' = 0.$$

If we make the substitution $w = y'$, then this becomes a third-order Euler equation that we can solve by the method of Exercise 19. Do so, and show that the general solution is

$$y(x) = c_1 x^2 + c_2 x^{-1} + c_3 x^{-3} + c_4.$$

(These ideas are part of the mathematical foundation of the work of Robert Millikan in his famous oil-drop experiment of 1909 for measuring the charge of an electron. He won the Nobel Prize for this work in 1923.)

21. In Example 2.7.7, find the fourth-order differential equation for x_1 by eliminating x_2, as described at the end of the example.

22. In Exercise 21, solve the fourth-order equation for x_1 if the masses are equal and the spring constants are equal, so that $m_1 = m_2 = m$ and $k_1 = k_2 = k_3 = k$. In this special case, show directly that x_2 satisfies the same differential equation as x_1. The two frequencies associated with these coupled harmonic oscillators are called the *normal frequencies* of the system. What are they?

Historical Note

Euler

Leonhard Euler (1707–1783), a Swiss by birth, was one of the foremost mathematicians of all time. He was also arguably the most prolific author of all time in any field. The publication of Euler's complete works was begun in 1911 and the end is still not in sight. The works were originally planned for 72 volumes, but new manuscripts have been discovered and the work continues.

Euler's interests were vast, and ranged over all parts of mathematics and science. He wrote effortlessly and fluently. When he was stricken with blindness at the age of 59, he enlisted the help of assistants to record his thoughts. Aided by his powerful memory and fertile imagination, Euler's output actually increased.

Euler was a native of Basel, Switzerland, and a student of the noted mathematician Johann Bernoulli (mentioned elsewhere in this text). His working life was spent as a member of the Academies of Science at Berlin and St. Petersburg. He was a man of broad culture, well-versed in the classical languages and literatures (he knew the *Aeneid* by heart), physiology, medicine, botany, geography, and the entire body of physical science.

Euler had 13 children. Even so, his personal life was uneventful and placid. It is said that he died while dandling a grandchild on his knee. He never taught, but his influence on the teaching of mathematics has been considerable. Euler wrote three great treatises: *Introductio in Analysin Infinitorum* (1748), *Institutiones Calculi Differentialis* (1755), and *Institutiones Calculi Integralis* (1768–1794). These works both assessed and codified the works of all Euler's predecessors, and they contained many of his own ideas as well. It has been said that all elementary and advanced calculus textbooks since 1748 are either copies of Euler or copies of copies of Euler.

Among many other important advances, Euler's work extended and perfected plane and solid analytic geometry, introduced the analytic approach to trigonometry, and was responsible for the modern treatment of the functions $\ln x$ and e^x. He created a consistent theory of logarithms of negative and imaginary numbers, and discovered that $\ln z$ has infinitely many values. Euler's work established the use of the symbols e, π, and i (for $\sqrt{-1}$). Euler linked these three important numbers together with the remarkable formula

$$e^{\pi i} = -1 \,.$$

Euler was also the one who established the notation $\sin x$ and $\cos x$, the use of $f(x)$ for an arbitrary function, and the use of $\sum$ to denote a series.

The distinction between pure and applied mathematics did not exist in Euler's day. For him, the entire physical universe was grist for his mill. The foundations of classical mechanics were laid by Isaac Newton, but Euler was the principal architect of the subject. In his treatise of 1736 he was the first to explicitly introduce the concept of a mass-point or particle, and he was also the first to study the acceleration of a particle moving along any curve and to use the notion of a vector in connection with velocity and accceleration. Euler's discoveries were so pervasive that many of them continue to be used without any explicit credit to Euler. Among his named contributions are Euler's equations of motion for the rotation of a rigid body, Euler's hydrodynamic equation for the flow of an ideal incompressible fluid, Euler's law for the bending of elastic beams, and Euler's critical load in the theory of the buckling of columns.

Few mathematicians have had the fluency, the clarity of thought, and the profound influence of Leonhard Euler. His ideas are an integral part of modern mathematics.

2.8   Bessel Functions and the Vibrating Membrane

Bessel functions arise typically in the solution of Bessel's differential equation

$$x^2 y'' + xy' + (x^2 - p^2)y = 0 \,.$$

They are among the most important special functions of mathematical physics. In the present discussion we shall explore the use of these functions to describe Euler's analysis of a vibrating circular membrane. The approach is similar to that for the vibrating string, which is treated elsewhere in the present chapter (Section 2.5).

We shall be considering a uniform thin sheet of flexible material (polyester, perhaps). The sheet will be clamped along a given closed curve (a circle, perhaps) in the x-y plane and pulled taut into a state of uniform tension. Think, for instance, of a drum.

When the membrane is displaced slightly from its equilibrium position and then released, the restoring forces created by the deformation cause it to vibrate. For instance, this is how a drum works. To simplify the mathematics, we shall consider only *small oscillations* of a *freely vibrating* membrane.

We shall assume that the membrane lies in the x-y plane and that the

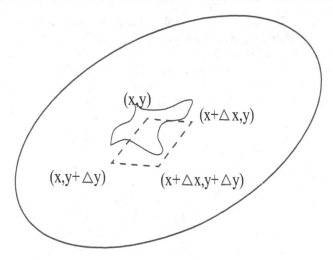

FIGURE 2.14

A segment of the vibrating membrane.

displacement of the membrane is so small that each point of the surface is moved only in the z direction (i.e., perpendicular to the plane of the membrane). The displacement is given by a function $z = z(x, y, t)$, where t is time. We shall consider a small, rectangular piece of the membrane with dimensions $\triangle x$ and $\triangle y$. The corners of this rectangle are the points (x, y), $(x + \triangle x, y)$, $(x, y + \triangle y)$, and $(x + \triangle x, y + \triangle y)$. This rectangle, and the portion of the displaced membrane that lies above it, are depicted in Figure 2.14.

If m is the constant mass per unit area of the membrane, then the mass of the rectangular piece is $m \triangle x \triangle y$. Newton's second law of motion then tells us that

$$F = m \triangle x \triangle y \frac{\partial^2 z}{\partial t^2} \tag{2.8.1}$$

is the force acting on the membrane in the z-direction.

When the membrane is in equilibrium position, the constant tension T in the surface has this physical meaning: Along any small line segment in the membrane of length $\triangle s$, the membrane material on one side of the segment exerts a force, normal to the segment and of magnitude $T \triangle s$, on the membrane material on the other side. In this case, because the membrane is in equilibrium, the forces on opposite sides of the segment are both parallel to the x-y plane and cancel one another. When the membrane is curved (i.e., displaced), however, as in the frozen moment depicted in Figure 2.14, we shall assume that the deformation is so small that the tension is still T but now acts parallel to the tangent plane, and therefore has a nontrivial vertical component. It is the curvature of the distorted membrane that produces different magnitudes for these vertical components on opposite edges, and this in turn is the source of the restoring forces that cause the vibrating motion.

We analyze these forces by assuming that the piece of the membrane indicated in the figure is only slightly tilted. This makes it possible to replace the sines of certain small angles by their tangents, thereby simplifying the calculations. We proceed as follows. Along the upper and lower edges in the figure, the forces are perpendicular to the x-axis and almost parallel to the y-axis, with small z-components approximately equal to

$$T \, \triangle x \left(\frac{\partial z}{\partial y} \right)_{y+\triangle y} \qquad \text{and} \qquad -T \, \triangle x \left(\frac{\partial z}{\partial y} \right)_{y} .$$

Hence their sum is approximately equal to

$$T \, \triangle x \left\{ \left(\frac{\partial z}{\partial y} \right)_{y+\triangle y} - \left(\frac{\partial z}{\partial y} \right)_{y} \right\} .$$

The subscripts on these partial derivatives indicate their values at the points $(x, y + \triangle y)$ and (x, y).

Performing the same type of analysis on the left and right edges, we find that the total force in the z-direction—coming from all four edges—is

$$T \, \triangle y \left\{ \left(\frac{\partial z}{\partial x} \right)_{x+\triangle x} - \left(\frac{\partial z}{\partial x} \right)_{x} \right\} + T \, \triangle x \left\{ \left(\frac{\partial z}{\partial y} \right)_{y+\triangle y} - \left(\frac{\partial z}{\partial y} \right)_{y} \right\} .$$

As a result, equation (2.8.1) can be rewritten as

$$T \frac{(\partial z/\partial x)_{x+\triangle x} - (\partial z/\partial x)_{x}}{\triangle x} + T \frac{(\partial z/\partial y)_{y+\triangle y} - (\partial z/\partial y)_{y}}{\triangle y} = m \frac{\partial^2 z}{\partial t^2} .$$

If we now set $a^2 = T/m$ and let $\triangle x \to 0$, $\triangle y \to 0$, then we find that

$$a^2 \left(\frac{\partial^2 z}{\partial x^2} + \frac{\partial^2 z}{\partial y^2} \right) = \frac{\partial^2 z}{\partial t^2} ; \qquad (2.8.2)$$

this is the *two-dimensional wave equation*. We note that this is a *partial differential equation*: it involves functions of three variables, and of course partial derivatives.

Now we shall consider the displacement of a circular membrane. So our study will be the model for a drum. Of course we shall use polar coordinates, with the origin at the center of the drum. The wave equation now has the form

$$a^2 \left(\frac{\partial^2 z}{\partial r^2} + \frac{1}{r} \frac{\partial z}{\partial r} + \frac{1}{r^2} \frac{\partial^2 z}{\partial \theta^2} \right) = \frac{\partial^2 z}{\partial t^2} . \qquad (2.8.3)$$

Here, naturally, $z = z(r, \theta, t)$ is a function of the polar coordinates r, θ and of time t. We assume, without loss of generality, that the membrane has radius 1. Thus it is clamped to its plane of equilibrium along the circle $r = 1$ in the polar coordinates. Thus our boundary condition is

$$z(1, \theta, t) = 0 , \qquad (2.8.4)$$

because the height of the membrane at the edge of the disc-shaped displaced region is 0. The problem, then, is to find a solution of (2.8.3) that satisfies the boundary condition (2.8.4) together with certain initial conditions that we shall not consider at the moment.

We shall apply the method of separation of variables. Thus we seek a solution of the form

$$z(r, \theta, t) = u(r)v(\theta)w(t). \tag{2.8.5}$$

We substitute (2.8.5) into (2.8.3) and perform a little algebra to obtain

$$\frac{u''(r)}{u(r)} + \frac{1}{r}\frac{u'(r)}{u(r)} + \frac{1}{r^2}\frac{v''(\theta)}{v(\theta)} = \frac{1}{a^2}\frac{w''(t)}{w(t)}. \tag{2.8.6}$$

Now our analysis follows familiar lines: Since the left-hand side of (2.8.6) depends on r and θ only and the right-hand side depends on t only, we conclude that both sides are equal to some constant K. In order for the membrane to vibrate, $w(t)$ must be periodic. Thus the constant K must be negative (if it is positive, then the solutions of $w'' - Ka^2w = 0$ will be real exponentials and hence *not* periodic). We thus equate both sides of (2.8.6) with $K = -\lambda^2$ for some $\lambda > 0$ and obtain the two ordinary differential equations

$$w''(t) + \lambda^2 a^2 w(t) = 0 \tag{2.8.7}$$

and

$$\frac{u''(r)}{u(r)} + \frac{1}{r}\frac{u'(r)}{u(r)} + \frac{1}{r^2}\frac{v''(\theta)}{v(\theta)} = -\lambda^2. \tag{2.8.8}$$

Now (2.8.7) is easy to solve, and its general solution is

$$w(t) = c_1 \cos \lambda at + c_2 \sin \lambda at. \tag{2.8.9}$$

We can rewrite (2.8.8) as

$$r^2\frac{u''(r)}{u(r)} + r\frac{u'(r)}{u(r)} + \lambda^2 r^2 = -\frac{v''(\theta)}{v(\theta)}. \tag{2.8.10}$$

Notice that in equation (2.8.10) we have a function of r only on the left and a function of θ only on the right. So, as usual, both sides must be equal to a constant L. Now we know, by the physical nature of our problem, that v must be 2π-periodic. Looking at the right-hand side of (2.8.10) then tells us that $L = n^2$ for $n \in \{0, 1, 2, \dots\}$.

With these thoughts in mind, equation (2.8.10) splits into

$$v''(\theta) + n^2 v(\theta) = 0 \tag{2.8.11}$$

and

$$r^2 u''(r) + ru'(r) + (\lambda^2 r^2 - n^2)u(r) = 0. \tag{2.8.12}$$

Of course equation (2.8.11) has, as its general solution,

$$v(\theta) = d_1 \cos n\theta + d_2 \sin n\theta.$$

(Note that this solution is not valid when $n = 0$. But, when $n = 0$ the equation has no nontrivial periodic solutions.) Also observe that equation (2.8.12) is a slight variant of Bessel's equation (in fact a change of variables of the form $r = w^b$, $u = v \cdot w^c$, for appropriately chosen b and c, will transform the Bessel's equation given at the beginning of this discussion to equation (2.8.12)). It turns out that, according to physical considerations, the solution that we want of equation (2.8.12) is

$$u(r) = k \cdot J_n(r).$$

Here k is a physical constant and J_n is the nth Bessel function, discussed in detail in Chapters 3 and 4 below. Note for now that the Bessel functions are transcendental functions, best described with power series.

Let us conclude this discussion by considering the boundary condition (2.8.4) for our problem. It can now be written simply as $u(1) = 0$ or

$$J_n(\lambda) = 0.$$

Thus the permissible values of λ are the positive zeros of the Bessel function J_n (see also the discussion in Chapter 3 below). It is known that there are infinitely many such zeros, and they have been studied intensively over the years. The reference [WAT] is a great source of information about Bessel functions.

Problems for Review and Discovery

A. Drill Exercises

1. Find the general solution of each of these differential equations.
   - (a) $y'' - 3y' + y = 0$
   - (b) $y'' + y' + y = 0$
   - (c) $y'' + 6y' + 9y = 0$
   - (d) $y'' - y' + 6y = 0$
   - (e) $y'' - 2y' - 5y = x$
   - (f) $y'' + y = e^x$
   - (g) $y'' + y' + y = \sin x$
   - (h) $y'' - y = e^{3x}$

2. Solve each of the following initial value problems.
   - (a) $y'' + 9y = 0$, $\quad y(0) = 1$, $y'(0) = 2$
   - (b) $y'' - y' + 4y = x$, $\quad y(1) = 2$, $y'(1) = 1$
   - (c) $y'' + 2y' + 5y = e^x$, $\quad y(0) = -1$, $y'(0) = 1$
   - (d) $y'' + 3y' + 4y = \sin x$, $\quad y(\pi/2) = 1$, $y'(\pi/2) = -1$

    (e) $y'' + y = e^{-x}$, $y(2) = 0$, $y'(2) = -2$
    (f) $y'' - y = \cos x$, $y(0) = 3$, $y'(0) = 2$
    (g) $y'' = \tan x$, $y(1) = 1$, $y'(1) = -1$
    (h) $y'' - 2y' = \ln x$, $y(1) = e$, $y'(1) = 1/e$

3. Solve each of these differential equations.

    (a) $y'' + 3y' + 2y = 2x - 1$
    (b) $y'' - 3y' + 2y = e^{-x}$
    (c) $y'' - y' - 2y = \cos x$
    (d) $y'' + 2y' - y = xe^x \sin x$
    (e) $y'' + 9y = \sec 2x$
    (f) $y'' + 4y' + 4y = x \ln x$
    (g) $x^2 y'' + 3xy' + y = 2/x$
    (h) $y'' + 4y = \tan^2 x$

4. Use the given solution of the differential equation to find the general solution.

    (a) $y'' - y = 3e^{2x}$, $y_1(x) = e^{2x}$
    (b) $y'' + y = -8 \sin 3x$, $y_1(x) = \sin 3x$
    (c) $y'' + y' + y = x^2 + 2x + 2$, $y_1(x) = x^2$
    (d) $y'' + y' = \frac{x-1}{x^2}$, $y_1(x) = \ln x$

B. Challenge Exercises

1. Consider the differential equation $y'' + 4y = 0$. Convert it to a system of first-order, linear ordinary differential equations by setting $v = y'$. Hence we have

$$y' = v$$
$$v' = -4y$$

Find solutions $y(x), v(x)$ for this system. If we think of x as a parameter, then the map

$$x \longmapsto (y(x), v(x))$$

describes a curve in the plane. Draw this curve for a variety of different sets of initial conditions. What sort of curve is it?

2. Explain why $y_1(x) = \sin x$ and $y_2(x) = 2x$ cannot both be solutions of the same ordinary differential equation

$$y'' = F(x, y, y')$$

for a smooth F.

3. Show that the Euler equation

$$x^2 y'' - 2xy' + 2y = 0$$

with initial conditions

$$y(0) = 0, \ y'(0) = 0$$

has infinitely many solutions. Why does this surprising result not contradict the general ideas that we learned in this chapter?

4. Does the differential equation

$$y'' + 9y = -3\cos 2x$$

have any periodic solutions? Say exactly what you mean by "periodic" as you explain your answer.

C. Problems for Discussion and Exploration

1. Show that the ordinary differential equation $y' + y = \cos x$ has a unique periodic solution.

2. Find the regions where the solution of the initial value problem

$$y'' = -3y, \quad y(0) = -1$$

is concave down. In what regions is it concave up? What do these properties tell us about the solution? Do you need to actually solve the differential equation in order to answer this question?

3. Consider solutions of the differential equation

$$\frac{d^2y}{dx^2} - c\frac{dy}{dx} + y = 0$$

for a constant c. Describe how the behavior of this solution changes as c varies.

4. Endeavor to find an approximate solution to the differential equation

$$\frac{d^2y}{dx^2} + \sin y = 0$$

by guessing that the solution is a polynomial. Think of this polynomial as the Taylor polynomial of the actual solution. Can you say anything about how accurately your polynomial solution approximates the true solution?

3

Power Series Solutions and Special Functions

- Power series basics

- Convergence of power series

- Series solutions of first-order equations

- Series solutions of second-order equations

- Ordinary points

- Regular singular points

- Frobenius's method

3.1 Introduction and Review of Power Series

It is useful to classify the functions that we know, or will soon know, in an informal way. The *polynomials* are functions of the form

$$a_0 + a_1 x + a_2 x^2 + \cdots + a_{k-1} x^{k-1} + a_k x^k,$$

where $a_0, a_1, \ldots, a_k$ are constants. This is a polynomial of degree k. A *rational function* is a quotient of polynomials. For example,

$$r(x) = \frac{3x^3 - x + 4}{x^2 + 5x + 1}$$

is a rational function.

A *transcendental function* is one that is not a polynomial or a rational function or a root. The *elementary transcendental functions* are the ones that we encounter in calculus class: sine, cosine, logarithm, exponential, and their inverses and combinations using arithmetic/algebraic operations.

The *higher transcendental functions* are ones that are not elementary and are defined using power series (although they often arise by way of integrals or asymptotic expansions or other means). These often are discovered as solutions of differential equations. These functions are a bit difficult to understand, just because they are not given by elementary formulas. But they are frequently very important because they come from fundamental problems of mathematical physics. As an example, solutions of Bessel's equation, which we saw at the end of the last chapter, are called Bessel functions and are studied intensively (see [WAT]).

Higher transcendental functions are frequently termed *special functions*. These important functions were studied extensively in the eighteenth and nineteenth centuries—by Gauss, Euler, Abel, Jacobi, Weierstrass, Riemann, Hermite, Poincaré, and other leading mathematicians of the day. Although many of the functions that they studied were quite recondite, and are no longer of much interest today, others (such as the Riemann zeta function, the gamma function, and elliptic functions) are still intensively studied.

In the present chapter we shall learn to solve differential equations using the method of power series, and we shall have a very brief introduction to how special functions arise from this process. It is a long chapter, with a number of new ideas. But there are many rewards along the way.

3.1.1   Review of Power Series

We begin our study with a quick review of the key ideas from the theory of power series.

I.   A series of the form

$$\sum_{j=0}^{\infty} a_j x^j = a_0 + a_1 x + a_2 x^2 + \cdots \tag{3.1.1}$$

is called a *power series in x*. Slightly more general is the series

$$\sum_{j=0}^{\infty} a_j (x - a)^j \,,$$

which is a *power series in $x - a$* (or *expanded about the point a*).

II.   The series (3.1.1) is said to *converge* at a point x if the limit

$$\lim_{N \to \infty} \sum_{j=0}^{N} a_j x^j$$

exists. The value of the limit is called the *sum* of the series. (This is just the familiar idea of defining the value of a series to be the limit of its partial sums.)

Obviously (3.1.1) converges when $x = 0$, since all terms but the first (or zeroeth) will then be equal to 0. The following three examples illustrate, in an informal way, what the convergence properties might be at other values of x.

(a)   The series

$$\sum_{j=0}^{\infty} j! x^j = 1 + x + 2! x^2 + 3! x^3 + \cdots$$

diverges[1] at every $x \neq 0$. This can be seen by using the ratio test from the theory of series. It of course converges at $x = 0$.

(b)   The series

$$\sum_{j=0}^{\infty} \frac{x^j}{j!} = 1 + x + \frac{x^2}{2!} + \frac{x^3}{3!} + \cdots$$

converges at every value of x, including $x = 0$. This can be seen by applying the ratio test from the theory of series.

(c)   The series

$$\sum_{j=0}^{\infty} x^j = 1 + x + x^2 + x^3 + \cdots$$

converges when $|x| < 1$ and diverges when $|x| \geq 1$.

These three examples are special instances of a general phenomenon that governs the convergence behavior of power series. There will *always* be a number R, $0 \leq R \leq \infty$, such that the series converges for $|x| < R$ and diverges for $|x| > R$. In the first example, $R = 0$; in the second example, $R = +\infty$; in the third example, $R = 1$. We call R the *radius of convergence* of the power series. The interval $(-R, R)$ is called the *interval of convergence*. In practice, we check convergence at the endpoints of the interval of convergence by hand in each example. We add those points to the interval of convergence as appropriate. The next three examples will illustrate how we calculate R in practice.

EXAMPLE 3.1.2 Calculate the interval of convergence of the series

$$\sum_{j=0}^{\infty} \frac{x^j}{j^2}.$$

Solution: We apply the ratio test:

$$\lim_{j \to \infty} \left| \frac{x^{j+1}/(j+1)^2}{x^j/j^2} \right| = \left| \lim_{j \to \infty} \frac{j^2}{(j+1)^2} \cdot x \right| = |x|.$$

---

[1]Here we use the notation $n! = n \cdot (n-1) \cdot (n-2) \cdots 3 \cdot 2 \cdot 1$. This is called the *factorial notation*. By convention, $1! = 1$ and $0! = 1$.

We know that the series will converge when this limit is less than 1, or $|x| < 1$. Likewise, it diverges when $|x| > 1$. Thus the radius of convergence is $R = 1$.

In practice, one has to check the *endpoints* of the interval of convergence by hand for each case. In this example, we see immediately that the series converges at $x = \pm 1$. Thus we may say that the interval of convergence is $[-1, 1]$.   ∎

EXAMPLE 3.1.3 Calculate the interval of convergence of the series

$$\sum_{j=0}^{\infty} \frac{x^j}{j}.$$

Solution: We apply the ratio test:

$$\lim_{j \to \infty} \left| \frac{x^{j+1}/(j+1)}{x^j/j} \right| = \left| \lim_{j \to \infty} \frac{j}{j+1} \cdot x \right| = |x|.$$

We know that the series will converge when this limit is less than 1, or $|x| < 1$. Likewise, it diverges when $|x| > 1$. Thus the radius of convergence is $R = 1$.

In this example, we see immediately that the series converges at -1 (by the alternating series test) and diverges at $+1$ (since this gives the harmonic series). Thus we may say that the interval of convergence is $[-1, 1)$.   ∎

EXAMPLE 3.1.4 Calculate the interval of convergence of the series

$$\sum_{j=0}^{\infty} \frac{x^j}{j^j}.$$

Solution: We use the root test:

$$\lim_{j \to \infty} \left| \frac{x^j}{j^j} \right|^{1/j} = \lim_{j \to \infty} \left| \frac{x}{j} \right| = 0.$$

Of course $0 < 1$, regardless of the value of x. So the series converges for all x. The radius of convergence is $+\infty$ and the interval of convergence is $(-\infty, +\infty)$. There is no need to check the endpoints of the interval of convergence, because there are none.   ∎

III. Suppose that our power series (3.1.1) converges for $|x| < R$ with $R > 0$. Denote its sum by $f(x)$, so

$$f(x) = \sum_{j=0}^{\infty} a_j x^j = a_0 + a_1 x + a_2 x^2 + \cdots. \tag{3.1.2}$$

Thus the power series defines a *function*, and we may consider differentiating it. In fact the function f is continuous and has derivatives of all orders. We may calculate the derivatives by differentiating the series termwise:

$$f'(x) = \sum_{j=1}^{\infty} j a_j x^{j-1} = a_1 + 2a_2 x + 3a_3 x^2 + \cdots, \qquad (3.1.3)$$

$$f''(x) = \sum_{j=2}^{\infty} j(j-1) x^{j-2} = 2a_2 + 3 \cdot 2a_3 x + \cdots, \qquad (3.1.4)$$

and so forth. Each of these series converges on the same interval $|x| < R$.

Observe that, if we evaluate (3.1.2) at $x = 0$, then we learn that

$$a_0 = f(0).$$

If instead we evaluate the series (3.1.3) at $x = 0$, then we obtain the useful fact that

$$a_1 = \frac{f'(0)}{1!}.$$

If we evaluate the series (3.1.4) at $x = 0$, then we obtain the analogous fact that

$$a_2 = \frac{f^{(2)}(0)}{2!}.$$

Here the superscript $^{(2)}$ denotes a second derivative.

In general, we can derive (by successive differentiations) the formula

$$a_j = \frac{f^{(j)}(0)}{j!}, \qquad (3.1.5)$$

which gives us an explicit way to determine the coefficients of the power series expansion of a function. It follows from these considerations that a power series is identically equal to 0 if and only if each of its coefficients is 0.

We may also note that a power series may be integrated termwise. If

$$f(x) = \sum_{j=0}^{\infty} a_j x^j = a_0 + a_1 x + a_2 x^2 + \cdots,$$

then

$$\int f(x)\, dx = \sum_{j=0}^{\infty} a_j \frac{x^{j+1}}{j+1} = a_0 x + a_1 \frac{x^2}{2} + a_2 \frac{x^3}{3} + \cdots + C.$$

If

$$f(x) = \sum_{j=0}^{\infty} a_j x^j = a_0 + a_1 x + a_2 x^2 + \cdots$$

and

$$g(x) = \sum_{j=0}^{\infty} b_j x^j = b_0 + b_1 x + b_2 x^2 + \cdots$$

for $|x| < R$, then these functions may be added or subtracted by adding or subtracting the series termwise:

$$f(x) \pm g(x) = \sum_{j=0}^{\infty} (a_j \pm b_j) x^j = (a_0 \pm b_0) + (a_1 \pm b_1) x + (a_2 \pm b_2) x^2 + \cdots .$$

Also f and g may be multiplied as if they were polynomials, to wit

$$f(x) \cdot g(x) = \sum_{j=0}^{\infty} c_n x^n ,$$

where

$$c_n = a_0 b_n + a_1 b_{n-1} + \cdots + a_n b_0 .$$

We shall say more about operations on power series below.

Finally, we note that if two different power series converge to the same function, then (3.1.5) tells us that the two series are precisely the same (i.e., have the same coefficients). In particular, if $f(x) \equiv 0$ for $|x| < R$ then all the coefficients of the power series expansion for f are equal to 0.

IV. Suppose that f is a function that has derivatives of all orders on $|x| < R$. We may calculate the coefficients

$$a_j = \frac{f^{(j)}(0)}{j!}$$

and then write the (formal) series

$$\sum_{j=0}^{\infty} a_j x^j . \tag{3.1.6}$$

It is then natural to ask whether the series (3.1.6) *converges to f*. When the function f is sine or cosine or logarithm or the exponential then the answer is "yes." But these are very special functions. Actually, the answer to our question is generically "no." Most infinitely differentiable functions do *not* have power series expansion that converges back to the function. In fact most have power series that do not converge at all; but even in the unlikely circumstance that the series does converge, it will generally *not* converge to the original function f.

This circumstance may seem rather strange, but it explains why mathematicians spent so many years trying to understand power series. The functions that *do* have convergent power series are called *real analytic* and they are very particular functions with remarkable properties. Even though the subject of real analytic functions is more than 300 years old, the first and only book written on the subject is [KRP1].

We do have a way of coming to grips with the unfortunate state of affairs that has just been described, and that is the theory of *Taylor expansions*. For a function with $(n+1)$ continuous derivatives, here is what is actually true:

$$f(x) = \sum_{j=0}^{n} \frac{f^{(j)}(0)}{j!} x^j + R_n(x), \qquad (3.1.7)$$

where the remainder term $R_n(x)$ is given by

$$R_n(x) = \frac{f^{(n+1)}(\xi)}{(n+1)!} x^{n+1}$$

for some number ξ between 0 and x. The power series converges to f precisely when the partial sums in (3.1.7) converge, and that happens precisely when the remainder term goes to zero. What is important for you to understand is that, generically, the remainder term *does not go to zero*. But formula (3.1.7) is still valid.

We can use formula (3.1.7) to obtain the familiar power series expansions for several important functions:

$$e^x = \sum_{j=0}^{\infty} \frac{x^j}{j!} = 1 + x + \frac{x^2}{2!} + \frac{x^3}{3!} + \cdots$$

$$\sin x = \sum_{j=0}^{\infty} (-1)^j \frac{x^{2j+1}}{(2j+1)!} = x - \frac{x^3}{3!} + \frac{x^5}{5!} - + \cdots$$

$$\cos x = \sum_{j=0}^{\infty} (-1)^j \frac{x^{2j}}{(2j)!} = 1 - \frac{x^2}{2!} + \frac{x^4}{4!} - + \cdots$$

Of course there are many others, including the logarithm and the other trigonometric functions. Just for practice, let us verify that the first of these formulas is actually valid.

First,

$$\frac{d^j}{dx^j} e^x = e^x \qquad \text{for every } j.$$

Thus

$$a_j = \frac{(d^j/dx^j)e^x\big|_{x=0}}{j!} = \frac{1}{j!}.$$

This confirms that the formal power series for e^x is just what we assert it to be. To check that it converges back to e^x, we must look at the remainder term, which is

$$R_n(x) = \frac{f^{(n+1)}(\xi)}{(n+1)!} x^{n+1} = \frac{e^\xi \cdot x^{n+1}}{(n+1)!}.$$

Of course, for x fixed, we have that $|\xi| < |x|$; also $n \to \infty$ implies that $(n+1)! \to \infty$ much faster than x^{n+1}. So the remainder term goes to zero and the series converges to e^x.

V. Operations on Series

Some operations on series, such as addition, subtraction, and scalar multiplication, are straightforward. Others, such as multiplication, entail subtleties.

Products of Series

In order to keep our discussion of multiplication of series as straightforward as possible, we deal at first with absolutely convergent series. It is convenient in this discussion to begin our sum at $j = 0$ instead of $j = 1$. If we wish to multiply

$$\sum_{j=0}^{\infty} a_j \quad \text{and} \quad \sum_{j=0}^{\infty} b_j,$$

then we need to specify what the partial sums of the product series should be. An obvious necessary condition that we wish to impose is that if the first series converges to α and the second converges to β then the product series $\sum_{j=0}^{\infty} c_j$, whatever we define it to be, should converge to $\alpha \cdot \beta$.

The Cauchy Product

Cauchy's idea was that the summands for the product series should be

$$c_m \equiv \sum_{j=0}^{m} a_j \cdot b_{m-j}.$$

This particular form for the summands can be easily motivated using polynomial considerations (which we shall provide later on). For now we concentrate on confirming that this "Cauchy product" of two series really works.

Theorem 3.1.5 *Let* $\sum_{j=0}^{\infty} a_j$ *and* $\sum_{j=0}^{\infty} b_j$ *be two absolutely convergent series which converge to limits* α *and* β, *respectively. Define the series* $\sum_{m=0}^{\infty} c_m$ *with summands*

$$c_m = \sum_{j=0}^{m} a_j \cdot b_{m-j}.$$

Then the series $\sum_{m=0}^{\infty} c_m$ *converges to* $\alpha \cdot \beta$.

EXAMPLE 3.1.6 Consider the Cauchy product of the two conditionally convergent series

$$\sum_{j=0}^{\infty} \frac{(-1)^j}{\sqrt{j+1}} \quad \text{and} \quad \sum_{j=0}^{\infty} \frac{(-1)^j}{\sqrt{j+1}}.$$

Observe that

$$\begin{aligned}
c_m &= \frac{(-1)^0(-1)^m}{\sqrt{1}\sqrt{m+1}} + \frac{(-1)^1(-1)^{m-1}}{\sqrt{2}\sqrt{m}} + \cdots \\
&\quad + \frac{(-1)^m(-1)^0}{\sqrt{m+1}\sqrt{1}} \\
&= \sum_{j=0}^{m} (-1)^m \frac{1}{\sqrt{(j+1)\cdot(m+1-j)}}.
\end{aligned}$$

However,

$$(j+1)\cdot(m+1-j) \le (m+1)\cdot(m+1) = (m+1)^2.$$

Thus

$$|c_m| \ge \sum_{j=0}^{m} \frac{1}{m+1} = 1.$$

We thus see that the terms of the series $\sum_{m=0}^{\infty} c_m$ do not tend to zero, so the series cannot converge. ∎

EXAMPLE 3.1.7 The series

$$A = \sum_{j=0}^{\infty} 2^{-j} \quad \text{and} \quad B = \sum_{j=0}^{\infty} 3^{-j}$$

are both absolutely convergent. We challenge the reader to calculate the Cauchy product and to verify that that product converges to 3. ∎

VI. We conclude by summarizing some properties of real analytic functions.

1. Polynomials and the functions e^x, $\sin x$, $\cos x$ are all real analytic at all points.

2. If f and g are real analytic at x_0 then $f \pm g$, $f \cdot g$, and f/g (provided $g(x_0) \neq 0$) are real analytic at x_0.

3. If f is real analytic at x_0 and if f^{-1} is a continuous inverse on an interval containing $f(x_0)$ and $f'(x_0) \neq 0$, then f^{-1} is real analytic at $f(x_0)$.

4. If g is real analytic at x_0 and f is real analytic at $g(x_0)$, then $f \circ g$ is real analytic at x_0.

5. A function defined by the sum of a power series is real analytic at all interior points of its interval of convergence.

VII.   It is worth recording that all the basic properties of *real* power series that we have discussed above are still valid for *complex* power series. Such a series has the form

$$\sum_{j=0}^{\infty} c_j z^j \quad \text{or} \quad \sum_{j=0}^{\infty} c_j (z - a)^j \,,$$

where the coefficients c_j are real or complex numbers, a is a complex number, and z is a complex variable. The series has a radius of convergence R, and the domain of convergence is now a disc $D(0, R)$ or $D(a, R)$ rather than an interval. In practice, a complex analytic function has radius of convergence about a point a that is equal to the distance of the nearest singularity to a. See [KNO] or [GRK] or [KRP1] for further discussion of these ideas.

Exercises

1.   Use the ratio test (for example) to verify that $R = 0$, $R = \infty$, and $R = 1$ for the series **(a)**, **(b)**, **(c)** that are discussed in the text.

2.   If $p \neq 0$ and p is not a positive integer then show that the series

$$\sum_{j=1}^{\infty} \frac{p(p - 1)(p - 2) \cdots (p - j + 1)}{j!} \cdot x^j$$

   converges for $|x| < 1$ and diverges for $|x| > 1$.

3.   Verify that $R = +\infty$ for the power series expansions of sine and cosine.

4. Use Taylor's formula to confirm the validity of the power series expansions for $\ln(1 + x)$, $\sin x$, and $\cos x$.

5. When we first encounter geometric series we learn that

$$1 + x + x^2 + \cdots + x^n = \frac{1 - x^{n+1}}{1 - x}$$

provided $x \neq 1$. Indeed, let S be the expression on the left and consider $(1 - x) \cdot S$. Use this formula to show that the expansions

$$\frac{1}{1 - x} = 1 + x + x^2 + x^3 + \cdots$$

and

$$\frac{1}{1 + x} = 1 - x + x^2 - x^3 + \cdots$$

are valid for $|x| < 1$. Use these formulas to show that

$$\ln(1 + x) = x - \frac{x^2}{2} + \frac{x^3}{3} - \frac{x^4}{4} + - \cdots$$

and

$$\arctan x = x - \frac{x^3}{3} + \frac{x^5}{5} - \frac{x^7}{7} + - \cdots$$

for $|x| < 1$.

6. Use the first expansion given in Exercise 5 to find the power series for $1/(1 - x)^2$

(a) by squaring;

(b) by differentiating.

7. (a) Show that the series

$$y = 1 - \frac{x^2}{2!} + \frac{x^4}{4!} - \frac{x^6}{6!} + - \cdots$$

satisfies $y'' = -y$.

(b) Show that the series

$$y = 1 - \frac{x^2}{2^2} + \frac{x^4}{2^2 \cdot 4^2} - \frac{x^6}{2^2 \cdot 4^2 \cdot 6^2} + - \cdots$$

converges for all x. Verify that it defines a solution of the equation

$$xy'' + y' + xy = 0.$$

This function is the *Bessel function of order 0*. It will be encountered later in other contexts.

3.2   Series Solutions of First-Order Differential Equations

Now we get our feet wet and use power series to solve first-order linear equations. This will turn out to be misleadingly straightforward to do, but it will show us the basic moves.

At first we consider the equation

$$y' = y.$$

Of course we know that the solution to this equation is $y = C \cdot e^x$, but let us pretend that we do not know this. We proceed by *guessing* that the equation has a solution given by a power series, and we proceed to solve for the coefficients of that power series.

So our guess is a solution of the form

$$y = a_0 + a_1 x + a_2 x^2 + a_3 x^3 + \cdots.$$

Then

$$y' = a_1 + 2a_2 x + 3a_3 x^2 + \cdots$$

and we may substitute these two expressions into the differential equation. Thus

$$a_1 + 2a_2 x + 3a_3 x^2 + \cdots = a_0 + a_1 x + a_2 x^2 + \cdots.$$

Now the powers of x must match up (i.e., the coefficients must be equal). We conclude that

$$\begin{aligned} a_1 &= a_0 \\ 2a_2 &= a_1 \\ 3a_3 &= a_2 \end{aligned}$$

and so forth. Let us take a_0 to be an unknown constant C. Then we see that

$$\begin{aligned} a_1 &= C \\ a_2 &= \frac{C}{2} \\ a_3 &= \frac{C}{3 \cdot 2} \end{aligned}$$

etc.

In general,

$$a_n = \frac{C}{n!}.$$

In summary, our power series solution of the original differential equation is

$$y = \sum_{j=0}^{\infty} a_j x^j = \sum_{j=0}^{\infty} \frac{C}{j!} x^j = C \cdot \sum_{j=0}^{\infty} \frac{x^j}{j!} = C \cdot e^x \, .$$

Thus we have a new way, using power series, of discovering the general solution of the differential equation $y' = y$.

The next example illustrates the point that, by running our logic a bit differently, we can *use* a differential equation to derive the power series expansion for a given function.

EXAMPLE 3.2.1 Let p be an arbitrary real constant. Use a differential equation to derive the power series expansion for the function

$$y = (1 + x)^p \, .$$

Solution: Of course the given y is a solution of the initial value problem

$$(1 + x) \cdot y' = py \, , \quad y(0) = 1 \, .$$

We assume that the equation has a power series solution

$$y = \sum_{j=0}^{\infty} a_j x^j = a_0 + a_1 x + a_2 x^2 + \cdots$$

with positive radius of convergence R. Then

$$y' = \sum_{j=1}^{\infty} j \cdot a_j x^{j-1} = a_1 + 2a_2 x + 3a_3 x^2 + \cdots$$

$$xy' = \sum_{j=1}^{\infty} j \cdot a_j x^j = a_1 x + 2a_2 x^2 + 3a_3 x^3 + \cdots$$

$$py = \sum_{j=0}^{\infty} p a_j x^j = p a_0 + p a_1 x + p a_2 x^2 + \cdots \, .$$

We rewrite the differential equation as

$$y' + xy' = py \, .$$

Now we see that the equation tells us that the sum of the first two of our series equals the third. Thus

$$\sum_{j=1}^{\infty} j a_j x^{j-1} + \sum_{j=1}^{\infty} j a_j x^j = \sum_{j=0}^{\infty} p a_j x^j \, .$$

We immediately see two interesting anomalies: the powers of x on the left-hand side do not match up, so the two series cannot be immediately added. Also the summations do not all begin in the same place. We address these two concerns as follows.

First, we can change the index of summation in the first sum on the left to obtain

$$\sum_{j=0}^{\infty}(j+1)a_{j+1}x^j + \sum_{j=1}^{\infty} ja_j x^j = \sum_{j=0}^{\infty} pa_j x^j.$$

Write out the first few terms of the changed sum, and the original sum, to see that they are just the same.

Now every one of our series has x^j in it, but they begin at different places. So we break off the extra terms as follows:

$$\sum_{j=1}^{\infty}(j+1)a_{j+1}x^j + \sum_{j=1}^{\infty} ja_j x^j - \sum_{j=1}^{\infty} pa_j x^j = -a_1 x^0 + pa_0 x^0. \qquad (3.2.1.1)$$

Notice that all we have done is to break off the zeroeth terms of the first and third series, and put them on the right.

The three series on the left-hand side of (3.2.1.1) are begging to be put together: they have the same form, they all involve matching powers of x, and they all begin at the same index. Let us do so:

$$\sum_{j=1}^{\infty}\left[(j+1)a_{j+1} + ja_j - pa_j\right]x^j = -a_1 + pa_0.$$

Now the powers of x that appear on the left are $1, 2, \ldots$, and there are none of these on the right. The right-hand side only contains x to the zeroeth power.

We conclude that each of the coefficients on the left is zero; by the same reasoning, the coefficient $(-a_1 + pa_0)$ on the right (i.e., the constant term) equals zero. Here are we are using the uniqueness property of the power series representation. It gives us infinitely many equations.

So we have the equations[2]

$$
\begin{aligned}
-a_1 + pa_0 &= 0 \\
(j+1)a_{j+1} + (j-p)a_j &= 0 \qquad \text{for } j = 1, 2, \ldots.
\end{aligned}
$$

Our initial condition $y(0) = 1$ tells us that $a_0 = 1$. Then our first equation implies that $a_1 = p$. The next equation, with $j = 1$, says that

$$2a_2 + (1-p)a_1 = 0.$$

Hence

$$a_2 = \frac{(p-1)a_1}{2} = \frac{p(p-1)}{2}.$$

---

[2] A set of equations like this is called a *recursion*. It expresses later indexed a_js in terms of earlier indexed a_js.

Continuing, we take $j = 2$ in the second equation to get

$$3a_3 + (2 - p)a_2 = 0$$

so

$$a_3 = \frac{(p-2)a_2}{3} = \frac{p(p-1)(p-2)}{3 \cdot 2}.$$

We may continue in this manner to obtain that

$$a_j = \frac{p(p-1)(p-2) \cdots (p-j+1)}{j!}.$$

Thus the power series expansion for our solution y is

$$
\begin{aligned}
y \ = \ & 1 + px + \frac{p(p-1)}{2!}x + \frac{p(p-1)(p-2)}{3!} + \cdots \\
& + \frac{p(p-1)(p-2) \cdots (p-j+1)}{j!}x^j + \cdots .
\end{aligned}
$$

Since we knew in advance that the solution of our initial value problem was

$$y = (1 + x)^p$$

(and this function is certainly analytic at 0), we find that we have derived Isaac Newton's general binomial theorem (or binomial series):

$$
\begin{aligned}
(1 + x)^p &= 1 + px + \frac{p(p-1)}{2!}x^2 + \frac{p(p-1)(p-2)}{3!} + \cdots \\
&\quad + \frac{p(p-1)(p-2) \cdots (p-j+1)}{j!}x^j + \cdots \\
&= \sum_{j=0}^{\infty} \frac{p(p-1)(p-2) \cdots \cdots (p-j+1)}{j!}x^j .
\end{aligned}
\qquad (3.2.1.2)
$$

■

Exercises

1. For each of the following differential equations, find a power series solution of the form $\sum_j a_j x^j$. Endeavor to recognize this solution as the series expansion of a familiar function. Now solve the equation directly, using a method from the earlier part of the book, to confirm your series solution.

    (a) $y' = 2xy$

    (b) $y' + y = 1$

    (c) $y' - y = 2$

    (d) $y' + y = 0$

    (e) $y' - y = 0$

    (f) $y' - y = x^2$

2. For each of the following differential equations, find a power series so-lution of the form $\sum_j a_j x^j$. Then solve the equation directly. Compare your two answers, and explain any discrepancies that may arise.

    (a) $xy' = y$

    (b) $x^2 y' = y$

    (c) $y' - (1/x)y = x^2$

    (d) $y' + (1/x)y = x$

3. Express the function $\arcsin x$ in the form of a power series $\sum_j a_j x^j$ by solving the differential equation $y' = (1 - x^2)^{-1/2}$ in two different ways. [**Hint**: Remember the binomial series.] Use this result to obtain the for-mula

$$\frac{\pi}{6} = \frac{1}{2} + \frac{1}{2} \cdot \frac{1}{3 \cdot 2^3} + \frac{1 \cdot 3}{2 \cdot 4} \cdot \frac{1}{5 \cdot 2^5} + \frac{1 \cdot 3 \cdot 5}{2 \cdot 4 \cdot 6} \cdot \frac{1}{7 \cdot 2^7} + \cdots .$$

4. The differential equations considered in the text and in the preceding exercises are all linear. By contrast, the equation

$$y' = 1 + y^2 \tag{$*$}$$

is nonlinear. One may verify directly that $y(x) = \tan x$ is the particular solution for which $y(0) = 0$. Show that

$$\tan x = x + \frac{1}{3}x^3 + \frac{2}{15}x^5 + \cdots$$

by assuming a solution for equation $(*)$ in the form of a power series $\sum_j a_j x^j$ and then finding the coefficients a_j in two ways:

    (a) by the method of the examples in the text (note particularly how the nonlinearity of the equation complicates the formulas);

    (b) by differentiating equation $(*)$ repeatedly to obtain

$$y'' = 2yy', \quad y''' = 2yy'' + 2(y')^2, \quad \text{etc.}$$

    and using the formula $a_j = f^{(j)}(0)/j!$.

5. Solve the equation

$$y' = x - y, \quad y(0) = 0$$

by each of the methods suggested in the last exercise. What familiar function does the resulting series represent? Verify your conclusion by solving the equation directly as a first-order linear equation.

6. Use your symbol manipulation software, such as `Maple` or `Mathematica`, to write a routine that will find the coefficients of the power series solution for a given first-order ordinary differential equation.

3.3 Second-Order Linear Equations: Ordinary Points

We have invested considerable effort in studying equations of the form

$$y'' + p \cdot y' + q \cdot y = 0 \,. \tag{3.3.1}$$

Here p and q could be constants or, more generally, functions. In some sense, our investigations thus far have been misleading; for we have only considered particular equations in which a closed-form solution can be found. These cases are really the exception rather than the rule. For most such equations, there is no "formula" for the solution. Power series then give us some extra flexibility. Now we may seek a power series solution; that solution is valid, and may be calculated and used in applications, even though it may not be expressed in a compact formula.

A number of the differential equations that arise in mathematical physics—Bessel's equation, Lagrange's equation, and many others—in fact fit the description that we have just presented. So it is worthwhile to develop techniques for studying (3.3.1). In the present section we shall concentrate on finding a power series solution to equation (3.3.1)—written in standard form—expanded about a point x_0, where x_0 has the property that p and q have convergent power series expansions about x_0. In this circumstance we call x_0 an *ordinary point* of the differential equation. Later sections will treat the situation where either p or q (or both) has a singularity at x_0.

We begin our study with a familiar equation, just to see the basic steps, and how the solution will play out.[3] Namely, we endeavor to solve

$$y'' + y = 0$$

by power series methods. As usual, we guess a solution of the form

$$y = \sum_{j=0}^{\infty} a_j x^j = a_0 + a_1 x + a_2 x^2 + \cdots \,.$$

Of course it follows that

$$y' = \sum_{j=1}^{\infty} j a_j x^{j-1} = a_1 + 2 a_2 x + 3 a_3 x^2 + \cdots$$

and

$$y'' = \sum_{j=2}^{\infty} j(j-1) a_j x^{j-2} = 2 \cdot 1 \cdot a_2 + 3 \cdot 2 \cdot a_3 x + 4 \cdot 3 \cdot a_4 x^2 \cdots \,.$$

---

[3] Of course this is an equation that we know how to solve by other means. Now we are learning a new solution technique.

Plugging the first and third of these into the differential equation gives

$$\sum_{j=2}^{\infty} j(j-1)a_j x^{j-2} + \sum_{j=0}^{\infty} a_j x^j = 0.$$

As in the last example of Section 3.2, we find that the series that occur have x raised to different powers, and that the summations begin in different places. We follow the standard procedure for repairing these matters.

First, we change the index of summation in the second series. So

$$\sum_{j=2}^{\infty} j(j-1)a_j x^{j-2} + \sum_{j=2}^{\infty} a_{j-2} x^{j-2} = 0.$$

We invite the reader to verify that the new second series is just the same as the original second series (merely write out a few terms of each to check). We are fortunate in that both series now begin at the same index, and they both involve the same powers of x. So we may add them together to obtain

$$\sum_{j=2}^{\infty} [j(j-1)a_j + a_{j-2}] x^{j-2} = 0.$$

The only way that such a power series can be identically zero is if each of the coefficients is zero. So we obtain the recursion equations

$$j(j-1)a_j + a_{j-2} = 0, \qquad j = 2, 3, 4, \ldots.$$

Then $j = 2$ gives us

$$a_2 = -\frac{a_0}{2 \cdot 1}.$$

It will be convenient to take a_0 to be an arbitrary constant A, so that

$$a_2 = -\frac{A}{2 \cdot 1}.$$

The recursion for $j = 4$ says that

$$a_4 = -\frac{a_2}{4 \cdot 3} = \frac{A}{4 \cdot 3 \cdot 2 \cdot 1}.$$

Continuing in this manner, we find that

$$\begin{aligned}
a_{2j} &= (-1)^j \cdot \frac{A}{2j \cdot (2j-1) \cdot (2j-2) \cdots 3 \cdot 2 \cdot 1} \\
&= (-1)^j \cdot \frac{A}{(2j)!}, \qquad j = 1, 2, \ldots.
\end{aligned}$$

Thus we have complete information about the coefficients with even index.

Now let us consider the odd indices. Look at the recursion for $j = 3$. This is

$$a_3 = -\frac{a_1}{3 \cdot 2}.$$

It is convenient to take a_1 to be an arbitrary constant B. Thus

$$a_3 = -\frac{B}{3 \cdot 2 \cdot 1}.$$

Continuing with $j = 5$, we find that

$$a_5 = -\frac{a_3}{5 \cdot 4} = \frac{B}{5 \cdot 4 \cdot 3 \cdot 2 \cdot 1}.$$

In general,

$$a_{2j+1} = (-1)^j \frac{B}{(2j+1)!}, \quad j = 1, 2, \ldots.$$

In summary, then, the general solution of our differential equation is given by

$$y = A \cdot \left(\sum_{j=0}^{\infty} (-1)^j \cdot \frac{1}{(2j)!} x^{2j} \right) + B \cdot \left(\sum_{j=0}^{\infty} (-1)^j \frac{1}{(2j+1)!} x^{2j+1} \right).$$

Of course we recognize the first power series as the cosine function and the second as the sine function. So we have rediscovered that the general solution of $y'' + y = 0$ is

$$y = A \cdot \cos x + B \cdot \sin x.$$

This is consistent with what we learned earlier in the book about solving the equation $y'' + y = 0$. ∎

EXAMPLE 3.3.1 Use the method of power series to solve the differential equation

$$(1 - x^2)y'' - 2xy' + p(p+1)y = 0. \tag{3.3.1.1}$$

Here p is an arbitrary real constant. This is called *Legendre's equation*.

Solution: First we write the equation in standard form:

$$y'' - \frac{2x}{1 - x^2} y' + \frac{p(p+1)}{1 - x^2} y = 0.$$

Observe that, near $x = 0$, division by 0 is avoided and the coefficients p and q are real analytic. So 0 is an ordinary point.

We therefore guess a solution of the form

$$y = \sum_{j=0}^{\infty} a_j x^j = a_0 + a_1 x + a_2 x^2 + \cdots$$

and calculate

$$y' = \sum_{j=1}^{\infty} j a_j x^{j-1} = a_1 + 2a_2 x + 3a_3 x^2 + \cdots$$

and

$$y'' = \sum_{j=2}^{\infty} j(j-1) a_j x^{j-2} = 2a_2 + 3 \cdot 2 \cdot a_3 x + \cdots .$$

It is most convenient to treat the differential equation in the form (3.3.1.1). We calculate

$$-x^2 y'' = -\sum_{j=2}^{\infty} j(j-1) a_j x^j$$

and

$$-2xy' = -\sum_{j=1}^{\infty} 2j a_j x^j .$$

Substituting into the differential equation now yields

$$\sum_{j=2}^{\infty} j(j-1) a_j x^{j-2} - \sum_{j=2}^{\infty} j(j-1) a_j x^j - \sum_{j=1}^{\infty} 2j a_j x^j + \sum_{j=0}^{\infty} p(p+1) a_j x^j = 0 .$$

We adjust the index of summation in the first sum so that it contains x^j rather than x^{j-2} and we break off spare terms and collect them. The result is

$$\sum_{j=2}^{\infty} (j+2)(j+1) a_{j+2} x^j - \sum_{j=2}^{\infty} j(j-1) a_j x^j - \sum_{j=2}^{\infty} 2j a_j x^j$$

$$+ \sum_{j=2}^{\infty} p(p+1) a_j x^j + \Big(2a_2 + 6a_3 x - 2a_1 x$$

$$+ p(p+1) a_0 + p(p+1) a_1 x \Big) = 0 .$$

In other words,

$$\sum_{j=2}^{\infty} \Big((j+2)(j+1) a_{j+2} - j(j-1) a_j - 2j a_j + p(p+1) a_j \Big) x^j$$

$$+ \Big(2a_2 + p(p+1) a_0 \Big) + \Big(6a_3 - 2a_1 + p(p+1) a_1 \Big) x = 0 .$$

As a result,

$$(j+2)(j+1) a_{j+2} - j(j-1) a_j - 2j a_j + p(p+1) a_j = 0 \qquad \text{for } j = 2, 3, \ldots$$

together with

$$2a_2 + p(p+1) a_0 = 0$$

and

$$6a_3 - 2a_1 + p(p+1)a_1 = 0.$$

We have arrived at the recursion relations

$$a_2 = -\frac{(p+1)p}{2 \cdot 1} \cdot a_0$$

$$a_3 = -\frac{(p+2)(p-1)}{3 \cdot 2 \cdot 1} \cdot a_1$$

$$a_{j+2} = -\frac{(p+j+1)(p-j)}{(j+2)(j+1)} \cdot a_j \qquad \text{for } j = 2, 3, \ldots. \qquad (3.3.1.2)$$

We recognize a familiar pattern: The coefficients a_0 and a_1 are unspecified, so we set $a_0 = A$ and $a_1 = B$. Then we may proceed to solve for the rest of the coefficients. Now

$$a_2 = -\frac{(p+1)p}{2!} \cdot A$$

$$a_3 = -\frac{(p+2)(p-1)}{3!} \cdot B$$

$$a_4 = -\frac{(p+3)(p-2)}{4 \cdot 3} a_2 = \frac{(p+3)(p+1)p(p-2)}{4!} \cdot A$$

$$a_5 = -\frac{(p+4)(p-3)}{5 \cdot 4} a_3 = \frac{(p+4)(p+2)(p-1)(p-3)}{5!} \cdot B$$

$$a_6 = -\frac{(p+5)(p-4)}{6 \cdot 5} a_4 = -\frac{(p+5)(p+3)(p+1)p(p-2)(p-4)}{6!} \cdot A$$

$$a_7 = -\frac{(p+6)(p-5)}{7 \cdot 6} a_5 = -\frac{(p+6)(p+4)(p+2)(p-1)(p-3)(p-5)}{7!} \cdot B,$$

and so forth. Putting these coefficient values into our supposed power series solution, we find that the general solution of our differential equation is

$$
\begin{aligned}
y = {} & A\Bigg(1 - \frac{(p+1)p}{2!}x^2 + \frac{(p+3)(p+1)p(p-2)}{4!}x^4 \\
& - \frac{(p+5)(p+3)(p+1)p(p-2)(p-4)}{6!}x^6 + - \cdots \Bigg) \\
& + B\Bigg(x - \frac{(p+2)(p-1)}{3!}x^3 + \frac{(p+4)(p+2)(p-1)(p-3)}{5!}x^5 \\
& - \frac{(p+6)(p+4)(p+2)(p-1)(p-3)(p-5)}{7!}x^7 + - \cdots \Bigg).
\end{aligned}
$$

We assure the reader that, when p is not an integer, then these are *not* familiar elementary transcendental functions. These are what we call *Legendre functions*. In the special circumstance that p is a positive even integer, the first function (that which is multiplied by A) terminates as a polynomial. In the special circumstance that p is a positive odd integer, the second function (that which is multiplied by B) terminates as a polynomial. These are called *Legendre polynomials* P_n, and they play an important role in mathematical physics, representation theory, and interpolation theory. We shall encounter the Legendre polynomials later in the chapter. ∎

It is actually possible, without much effort, to check the radius of convergence of the functions we discovered as solutions in the last example. In fact we use the recursion relation (3.3.1.2) to see that

$$\left| \frac{a_{2j+2}x^{2j+2}}{a_{2j}x^{2j}} \right| = \left| -\frac{(p-2j)(p+2j+1)}{(2j+1)(2j+2)} \right| \cdot |x|^2 \to |x|^2$$

as $j \to \infty$. Thus the series expansion of the first Legendre function converges when $|x| < 1$, so the radius of convergence is 1. A similar calculation shows that the radius of convergence for the second Legendre function is 1.

We now enunciate a general result about the power series solution of an ordinary differential equation at an ordinary point. The proof is omitted.

Theorem 3.3.2 *Let x_0 be an ordinary point of the differential equation*

$$y'' + p \cdot y' + q \cdot y = 0, \qquad (3.3.2.1)$$

and let α and β be arbitrary real constants. Then there exists a unique real analytic function $y = y(x)$ that has a power series expansion about x_0 and so that

(a) *The function y solves the differential equation (3.3.2.1).*

(b) *The function y satisfies the initial conditions $y(x_0) = \alpha$, $y'(x_0) = \beta$.*

If the functions p and q have power series expansions about x_0 with radius of convergence R then so does y.

We conclude with this remark. The examples that we have worked in detail resulted in solutions with *two-term* (or *binary*) recursion formulas: a_2 was expressed in terms of a_0 and a_3 was expressed in terms of a_1, etc. In general, the recursion formulas that arise in solving an ordinary differential equation at an ordinary point may result in more complicated recursion relations.

Exercises

1. In each of the following problems, verify that 0 is an ordinary point. Then find the power series solution of the given equation.

   (a) $y'' + xy' + y = 0$

   (b) $y'' - y' + xy = 0$

   (c) $y'' + 2xy' - y = x$

   (d) $y'' + y' - x^2 y = 1$

   (e) $(1 + x^2)y'' + xy' + y = 0$

   (f) $y'' + (1 + x)y' - y = 0$

2. Find the general solution of

$$(1 + x^2)y'' + 2xy' - 2y = 0$$

   in terms of power series in x. Can you express this solution by means of elementary functions?

3. Consider the equation $y'' + xy' + y = 0$.

   (a) Find its general solution $y = \sum_j a_j x^j$ in the form $y = c_1 y_1(x) + c_2 y_2(x)$, where y_1, y_2 are power series.

   (b) Use the ratio test to check that the two series y_1 and y_2 from part (a) converge for all x (as Theorem 3.3.2 actually asserts).

   (c) Show that one of these two solutions, say y_1, is the series expansion of $e^{-x^2/2}$ and use this fact to find a second independent solution by the method discussed in Section 2.4. Convince yourself that this second solution is the function y_2 found in part (a).

4. Verify that the equation $y'' + y' - xy = 0$ has a three-term recursion formula and find its series solutions y_1 and y_2 such that

   (a) $y_1(0) = 1, \quad y_1'(0) = 0$

   (b) $y_2(0) = 0, \quad y_2'(0) = 1$

   Theorem 3.3.2 guarantees that both series converge at every $x \in \mathbf{R}$. Notice how difficult it would be to verify this assertion by working directly with the series themselves.

5. The equation $y'' + (p + 1/2 - x^2/4)y = 0$, where p is a constant, certainly has a series solution of the form $y = \sum_j a_j x^j$.

   (a) Show that the coefficients a_j are related by the three-term recursion formula

$$(n+1)(n+2)a_{n+2} + \left(p + \frac{1}{2}\right)a_n - \frac{1}{4}a_{n-2} = 0.$$

   (b) If the dependent variable is changed from y to w by means of $y = we^{-x^2/4}$, then show that the equation is transformed into $w'' - xw' + pw = 0$.

   (c) Verify that the equation in (b) has a two-term recursion formula and find its general solution.

6. *Chebyshev's equation* is

$$(1 - x^2)y'' - xy' + p^2 y = 0,$$

where p is constant.

(a) Find two linearly independent series solutions valid for $|x| < 1$.

(b) Show that, if $p = n$ where n is an integer ≥ 0, then there is a polynomial solution of degree n. When these polynomials are multiplied by suitable constants, then they are called the *Chebyshev polynomials*. We shall see this topic again later in the book.

7. *Hermite's equation* is

$$y'' - 2xy' + 2py = 0,$$

where p is a constant.

(a) Show that its general solution is $y(x) = a_1 y_1(x) + a_2 y_2(x)$, where

$$y_1(x) = 1 - \frac{2p}{2!}x^2 + \frac{2^2 p(p-2)}{4!}x^4$$
$$- \frac{2^3 p(p-2)(p-4)}{6!}x^6 + - \cdots$$

and

$$y_2(x) = x - \frac{2(p-1)}{3!}x^3 + \frac{2^2(p-1)(p-3)}{5!}x^5$$
$$- \frac{2^3(p-1)(p-3)(p-5)}{7!}x^7 + - \cdots .$$

By Theorem 3.3.2, both series converge for all x. Verify this assertion directly.

(b) If p is a nonnegative integer, then one of these series terminates and is thus a polynomial—y_1 if p is even and y_2 if p is odd—while the other remains an infinite series. Verify that for $p = 0, 1, 2, 3, 4, 5$, these polynomials are $1, x, 1 - 2x^2, x - 2x^3/3, 1 - 4x^2 + 4x^4/3, x - 4x^3/3 + 4x^5/15$.

(c) It is clear that the only polynomial solutions of Hermite's equation are constant multiples of the polynomials described in part **(b)**. Those constant multiples which have the additional property that the terms containing the highest powers of x are of the form $2^n x^n$ are denoted by $H_n(x)$ and called the *Hermite polynomials*. Verify that $H_0(x) = 1$, $H_1(x) = 2x$, $H_2(x) = 4x^2 - 2$, $H_3(x) = 8x^3 - 12x$, $H_4(x) = 16x^4 - 48x^2 + 12$, and $H_5(x) = 32x^5 - 160x^3 + 120x$.

(d) Verify that the polynomials listed in **(c)** are given by the general formula

$$H_n(x) = (-1)^n e^{x^2} \frac{d^n}{dx^n} e^{-x^2} .$$

8. Use your symbol manipulation software, such as `Maple` or `Mathematica`, to write a routine that will find the coefficients of the power series solution, expanded about an ordinary point, for a given second-order ordinary differential equation.

3.4   Regular Singular Points

Consider a second-order, linear differential equation in the standard form

$$y'' + p \cdot y' + q \cdot y = 0.$$

Let us examine a solution about a point x_0. If either of the coefficient functions p or q fails to be analytic at x_0, then x_0 is not an ordinary point for the differential equation and the methods of the previous section do not apply. In this circumstance we call x_0 a *singular point*.

There is some temptation to simply avoid the singular points. But often these are the points that are of the greatest physical interest. We must learn techniques for analyzing singular points. A simple example begins to suggest what the behavior near a singular point might entail. Consider the differential equation

$$y'' + \frac{2}{x}y' - \frac{2}{x^2}y = 0. \tag{3.4.1}$$

Obviously the point $x_0 = 0$ is a singular point for this equation. One may verify directly that the functions $y_1(x) = x$ and $y_2(x) = x^{-2}$ are solutions of this differential equation. Thus the general solution is

$$y = Ax + Bx^{-2}. \tag{3.4.2}$$

If we are interested only in solutions that are bounded near the origin, then we must take $B = 0$ and the solutions will have the form $y = Ax$. Most likely, the important physical behavior will take place when $B \neq 0$; we want to consider (3.4.2) in full generality.

The solution of ordinary differential equations near singular points of arbitrary type is extremely difficult. Many equations are intractable. Fortunately, many equations that arise in practice, or from physical applications, are of a particularly tame sort. We say that a singular point x_0 for the differential equation

$$y'' + p \cdot y' + q \cdot y = 0$$

is a *regular singular point* if

$$(x - x_0) \cdot p(x) \qquad \text{and} \qquad (x - x_0)^2 q(x)$$

are analytic at x_0. As an example, equation (3.4.1) has a regular singular point at 0 because

$$x \cdot p(x) = x \cdot \frac{2}{x} = 2$$

is analytic at 0 and

$$x^2 \cdot \left(-\frac{2}{x^2}\right) = -2$$

is analytic at 0.

Let us now consider some important differential equations from mathematical physics, just to illustrate how regular singular points arise in practice. Recall Legendre's equation

$$y'' - \frac{2x}{1 - x^2} y' + \frac{p(p+1)}{1 - x^2} y = 0.$$

We see immediately that ± 1 are singular points. The point $x_0 = 1$ is a regular singular point because

$$(x - 1) \cdot p(x) = (x - 1) \cdot \left(-\frac{2x}{1 - x^2} \right) = \frac{2x}{x + 1}$$

and

$$(x - 1)^2 q(x) = (x - 1)^2 \cdot \left(\frac{p(p+1)}{1 - x^2} \right) = -\frac{(x - 1)p(p+1)}{x + 1}$$

are both real analytic at $x = 1$ (namely, we *avoid* division by 0). A similar calculation shows that $x_0 = -1$ is a regular singular point.

As a second example, consider *Bessel's equation of order p*, where $p \geq 0$ is a constant:

$$x^2 y'' + xy' + (x^2 - p^2)y = 0.$$

Written in the form

$$y'' + \frac{1}{x} y' + \frac{x^2 - p^2}{x^2} y = 0,$$

the equation is seen to have a singular point at $x_0 = 0$. But it is regular because

$$x \cdot p(x) = 1 \quad \text{and} \quad x^2 \cdot q(x) = x^2 - p^2$$

are both real analytic at 0.

Let us assume for the rest of this discussion, for simplicity, that the regular singular point is at $x_0 = 0$.

The key idea in solving a differential equation at a regular singular point is to guess a solution of the form

$$y = y(x) = x^m \cdot \left(a_0 + a_1 x + a_2 x^2 + \cdots \right). \tag{3.4.3}$$

We see that we have modified the guess used in the last section by adding a factor of x^m in front. Here the exponent m can be positive or negative or zero—and m *need not be an integer*. In practice—and this is conceptually important to avoid confusion—we assume that we have factored out the greatest possible power of x; thus the coefficient a_0 will always be a nonzero constant.

We call a series of the type (3.4.3) a *Frobenius series*. We now solve the differential equation at a regular singular point just as we did in the last section, except that now our recursion relations will be more complicated—as they will involve both the coefficients a_j and also the new exponent m. The method is best understood by examining an example.

EXAMPLE 3.4.4 Use the method of Frobenius series to solve the differential equation

$$2x^2 y'' + x(2x + 1)y' - y = 0 \qquad (3.4.4.1)$$

about the regular singular point 0.

Solution: Writing the equation in the standard form

$$y'' + \frac{x(2x + 1)}{2x^2} y' - \frac{1}{2x^2} y = 0,$$

we readily check that

$$x \cdot \left(\frac{x(2x + 1)}{2x^2} \right) = \frac{2x + 1}{2} \quad \text{and} \quad x^2 \cdot \left(-\frac{1}{2x^2} \right) = -\frac{1}{2}$$

are both real analytic at $x_0 = 0$. So $x_0 = 0$ is a regular singular point for this differential equation.

We guess a solution of the form

$$y = x^m \cdot \sum_{j=0}^{\infty} a_j x^j = \sum_{j=0}^{\infty} a_j x^{m+j}$$

and therefore calculate that

$$y' = \sum_{j=0}^{\infty} (m + j) a_j x^{m+j-1}$$

and

$$y'' = \sum_{j=0}^{\infty} (m + j)(m + j - 1) a_j x^{m+j-2}.$$

Notice that we do not begin the series for y' at 1 and we do not begin the series for y'' at 2, just because m may not be an integer so none of the summands may vanish (as in the case of an ordinary point).

Plugging these calculations into the differential equation, written in the form (3.4.4.1), yields

$$2 \sum_{j=0}^{\infty} (m + j)(m + j - 1) a_j x^{m+j}$$

$$+2 \sum_{j=0}^{\infty} (m + j) a_j x^{m+j+1} + \sum_{j=0}^{\infty} (m + j) a_j x^{m+j} - \sum_{j=0}^{\infty} a_j x^{m+j} = 0.$$

We make the usual adjustments in the indices so that all powers of x are x^{m+j}, and break off the odd terms so that all the series begin at the same place. We obtain

$$2\sum_{j=1}^{\infty}(m+j)(m+j-1)a_jx^{m+j}+2\sum_{j=1}^{\infty}(m+j-1)a_{j-1}x^{m+j}$$

$$+\sum_{j=1}^{\infty}(m+j)a_jx^{m+j}-\sum_{j=1}^{\infty}a_jx^{m+j}+\left(2m(m-1)a_0x^m+ma_0x^m-a_0x^m\right)=0.$$

The result is

$$\left(2(m+j)(m+j-1)a_j+2(m+j-1)a_{j-1}+(m+j)a_j-a_j\right)=0 \quad \text{for } j=1,2,3,\ldots$$
$$(3.4.4.2)$$

together with

$$[2m(m-1)+m-1]a_0=0.$$

It is clearly not to our advantage to let $a_0=0$. Thus

$$2m(m-1)+m-1=0.$$

This is the *indicial equation*.

The roots of this quadratic equation are $m=-1/2, 1$. We put each of these values into (3.4.4.2) and solve the two resulting recursion relations.

Now (3.4.4.2) says that

$$(2m^2+2j^2+4mj-j-m-1)a_j=(-2m-2j+2)a_{j-1}.$$

For $m=-1/2$ this is

$$a_j=\frac{3-2j}{-3j+2j^2}a_{j-1}=-\frac{1}{j}a_{j-1}.$$

We set $a_0=A$, so that

$$a_1=-a_0=-A, \quad a_2=-\frac{1}{2}a_1=\frac{1}{2}a_0=\frac{1}{2}A, \text{ etc.}$$

For $m=1$ we have

$$a_j=\frac{-2j}{3j+2j^2}a_{j-1}=\frac{-2}{3+2j}a_{j-1}.$$

We set $a_0=B$, so that

$$a_1=-\frac{2}{5}a_0=-\frac{2}{5}B, \quad a_2=-\frac{2}{7}a_1=\frac{4}{35}B, \text{ etc.}$$

Thus we have found the linearly independent solutions

$$Ax^{-1/2}\cdot\left(1-x+\frac{1}{2}x^2-+\cdots\right)$$

and

$$Bx \cdot \left(1 - \frac{2}{5}x + \frac{4}{35}x^2 - + \cdots\right).$$

The general solution of our differential equation is then

$$y = Ax^{-1/2} \cdot \left(1 - x + \frac{1}{2}x^2 - + \cdots\right) + Bx \cdot \left(1 - \frac{2}{5}x + \frac{4}{35}x^2 - + \cdots\right).$$

■

There are some circumstances (such as when the indicial equation has a repeated root) that the method we have presented will not yield two linearly independent solutions. We explore these circumstances in the next section.

Exercises

1. For each of the following differential equations, locate and classify its singular points on the x-axis.

   (a) $x^3(x-1)y'' - 2(x-1)y' + 3xy = 0$
   (b) $x^2(x^2-1)y'' - x(1-x)y' + 2y = 0$
   (c) $x^2y'' + (2-x)y' = 0$
   (d) $(3x+1)xy'' - (x+1)y' + 2y = 0$

2. Determine the nature of the point $x = 0$ (i.e., what type of singular point it is) for each of the following differential equations.

   (a) $y'' + (\sin x)y = 0$ (d) $x^3y'' + (\sin x)y = 0$
   (b) $xy'' + (\sin x)y = 0$ (e) $x^4y'' + (\sin x)y = 0$
   (c) $x^2y'' + (\sin x)y = 0$

3. Find the indicial equation and its roots (for a series solution in powers of x) for each of the following differential equations.

   (a) $x^3y'' + (\cos 2x - 1)y' + 2xy = 0$
   (b) $4x^2y'' + (2x^4 - 5x)y' + (3x^2 + 2)y = 0$
   (c) $x^2y'' + 3xy' + 4xy = 0$
   (d) $x^3y'' - 4xY2y' + 3xy = 0$

4. For each of the following differential equations, verify that the origin is a regular singular point and calculate two independent Frobenius series solutions:

   (a) $4x^2y'' + 3y' + y = 0$ (c) $2xy'' + (x+1)y' + 3y = 0$
   (b) $2xy'' + (3-x)y' - y = 0$ (d) $2x^2y'' + xy' - (x+1)y = 0$

5. When $p = 0$, Bessel's equation becomes

$$x^2 y'' + x y' + x^2 y = 0.$$

Show that its indicial equation has only one root, and use the method of this section to deduce that

$$y(x) = \sum_{j=0}^{\infty} \frac{(-1)^j}{2^{2j} (j!)^2} x^{2j}$$

is the corresponding Frobenius series solution.

6. Consider the differential equation

$$y'' + \frac{1}{x^2} y' - \frac{1}{x^3} y = 0.$$

(a) Show that $x = 0$ is an irregular singular point.

(b) Use the fact that $y_1 = x$ is a solution to find a second independent solution y_2 by the method discussed earlier in the book.

(c) Show that the second solution y_2 that we found in part **(b)** cannot be expressed as a Frobenius series.

7. Consider the differential equation

$$y'' + \frac{p}{x^b} y' + \frac{q}{x^c} = 0,$$

where p and q are nonzero real numbers and b, c are positive integers. It is clear that, if $b > 1$ or $c > 2$, then $x = 0$ is an irregular singular point.

(a) If $b = 2$ and $c = 3$, then show that there is only one possible value of m (from the indicial equation) for which there might exist a Frobenius series solution.

(b) Show similarly that m satisfies a quadratic equation—and hence we can hope for two Frobenius series solutions, corresponding to the roots of this equation—if and only if $b = 1$ and $c \leq 2$. Observe that these are exactly the conditions that characterize $x = 0$ as a "weak" or regular singular point as opposed to a "strong" or irregular singular point.

8. The differential equation

$$x^2 y'' + (3x - 1) y' + y = 0 \qquad\qquad (\star)$$

has $x = 0$ as an irregular singular point. If

$$\begin{aligned}
y &= x^m (a_0 + a_1 x + a_2 x^2 + \cdots) \\
&= a_0 x^m + a_1 x^{m+1} + a_2 x^{m+2} + \cdots
\end{aligned}$$

is inserted into $(\star)$, then show that $m = 0$ and the corresponding Frobenius series "solution" is the power series

$$y = \sum_{j=0}^{\infty} j! x^j,$$

which converges only at $x = 0$. This demonstrates that, even when a Frobenius series formally satisfies such an equation, it is not necessarily a valid solution.

3.5   More on Regular Singular Points

We now look at the Frobenius series solution of

$$y'' + p \cdot y' + q \cdot y = 0$$

at a regular singular point from a theoretical point of view.

Assuming that 0 is regular singular, we may suppose that

$$x \cdot p(x) = \sum_{j=0}^{\infty} p_j x^j \qquad \text{and} \qquad x^2 \cdot q(x) = \sum_{j=0}^{\infty} q_j x^j \,,$$

valid on a nontrivial interval $(-R, R)$. We guess a solution of the form

$$y = x^m \sum_{j=0}^{\infty} a_j x^j = \sum_{j=0}^{\infty} a_j x^{j+m}$$

and calculate

$$y' = \sum_{j=0}^{\infty} (j + m) a_j x^{j+m-1}$$

and

$$y'' = \sum_{j=0}^{\infty} (j + m)(j + m - 1) a_j x^{j+m-2} \,.$$

Then

$$
\begin{aligned}
p(x) \cdot y' &= \frac{1}{x} \left(\sum_{j=0}^{\infty} p_j x^j \right) \left(\sum_{j=0}^{\infty} a_j (m+j) x^{m+j-1} \right) \\
&= x^{m-2} \left(\sum_{j=0}^{\infty} p_j x^j \right) \left(\sum_{j=0}^{\infty} a_j (m+j) x^j \right) \\
&= x^{m-2} \sum_{j=0}^{\infty} \left(\sum_{k=0}^{j} p_{j-k} a_k (m+k) \right) x^j \,,
\end{aligned}
$$

where we have used the formula (Section 3.1) for the product of two power series. Now, breaking off the summands corresponding to $k = j$, we find that this last is equal to

$$x^{m-2} \sum_{j=0}^{\infty} \left(\sum_{k=0}^{j-1} p_{j-k} a_k (m+k) + p_0 a_j (m+j) \right) x^j \,.$$

Similarly,

$$
\begin{aligned}
q(x) \cdot y &= \frac{1}{x^2}\left(\sum_{j=0}^{\infty} q_j x^j\right)\left(\sum_{j=0}^{\infty} a_j x^{m+j}\right) \\
&= x^{m-2}\left(\sum_{j=0}^{\infty} q_j x^j\right)\left(\sum_{j=0}^{\infty} a_j x^j\right) \\
&= x^{m-2}\sum_{j=0}^{\infty}\left(\sum_{k=0}^{j} q_{j-k}a_k\right)x^j \\
&= x^{m-2}\sum_{j=0}^{\infty}\left(\sum_{k=0}^{j-1} q_{j-k}a_k + q_0 a_j\right)x^j \,.
\end{aligned}
$$

We put the series expressions for y'', $p \cdot y'$, and $q \cdot y$ into the differential equation, and cancel the common factor of x^{m-2}. The result is

$$
\sum_{j=0}^{\infty}\Bigg\{a_j[(m+j)(m+j-1)+(m+j)p_0+q_0] \\
+ \sum_{k=0}^{j-1} a_k[(m+k)p_{j-k}+q_{j-k}]\Bigg\}x^j = 0\,.
$$

Now of course each coefficient of x^j must be 0, so we obtain the following recursion relations:

$$
a_j[(m+j)(m+j-1)+(m+j)p_0+q_0] \\
+ \sum_{k=0}^{j-1} a_k[(m+k)p_{j-k}+q_{j-k}] = 0
$$

for $j = 0, 1, 2, \ldots$. (Incidentally, this illustrates a point we made earlier, in Section 3.3: That recursion relations need not be binary.)

It is convenient now to isolate the preceding formula for $j = 0$. This gives us the *indicial equation*

$$
f(m) = m(m-1) + mp_0 + q_0\,.
$$

This is an important and useful formula for the indicial equation. It is the equation that we solve to determine the values of m for any particular problem.

Then the recursion relation for $j = 0$ is

$$
a_0 f(m) = 0
$$

(because the sum in the recursion is vacuous when $j = 0$).

The successive recursion relations are

$$a_1 f(m+1) + a_0(mp_1 + q_1) = 0$$

$$a_2 f(m+2) + a_0[mp_2 + q_2] + a_1[(m+1)p_1 + q_1] = 0$$

$$\cdots$$

$$a_j f(m+j) + a_0[mp_j + q_j] + a_1[(m+1)p_{j-1} + q_{j-1}] + \cdots + a_{j-1}[(m+j-1)p_1 + q_1] = 0$$

etc.

The first recursion formula tells us, since $a_0 \neq 0$, that

$$f(m) = m(m-1) + mp_0 + q_0 = 0.$$

This is of course the indicial equation. The roots m_1, m_2 of this equation are called the *exponents* of the differential equation at the regular singular point.

Theorem 3.5.1 *Suppose that $x_0 = 0$ is a regular singular point for the differential equation*

$$y'' + p \cdot y' + q \cdot y = 0.$$

Assume that the power series for $x \cdot p(x)$ and $x^2 \cdot q(x)$ have radius of convergence $R > 0$. Suppose that the indicial equation $m(m-1) + mp_0 + q_0 = 0$ has roots m_1, m_2 with $m_1 \leq m_2$. Then the differential equation has at least one solution of the form

$$y_1 = x^{m_1} \sum_{j=0}^{\infty} a_j x^j$$

on the interval $(-R, R)$.

In case $m_2 - m_1$ is not zero or a positive integer, then the differential equation has a second independent solution

$$y_2 = x^{m_2} \sum_{j=0}^{\infty} b_j x^j$$

on the interval $(-R, R)$.

Now let us explain, and put this theorem in context. If the roots m_1 and m_2 are distinct and *do not differ by an integer*, then our procedures will produce two linearly independent solutions for the differential equation. If

$m_1 \leq m_2$ differ by an integer, say $m_2 = m_1 + k$ for some integer $k \geq 1$, then the recursion procedure breaks down because the coefficient of a_j in the jth recursion relation for m_1 will be 0—so that we cannot solve for a_j. The case $m_1 = m_2$ also leads to difficulties—because then our methods only generate one solution. *When m_1 and m_2 differ by a positive integer, then it is only necessary to do the analysis (leading to a solution of the differential equation) for the exponent m_1. The exponent m_2 is not guaranteed to lead to anything new.* The exponent m_2 *could* lead to something new, as the next example shows. But we cannot depend on it.

EXAMPLE 3.5.2 Find two independent Frobenius series solutions of

$$xy'' + 2y' + xy = 0.$$

Solution: We write the differential equation as

$$y'' + \frac{2}{x} \cdot y' + 1 \cdot y = 0.$$

Then $x \cdot p(x) = 2$ and $x^2 \cdot q(x) = x^2$. Notice that the constant term p_0 of $x \cdot p$ is 2 and the constant term q_0 of $x^2 \cdot q$ is 0. Thus the indicial equation is

$$m(m-1) + 2m + 0 = 0.$$

The exponents for the regular singular point 0 are then $0, -1$.
    Corresponding to $m_2 = 0$, we guess a solution of the form

$$y = \sum_{j=0}^{\infty} a_j x^j$$

which entails

$$y' = \sum_{j=1}^{\infty} j a_j x^{j-1}$$

and

$$y'' = \sum_{j=2}^{\infty} j(j-1) a_j x^{j-2}.$$

Putting these expressions into the differential equation yields

$$x \sum_{j=2}^{\infty} j(j-1) a_j x^{j-2} + 2 \sum_{j=1}^{\infty} j a_j x^{j-1} + x \sum_{j=0}^{\infty} a_j x^j = 0.$$

We adjust the indices, so that all powers are x^j, and break off the lower indices so that all sums begin at $j = 1$. The result is

$$\sum_{j=1}^{\infty} \left(a_{j+1}[j(j+1) + 2(j+1)] + a_{j-1} \right) x^j + 2a_1 = 0.$$

We read off the recursion relations

$$a_1 = 0$$

and

$$a_{j+1} = -\frac{a_{j-1}}{(j+2)(j+1)}$$

for $j \geq 1$. Thus all the coefficients with odd index are 0 and we calculate that

$$a_2 = -\frac{a_0}{3 \cdot 2}, \quad a_4 = \frac{a_0}{5 \cdot 4 \cdot 3 \cdot 2}, \quad a_6 = -\frac{a_0}{7 \cdot 6 \cdot 5 \cdot 4 \cdot 3 \cdot 2},$$

etc. Thus we have found one Frobenius solution

$$
\begin{aligned}
y_1 &= a_0 \left(1 - \frac{1}{3!}x^2 + \frac{1}{5!}x^4 - \frac{1}{7!}x^6 - + \cdots \right) \\
&= a_0 \cdot \frac{1}{x} \cdot \left(x - \frac{x^3}{3!} + \frac{x^5}{5!} - + \cdots \right) \\
&= a_0 \cdot \frac{1}{x} \cdot \sin x.
\end{aligned}
$$

Corresponding to $m_1 = -1$ we guess a solution of the form

$$y = x^{-1} \cdot \sum_{j=0}^{\infty} b_j x^j = \sum_{j=0}^{\infty} b_j x^{j-1}.$$

Thus we calculate that

$$y' = \sum_{j=0}^{\infty} (j-1) b_j x^{j-2}$$

and

$$y'' = \sum_{j=0}^{\infty} (j-1)(j-2) b_j x^{j-3}.$$

Putting these calculations into the differential equation gives

$$\sum_{j=0}^{\infty} (j-1)(j-2) b_j x^{j-2} + \sum_{j=0}^{\infty} 2(j-1) b_j x^{j-2} + \sum_{j=0}^{\infty} b_j x^j = 0.$$

We adjust the indices so that all powers that occur are x^{j-2} and break off extra terms so that all sums begin at $j = 2$. The result is

$$\sum_{j=2}^{\infty} [(j-1)(j-2) b_j + 2(j-1) b_j + b_{j-2}] x^{j-2}$$

$$+ \left((-1)(-2) b_0 x^{-2} + (0)(-1) b_1 x^{-1} + 2(-1) b_0 x^{-2} + 2(0) b_1 x^{-1} \right) = 0.$$

The four terms at the end cancel out. Thus our recursion relation is

$$b_j = -\frac{b_{j-2}}{j(j-1)}$$

for $j \geq 2$.

It is now easy to calculate that

$$b_2 = -\frac{b_0}{2 \cdot 1} \,, \quad b_4 = \frac{b_0}{4 \cdot 3 \cdot 2 \cdot 1} \,, \quad b_6 = -\frac{b_0}{6 \cdot 5 \cdot 4 \cdot 3 \cdot 2 \cdot 1} \,, \quad \text{etc.}$$

Also

$$b_3 = -\frac{b_1}{3 \cdot 2} \,, \quad b_5 = \frac{b_1}{5 \cdot 4 \cdot 3 \cdot 2 \cdot 1} \,, \quad b_7 = -\frac{b_1}{7 \cdot 6 \cdot 5 \cdot 4 \cdot 3 \cdot 2 \cdot 1} \,, \quad \text{etc.}$$

This gives the solution

$$
\begin{aligned}
y &= b_0 \cdot \frac{1}{x} \cdot \left(1 - \frac{1}{2!}x^2 + \frac{1}{4!}x^4 - + \cdots \right) + b_1 \cdot \frac{1}{x} \left(x - \frac{1}{3!}x^3 + \frac{1}{5!}x^5 - + \cdots \right) \\
&= b_0 \cdot \frac{1}{x} \cdot \cos x + b_1 \cdot \frac{1}{x} \cdot \sin x \,.
\end{aligned}
$$

Now we already discovered the solution $(1/x) \sin x$ in our first calculation with $m_2 = 0$. Our second calculation, with $m_1 = -1$, reproduces that solution and discovers the new, linearly independent solution $(1/x) \cos x$. ∎

EXAMPLE 3.5.3 The equation

$$4x^2 y'' - 8x^2 y' + (4x^2 + 1)y = 0$$

has only one Frobenius series solution. Find the general solution.

Solution: The indicial equation is

$$m(m-1) + m \cdot 0 + \frac{1}{4} = 0 \,.$$

The roots are $m_1 = 1/2, m_2 = 1/2$. This is a repeated indicial root, and it will lead to complications.

We guess a solution of the form

$$y = x^{1/2} \sum_{j=0}^{\infty} a_j x^j = \sum_{j=0}^{\infty} a_j x^{j+1/2} \,.$$

Thus

$$y' = \sum_{j=0}^{\infty} (j + 1/2) a_j x^{j-1/2}$$

and

$$y'' = \sum_{j=0}^{\infty} (j + 1/2)(j - 1/2) a_j x^{j-3/2}.$$

Putting these calculations into the differential equation yields

$$4x^2 \cdot \sum_{j=0}^{\infty} (j + 1/2)(j - 1/2) x^{j-3/2}$$

$$-8x^2 \cdot \sum_{j=0}^{\infty} (j + 1/2) a_j x^{j-1/2} + (4x^2 + 1) \cdot \sum_{j=0}^{\infty} a_j x^{j+1/2} = 0.$$

We may rewrite this as

$$\sum_{j=0}^{\infty} (2j + 1)(2j - 1) a_j x^{j+1/2}$$

$$-\sum_{j=0}^{\infty} (8j + 4) a_j x^{j+3/2} + \sum_{j=0}^{\infty} 4a_j x^{j+5/2} + \sum_{j=0}^{\infty} a_j x^{j+1/2}.$$

We adjust the indices so that all powers of x are $x^{j+1/2}$ and put extra terms on the right so that all sums begin at $j = 2$. The result is

$$\sum_{j=2}^{\infty} \left[(2j + 1)(2j - 1) a_j - (8j - 4) a_{j-1} + 4a_{j-2} + a_j \right] x^{j+1/2}$$

$$= -1 \cdot (-1) a_0 x^{1/2} - 3 \cdot 1 \cdot a_1 x^{3/2} + 4a_0 x^{3/2} - a_0 x^{1/2} - a_1 x^{3/2}$$

or

$$\sum_{j=2}^{\infty} [4j^2 a_j - (8j - 4) a_{j-1} + 4a_{j-2}] x^{j+1/2} = x^{3/2} (4a_0 - 4a_1).$$

We thus discover the recursion relations

$$a_1 = a_0$$

$$a_j = \frac{(2j - 1) a_{j-1} - a_{j-2}}{j^2} \qquad \text{for } j \geq 2.$$

Therefore

$$a_2 = \frac{a_0}{2!}, \quad a_3 = \frac{a_0}{3!}, \quad a_4 = \frac{a_0}{4!}, \quad a_5 = \frac{a_0}{5!}, \quad \text{etc.}$$

We thus have found the Frobenius series solution to our differential equation given by

$$y_1(x) = x^{1/2} \cdot \left(1 + \frac{x}{1!} + \frac{x^2}{2!} + \frac{x^3}{3!} + \frac{x^4}{4!} + \cdots \right) = x^{1/2} \cdot e^x.$$

But now we are stuck because $m = 1/2$ is a repeated root of the indicial equation. There is no other. We can rescue the situation by thinking back to how we handled matters for second-order linear equations with constant coefficients. In that circumstance, if the associated polynomial had distinct roots r_1, r_2 then the differential equation had solutions $y_1 = e^{r_1 x}$ and $y_2 = e^{r_2 x}$. But if the associated polynomial had repeated roots r and r, then solutions of the differential equation were given by $y_1 = e^{r x}$ and $y_2 = x \cdot e^{r x}$. Reasoning by analogy, and considering that x (or its powers) is a *logarithm* of e^x, we might hope that when m is a repeated root of the indicial equation, and $y_1 = y_1(x)$ is a solution corresponding to the root m, then $y_2(x) = \ln x \cdot y_1(x)$ is a second solution. We in fact invite the reader to verify that

$$y_2(x) = \ln x \cdot x^{1/2} \cdot e^x$$

is a second, linearly independent solution of the differential equation. So its general solution is

$$y = A x^{1/2} \cdot e^x + B \ln x \cdot x^{1/2} \cdot e^x . \qquad \blacksquare$$

Exercises

1. The equation
$$x^2 y'' - 3xy' + (4x + 4)y = 0$$
   has only one Frobenius series solution. Find it.

2. The equation
$$4x^2 y'' - 8x^2 y' + (4x^2 + 1)y = 0$$
   has only one Frobenius series solution. Find the general solution.

3. Find two independent Frobenius series solutions of each of the following equations.
   (a) $xy'' + 2y' + xy = 0$
   (b) $x^2 y'' - x^2 y' + (x^2 - 2)y = 0$
   (c) $xy'' - y' + 4x^3 y = 0$

4. Verify that the point 1 is a regular singular point for the equation
$$(x - 1)^2 y'' - 3(x - 1)y' + 2y = 0 .$$
   Now use Frobenius's method to solve the equation.

5. Verify that the point -1 is a regular singular point for the equation
$$3(x + 1)^2 y'' - (x + 1)y' - y = 0 .$$
   Now use Frobenius's method to solve the equation.

6. Bessel's equation of order $p = 1$ is

$$x^2 y'' + xy' + (x^2 - 1)y = 0.$$

We see that $x_0 = 0$ is a regular singular point. Show that $m_1 - m_2 = 2$ and that the equation has only one Frobenius series solution. Then find it.

7. Bessel's equation of order $p = 1/2$ is

$$x^2 y'' + xy' + \left(x^2 - \frac{1}{4}\right) y = 0.$$

Show that $m_2 - m_1 = 1$, but that the equation still has two independent Frobenius series solutions. Then find them.

Historical Note

Gauss

Often called the "prince of mathematicians," Carl Friedrich Gauss (1777–1855) left a legacy of mathematical genius that exerts considerable influence even today.

Gauss was born in the city of Brunswick in northern Germany. He showed an early aptitude with arithmetic and numbers. He was finding errors in his father's bookkeeping at the age of 3, and his facility with calculations soon became well known. He came to the attention of the Duke of Brunswick, who supported the young man's further education.

Gauss attended the Caroline College in Brunswick (1792–1795), where he completed his study of the classical languages and explored the works of Newton, Euler, and Lagrange. Early in this period he discovered the prime number theorem—legend has it by staring for hours at tables of primes. Gauss did not prove the theorem (it was finally proved in 1896 by Hadamard and de la Vallee Poussin). He also, at this time, invented the method of least squares for minimizing errors—a technique that is still widely used today. He also conceived the Gaussian (or normal) law of distribution in the theory of probability.

At the university, Gauss was at first attracted by philology and put off by the mathematics courses. But at the age of eighteen he made a remarkable discovery—of which regular polygons can be constructed by ruler and compass—and that set his future for certain. During these years Gauss was flooded, indeed nearly overwhelmed, by mathematical ideas. In 1795, just as an instance, Gauss discovered the fundamental law of quadratic reciprocity. It took a year of concerted effort for him to prove it. It is the core of his celebrated treatise *Disquisitiones Arithmeticae*, published in 1801. That book is arguably the cornerstone of modern number theory, as it contains the fundamental theorem of arithmetic as well as foundational ideas on congruences, forms, residues, and the theory of cyclotomic equations. The hallmark of Gauss's *Disquisitiones* is a strict adherence to rigor (unlike much of the mathematics of Gauss's day) and a chilling formality and lack of motivation.

Gauss's work touched all parts of mathematics—not just number theory. He discovered what is now known as the Cauchy integral formula, developed the intrinsic geometry of surfaces, discovered the mean-value property for harmonic functions, proved a version of Stokes's theorem, developed the theory of elliptic functions, and he anticipated Bolyai and Lobachevsky's ideas on non-Euclidean geometry. With regard to the latter—which was really an earth-shaking breakthrough—Gauss said that he did not publish his ideas because nobody (i.e., none of the mathematicians of the time) would appreciate or understand them.

Beginning in the 1830s, Gauss was increasingly occupied by physics. He had already had a real *coup* in helping astronomers to locate the planet Ceres using strictly mathematical reasoning. Now he turned his attention to conservation of energy, the calculus of variations, optics, geomagnetism, and potential theory.

Carl Friedrich Gauss was an extraordinarily powerful and imaginative mathematician who made fundamental contributions to all parts of the subject. He had a long and productive scientific career. When, one day, a messenger came to him with the news that his wife was dying, Gauss said, "Tell her to wait a bit until I am done with this theorem." Such is the life of a master of mathematics.

Historical Note

Abel

Niels Henrik Abel (1802–1829) was one of the foremost mathematicians of the nineteenth century, and perhaps the greatest genius ever produced by the Scandinavian countries. Along with his contemporaries Gauss and Cauchy, Abel helped to lay the foundations for the modern mathematical method.

Abel's genius was recognized when he was still young. In spite of grinding poverty, he managed to attend the University of Oslo. When only 21 years old, Abel produced a proof that the fifth-degree polynomial cannot be solved by an elementary formula (involving only arithmetic operations and radicals). Recall that the quadratic equation can be solved by the quadratic formula, and cubic and quartic equations can be solved by similar but more complicated formulas. This was an age-old problem, and Abel's solution was a personal triumph. He published the proof in a small pamphlet at his own expense. This was typical of the poor luck and lack of recognition that plagued Abel's short life.

Abel desired to spend time on the Continent and commune with the great mathematicians of the day. He finally got a government fellowship. His first year was spent in Berlin, where he became a friend and colleague of August Leopold Crelle. He helped Crelle to found the famous *Journal für die Reine und Angewandte Mathematik*, now the oldest extant mathematics journal.

There are many parts of modern mathematics that bear Abel's name. These include Abel's integral equation, Abelian integrals and functions, Abelian groups, Abel's series, Abel's partial summation formula, Abel's limit theorem, and Abel summability. The basic theory of elliptic functions is due to Abel (the reader may recall that these played a decisive role in Andrew Wiles's solution of Fermat's Last Problem).

Like Riemann (discussed elsewhere in this book), Abel lived in penury. He never held a proper academic position (although, shortly before Abel's death, Crelle finally secured for him a professorship in Berlin). The young genius contracted tuberculosis at the age of 26 and died soon thereafter.

Crelle eulogized Abel in his *Journal* with these words:

All of Abel's works carry the imprint of an ingenuity and force of thought which is amazing. One may say that he was able to penetrate all obstacles down to the very foundation of the problem, with a force which appeared irresistible ... He distinguished himself equally by the purity and nobility of his character and by a rare modesty which made his person cherished to the same unusual degree as was his genius.

It is difficult to imagine what Abel might have accomplished had he lived a normal lifespan and had an academic position with adequate financial support. His was one of the great minds of mathematics.

Today one may see a statue of Abel in the garden of the Norwegian King's palace—he is depicted stamping out the serpents of ignorance. Also the Abel Prize, one of the most distinguished of mathematical encomia, is named for this great scientist.

3.6 Steady-State Temperature in a Ball

Let us show how to reduce the analysis of the heat equation in a three-dimensional example to the study of Legendre's equation. Imagine a solid

ball of radius 1. We shall work in spherical coordinates (r, θ, ϕ) on this
ball. We hypothesize that the surface temperature at the boundary is held
at $g(\theta) = T_0 \sin^4 \theta$. Of course the steady-state temperature T will satisfy
Laplace's equation

$$\nabla^2 T = \left(\frac{\partial^2}{\partial x^2} + \frac{\partial^2}{\partial y^2} + \frac{\partial^2}{\partial z^2} \right) T = 0.$$

The Laplace operator is rotationally invariant and the boundary condition
does not depend on the azimuthal angle ϕ, hence the solution will not depend
on ϕ either. One may calculate that Laplace's equation thus takes the form

$$\frac{1}{r^2} \frac{\partial}{\partial r} \left(r^2 \frac{\partial T}{\partial r} \right) + \frac{1}{r^2 \sin \theta} \frac{\partial}{\partial \theta} \left(\sin \theta \frac{\partial T}{\partial \theta} \right) = 0. \tag{3.6.1}$$

We seek a solution by the method of separation of variables. Thus we set
$T(r, \theta) = A(r) \cdot B(\theta)$. Substituting into (3.6.1) gives

$$B \cdot \frac{d}{dr} \left(r^2 \frac{dA}{dr} \right) + \frac{A}{\sin \theta} \cdot \frac{d}{d\theta} \left(\sin \theta \frac{dB}{d\theta} \right) = 0.$$

Thus we find that

$$\frac{1}{A} \frac{d}{dr} \left(r^2 \frac{dA}{dr} \right) = - \frac{1}{B \sin \theta} \frac{d}{d\theta} \left(\sin \theta \frac{dB}{d\theta} \right).$$

The left-hand side depends only on r and the right-hand side depends only
on θ. We conclude that both sides are equal to a constant. Looking ahead to
the use of the Legendre equation, we are going to suppose that the constant
has the form $c = n(n+1)$ for n a nonnegative integer. This rather surprising
hypothesis will be justified later. Also refer back to the discussion in Example
3.3.1.

Now we have this ordinary differential equation for B:

$$\frac{1}{B} \cdot \frac{1}{\sin \theta} \frac{d}{d\theta} \left(\sin \theta \frac{dB}{d\theta} \right) + n(n+1) = 0.$$

We make the change of variable

$$\nu = \cos \theta, \quad y(\nu) = B(\theta).$$

With the standard identities

$$\sin^2 \theta = 1 - \nu^2 \quad \text{and} \quad \frac{d}{d\nu} = \frac{d\theta}{d\nu} \frac{d}{d\theta} = - \frac{1}{\sin \theta} \frac{d}{d\theta},$$

we find our differential equation converted to

$$\frac{d}{d\nu} \left((1 - \nu^2) \frac{dy}{d\nu} \right) + n(n+1)y = 0.$$

This is equivalent to

$$(1 - \nu^2)y'' - 2\nu y' + n(n+1)y = 0. \tag{3.6.2}$$

This is Legendre's equation.

Observe that $\nu = \pm 1$ corresponds to $\theta = 0, \pi$, i.e., the poles of the sphere. A physically meaningful solution will certainly be finite at these points. We conclude that the solution of our differential equation (3.6.2) is the Legendre polynomial $y(\nu) = P_n(\nu)$. Therefore the solution to our original problem is $B_n(\theta) = P_n(\cos \theta)$.

Our next task is to solve for $A(r)$. The differential equation (resulting from our separation of variables) is

$$\frac{d}{dr}\left(r^2 \frac{dA}{dr}\right) = n(n+1)A.$$

One can use the method of power series to solve this equation. Or one can take a shortcut and simply guess that the solution will be a power of r. Using the latter method, we find that

$$A_n(r) = c_n r^n + d_n r^{-1-n}.$$

Again, physical considerations can guide our thinking. We know that the temperature must be finite at the center of the sphere. Thus d_n must equal 0. Thus $A_n(r) = c_n r^n$. Putting this information together with our solution in θ, we find the solution of Laplace's equation to be

$$T = c_n r^n P_n(\cos \theta).$$

Here, of course, c_n is an arbitrary real constant.

Now we invoke the familiar idea of taking a linear combination of the solutions we have found to produce more solutions. We write our general solution as

$$T = \sum_{n=0}^{\infty} c_n r^n P_n(\cos \theta).$$

Recall that we specified the initial temperature on the sphere (the boundary of the ball) to be $T = T_0 \sin^4 \theta$ when $r = 1$. Thus we know that

$$T_0 \sin^4 \theta = f(\theta) = \sum_{n=0}^{\infty} c_n P_n(\cos \theta).$$

It is then possible to use the theory of Fourier–Legendre expansions to solve for the c_n. Since we have not developed that theory in the present book, we shall not carry out these calculations. We merely record the fact that the solution turns out to be

$$T = T_0 \left(\frac{8}{15} P_0(\cos \theta) - \frac{16}{21} r^2 P_2(\cos \theta) + \frac{8}{35} r^4 P_4(\cos \theta)\right).$$

Problems for Review and Discovery

A. Drill Exercises

1. Use the method of power series to find a solution of each of these differential equations.

    (a) $y'' + 2xy = x^2$
    (b) $y'' - xy' + y = x$
    (c) $y'' + y' + y = x^3 - x$
    (d) $2y'' + xy' + y = 0$
    (e) $(4 + x^2)y'' - y' + y = 0$
    (f) $(x^2 + 1)y'' - xy' + y = 0$
    (g) $y'' - (x + 1)y' - xy = 0$
    (h) $(x - 1)y'' + (x + 1)y' + y = 0$

2. For each of the following differential equations, verify that the origin is a regular singular point and calculate two independent Frobenius series solutions.

    (a) $(x^2 + 1)x^2y'' - xy' + (2 + x)y = 0$
    (b) $x^2y'' + xy' + (1 + x)y = 0$
    (c) $xy'' - 4y' + xy = 0$
    (d) $4x^2y'' + 4x^2y' + 2y = 0$
    (e) $2xy'' + (1 - x)y' + y = 0$
    (f) $xy'' - (x - 1)y' + 2y = 0$
    (g) $x^2y'' + x(1 - x)y' + y = 0$
    (h) $xy'' + (x + 1)y' + y = 0$

3. In each of these problems, use the method of Frobenius to find the first four nonzero terms in the series solution about $x = 0$ for a solution to the given differential equation.

    (a) $x^3y''' + 2x^2y'' + (x + x^2)y' + xy = 0$
    (b) $x^3y''' + x^2y'' - 3xy' + (x - 1)y = 0$
    (c) $x^3y''' - 2x^2y'' + (x^2 + 2x)y' - xy = 0$
    (d) $x^3y''' + (2x^3 - x^2)y'' - xy' + y = 0$

B. Challenge Problems

1. For some applications it is useful to have a series expansion about the point at infinity. In order to study such an expansion, we institute the change of variables $z = 1/x$ (of course we must remember to use the chain rule to transform the derivative as well) and expand about $z = 0$. In each of the following problems, use this idea to verify that ∞ is a regular singular point for the given differential equation by checking that $z = 0$ is a regular singular point for the transformed equation. Then find the first four nonzero terms in the series expansion about ∞ of a solution to the original differential equation.

    (a) $x^3 y'' + x^2 y' + y = 0$

    (b) $9(x-2)^2(x-3)y'' + 6x(x-2)y' + 16y = 0$

    (c) $(1-x^2)y'' - 2xy' + p(p+1)y = 0$ (Legendre's equation)

    (d) $x^2 y'' + xy' + (x^2 - p^2)y = 0$ (Bessel's equation)

2. *Laguerre's equation* is

$$xy'' + (1-x)y' + py = 0,$$

where p is a constant. Show that the only solutions that are bounded near the origin are series of the form

$$1 + \sum_{n=1}^{\infty} \frac{-p(-p+1)\cdots(-p+n-1)}{(n!)^2} x^n.$$

This is the series representation of a *confluent hypergeometric function*, and is often denoted by the symbol $F(-p, 1, x)$. In case $p \geq 0$ is an integer, then show that this solution is in fact a polynomial. These solutions are called *Laguerre polynomials*, and they play an important role in the study of the quantum mechanics of the hydrogen atom.

3. The ordinary differential equation

$$x^4 \frac{d^2 y}{dx^2} + \lambda^2 y = 0, \quad x > 0$$

is the mathematical model for the buckling of a column in the shape of a truncated cone. The positive constant λ depends on the rigidity of the column, the moment of inertia at the top of the column, and the load. Use the substitution $x = 1/z$ to reduce this differential equation to the form

$$\frac{d^2 y}{dz^2} + \frac{2}{z}\frac{dy}{dz} + \lambda^2 y = 0.$$

Find the first five terms in the series expansion about the origin of a solution to this new equation. Convert it back to an expansion for the solution of the original equation.

C. Problems for Discussion and Exploration

1. Consider a nonlinear ordinary differential equation such as

$$[\sin y]y'' + e^y y' - y^2 = 0.$$

Why would it be neither efficient nor useful to guess a power series solution for this equation?

2. A celebrated theorem of Cauchy and Kowalewska guarantees that a non-singular ordinary differential equation with real analytic coefficients will have a real analytic solution. What will happen if you seek a real analytic (i.e., a power series) solution to a differential equation that does *not* have real analytic coefficients?

3. Show that if y is a solution of *Bessel's equation*

$$x^2 y'' + x y' + (x^2 - p^2)y = 0$$

of order p, then $u(x) = x^{-c} y(ax^b)$ is a solution of

$$x^2 u'' + (2c+1)xu' + [a^2 b^2 x^{2b} + (c^2 - p^2 b^2)]u = 0.$$

Use this result to show that the general solution of *Airy's equation* $y'' - xy = 0$ is

$$y = |x|^{1/2} \left(A J_{1/3} \left(\frac{2|x|^{3/2}}{3} \right) + B J_{-1/3} \left(\frac{2|x|^{3/2}}{3} \right) \right).$$

Here J_n is the *Bessel function* defined by

$$J_n(x) = \left(\frac{x}{2} \right)^n \sum_{k=0}^{\infty} \frac{(-1)^k}{k!(k+n)!} \left(\frac{x}{2} \right)^{2k}$$

as long as $n \geq 0$ is an integer. In case n is replaced by p not an integer, then we replace $(k+n)!$ by $\Gamma(k+p+1)$, where

$$\Gamma(z) \equiv \int_0^{\infty} x^{z-1} e^{-x} \, dx.$$

4

Sturm–Liouville Problems and Boundary Value Problems

- The concept of a Sturm–Liouville problem

- How to solve a Sturm–Liouville problem

- Eigenvalues and eigenfunctions

- Orthogonal expansions

- Singular Sturm–Liouville

- Separation of variables

4.1   What Is a Sturm–Liouville Problem?

We wish to introduce the idea of eigenvalues and eigenfunctions. We can motivate the idea with the fairly extensive and far-reaching subject of Sturm–Liouville problems.

A sequence y_j of functions such that

$$\int_a^b y_m(x) y_n(x)\, dx = 0 \qquad \text{for } m \neq n$$

is said to be an *orthogonal system* on the interval $[a, b]$. If

$$\int_a^b y_j^2(x)\, dx = 1$$

for each j then we call this an *orthonormal system* or *orthonormal sequence*. It turns out (see below) that the sequence of eigenfunctions associated with a wide variety of boundary value problems enjoys the orthogonality property.

Now consider a differential equation of the form

$$\frac{d}{dx}\left(p(x)\frac{dy}{dx}\right) + [\lambda q(x) + r(x)]y = 0 ; \tag{4.1.1}$$

171

we shall be interested in solutions valid on a bounded interval $[a, b]$. We know that, under suitable conditions on the coefficients, a solution of equation (4.1.1) that takes a prescribed value and a prescribed derivative value at a fixed point $x_0 \in [a, b]$ will be uniquely determined. In other circumstances, we may wish to prescribe the values of y at two distinct points, say at a and at b. We now begin to examine the conditions under which such a *boundary value problem* has a nontrivial solution.

EXAMPLE 4.1.2 Consider equation (4.1.1) with $p(x) \equiv q(x) \equiv 1$ and $r(x) \equiv 0$. Then the differential equation becomes

$$y'' + \lambda y = 0.\qquad\qquad(4.1.2.1)$$

We take the domain interval to be $[0, \pi]$ and the boundary conditions to be

$$y(0) = 0, \quad y(\pi) = 0.\qquad\qquad(4.1.2.2)$$

Let us determine the eigenvalues and eigenfunctions for this problem.

Solution: The situation with boundary conditions is quite different from that for initial conditions. The latter is a sophisticated variation of the fundamental theorem of calculus. The former is rather more subtle. So let us begin to analyze.

First, if $\lambda < 0$ then the solutions of the differential equation are exponentials. So no nontrivial linear combination of these can satisfy the boundary conditions (4.1.2.2).

If $\lambda = 0$ then the general solution of (4.1.2.1) is the linear function $y = Ax + B$. Such a function cannot vanish at two points unless it is identically zero.

So the only interesting case is $\lambda > 0$. In this situation, the general solution of (4.1.2.1) is

$$y = A \sin \sqrt{\lambda} x + B \cos \sqrt{\lambda} x.$$

Since $y(0) = 0$, this in fact reduces to

$$y = A \sin \sqrt{\lambda} x.$$

In order for $y(\pi) = 0$, we must have $\sqrt{\lambda} \pi = n\pi$ for some positive integer n, thus $\lambda = n^2$. These values of λ are termed the *eigenvalues* of the problem, and the corresponding solutions

$$\sin x, \quad \sin 2x, \quad \sin 3x \ \ldots$$

are called the *eigenfunctions* of the problem (4.1.2.1), (4.1.2.2).   ∎

We note these immediate properties of the eigenvalues and eigenfunctions for our problem:

(i) If ϕ is an eigenfunction for eigenvalue λ, then so is $c \cdot \phi$ for any constant c.

(ii) The eigenvalues $1, 4, 9, \ldots$ form an increasing sequence that approaches $+\infty$.

(iii) The nth eigenfunction $\sin nx$ vanishes at the endpoints $0, \pi$ (as we originally mandated) and has exactly $n - 1$ zeros in the interval $(0, \pi)$.

It will turn out—and this is the basis for the Sturm–Liouville theory—that, if $p, q > 0$ on $[a, b]$, then equation (4.1.1) will have a solvable boundary value problem—for a certain discrete set of values of λ—with data specified at points a and b. These special values of λ will of course be the eigenvalues for the boundary value problem. They are real numbers that we shall arrange in their natural order

$$\lambda_1 < \lambda_2 < \cdots < \lambda_n < \cdots,$$

and we shall learn that $\lambda_j \to +\infty$. The corresponding eigenfunctions will then be ordered as $y_1, y_2, \ldots$.

Now let us examine possible orthogonality properties for the eigenfunctions of the boundary value problem for equation (4.1.1). Consider the differential equation (4.1.1) with two different eigenvalues λ_m and λ_n and y_m and y_n the corresponding eigenfunctions:

$$\frac{d}{dx}\left(p(x)\frac{dy_m}{dx}\right) + [\lambda_m q(x) + r(x)]y_m = 0$$

and

$$\frac{d}{dx}\left(p(x)\frac{dy_n}{dx}\right) + [\lambda_n q(x) + r(x)]y_n = 0.$$

We convert to the more convenient prime notation for derivatives, multiply the first equation by y_n and the second by y_m, and subtract. The result is

$$y_n(py_m')' - y_m(py_n')' + (\lambda_m - \lambda_n)qy_my_n = 0.$$

We move the first two terms to the right-hand side of the equation and integrate from a to b. Hence

$$(\lambda_m - \lambda_n)\int_a^b qy_my_n \, dx \quad = \quad \int_a^b y_m(py_n')' \, dx - \int_a^b y_n(py_m')' \, dx$$

$$\stackrel{\text{(parts)}}{=} \quad [y_m(py_n')]_a^b - \int_a^b y_m'(py_n') \, dx$$

$$- [y_n(py_m')]_a^b + \int_a^b y_n'(py_m') \, dx$$

$$= \quad p(b)[y_m(b)y_n'(b) - y_n(b)y_m'(b)]$$
$$- p(a)[y_m(a)y_n'(a) - y_n(a)y_m'(a)].$$

Notice that the two integrals on the right have cancelled.

Let us denote by $W(x)$ the Wronskian determinant of the two solutions y_m, y_n. Thus

$$W(x) = y_m(x)y_n'(x) - y_n(x)y_m'(x) \,.$$

Then our last equation can be written in the more compact form

$$(\lambda_m - \lambda_n) \int_a^b q y_m y_n \, dx = p(b)W(b) - p(a)W(a) \,.$$

Notice that things have turned out so nicely, and certain terms have cancelled, just because of the special form of the original differential equation.

We want the right-hand side of this last equation to vanish. This will certainly be the case if we require the familiar boundary condition

$$y(a) = 0 \qquad \text{and} \qquad y(b) = 0$$

or instead we require that

$$y'(a) = 0 \qquad \text{and} \qquad y'(b) = 0 \,.$$

Either of these will guarantee that the Wronskian vanishes, and therefore

$$\int_a^b y_m \cdot y_n \cdot q \, dx = 0 \,.$$

This is called an *orthogonality condition with weight q*.

With such a condition in place, we can consider representing an arbitrary function f as a linear combination of the y_j:

$$f(x) = a_1 y_1(x) + a_2 y_2(x) + \cdots + a_j y_j(x) + \cdots . \qquad (4.1.3)$$

We may determine the coefficients a_j by multiplying both sides of this equation by $y_k \cdot q$ and integrating from a to b. Thus

$$\int_a^b f(x)y_k(x)q(x)\,dx \;=\; \int_a^b \Big(a_1 y_1(x) + a_2 y_2(x) + \cdots $$
$$+ a_j y_j(x) + \cdots \Big) y_k(x)q(x)\,dx$$
$$=\; \sum_j a_j \int_a^b y_j(x)y_k(x)q(x)\,dx$$
$$=\; a_k \int_a^b y_k^2(x)q(x)\,dx$$

since all but one of the integrals vanishes (by orthogonality). Thus

$$a_k = \frac{\int_a^b f(x)y_k(x)q(x)\,dx}{\int_a^b y_k^2(x)q(x)\,dx} \,.$$

There is an important question that now must be asked. Namely, are there *enough* of the eigenfunctions y_j so that virtually any function f can be expanded as in (4.1.3)? For instance, the functions $y_1(x) = \sin x, y_3(x) = \sin 3x, y_7(x) = \sin 7x$ are orthogonal on $[-\pi, \pi]$, and for any function f one can calculate coefficients a_1, a_3, a_7. But there is no hope that a large class of functions f can be spanned by just y_1, y_3, y_7. We need to know that our y_j's "fill out the space." The study of this question is beyond the scope of the present text, as it involves ideas from Hilbert space (see [RUD1], [RUD2]). Our intention here has been merely to acquaint the reader with some of the language of Sturm–Liouville problems.

Math Nugget

Charles Hermite (1822–1901) was one of the most eminent French mathematicians of the nineteenth century. He was particularly noted for the elegance, indeed the artistry, of his work. As a student, he courted disaster by neglecting his routine assignments in order to study the classic masters of mathematics. Although he nearly failed his examinations, he became a first-rate and highly creative mathematician while still in his early twenties. In 1870 he was appointed to a professorship at the Sorbonne, where he trained a whole generation of important French mathematicians; these included Picard, Borel, and Poincaré.

The unusual character of Hermite's mind is suggested by the following remark of Poincaré: "Talk with M. Hermite. He never evokes a concrete image, yet you soon perceive that the most abstract entities are to him like living creatures." He disliked geometry, but was strongly attracted to number theory and analysis; his favorite subject was elliptic functions, where these two subjects interact in remarkable ways.

Several of Hermite's purely mathematical discoveries had unexpected applications many years later to mathematical physics. For example, the Hermite forms and matrices, which he invented in connection with certain problems of number theory, turned out to be crucial for Heisenberg's 1925 formulation of quantum mechanics. Also Hermite polynomials and Hermite functions are useful in solving Schrödinger's wave equation.

Exercises

1. The differential equation

$$P(x)y'' + Q(x)y' + R(x)y = 0$$

is called *exact* if it can be written in the form $[P(x)y']' + [S(x)y]' = 0$ for some function $S(x)$. If such an equation is not exact, it can often be made exact by multiplying through by a suitable integrating factor $\mu(x)$ (actually, there is always an integrating factor—but you may have trouble finding it). The function $\mu(x)$ must satisfy the condition that the equation

$$\mu(x)P(x)y'' + \mu(x)Q(x)y' + \mu(x)R(x)y = 0$$

can be expressed in the form

$$\left[\mu(x)P(x)y'\right]' + [S(x)y]' = 0$$

for some appropriate function S. Such an equation can be solved by the method of first-order linear equations.

Show that this μ must be the solution of the *adjoint equation*

$$P(x)\mu''(x) + \left[2P'(x) - Q(x)\right]\mu'(x) + \left[P''(x) - Q'(x) + R(x)\right]\mu(x) = 0.$$

Often the adjoint equation is just as difficult to solve as the original differential equation. But not always. Find the adjoint equation in each of the following instances.

(a)   **Legendre's equation:** $(1 - x^2)y'' - 2xy' + p(p+1)y = 0$

(b)   **Bessel's equation:** $x^2y'' + xy' + (x^2 - p^2)y = 0$

(c)   **Hermite's equation:** $y'' - 2xy' + 2py = 0$

(d)   **Laguerre's equation:** $xy'' + (1 - x)y' + py = 0$

2. Refer to Exercise 1. Consider the Euler equidimensional equation,

$$x^2y'' + xy' - n^2y = 0,$$

which we have seen before. Here n is a positive integer. Find the values of n for which this equation is exact, and for these values find the general solution by the method suggested in Exercise 1.

3. Refer to Exercise 1 for terminology. Solve the equation

$$y'' - \left(2x + \frac{3}{x}\right)y' - 4y = 0$$

by finding a simple solution of the adjoint equation by inspection.

4. Refer to Exercise 1. Show that the adjoint of the adjoint of the equation $P(x)y'' + Q(x)y' + R(x)y = 0$ is just the original equation.

5. Refer to Exercise 1 for terminology. The equation $P(x)y'' + Q(x)y' + R(x)y = 0$ is called *self-adjoint* if its adjoint is just the same equation (after a possible change of notation).

   (a) Show that this equation is self-adjoint if and only if $P'(x) = Q(x)$. In this case the equation becomes

   $$P(x)y'' + P'(x)y' + R(x) = 0$$

   or

   $$[P(x)y']' + R(x)y = 0.$$

   This is the standard form for a self-adjoint equation.

   (b) Which of the equations in Exercise 1 are self-adjoint?

6. Show that any equation $P(x)y'' + Q(x)y' + R(x)y = 0$ can be made self-adjoint by multiplying through by

   $$\frac{1}{P} \cdot e^{\int (Q/P)\, dx}.$$

7. Using Exercise 5 when appropriate, put each equation in Exercise 1 into the standard self-adjoint form described in Exercise 6.

4.2   Analyzing a Sturm–Liouville Problem

Let us now formulate and analyze some Sturm–Liouville problems.

EXAMPLE 4.2.1 For fixed n, the Bessel equation

$$\frac{d}{dx}\left[x\frac{dy}{dx}\right] + \left(k^2 x - \frac{n^2}{x}\right)y = 0, \quad a \le r \le b, \qquad (4.2.1.1)$$

is a Sturm–Liouville problem. Here $p = x$, $q = x$, $\lambda = k^2$, and $r = n^2/x$. We impose the endpoint conditions $y(a) = 0$ and $y(b) = 0$. ∎

EXAMPLE 4.2.2 Consider the differential equation

$$y'' + \lambda y = 0, \quad -\pi \le x \le \pi.$$

We impose the period endpoint conditions

$$\begin{aligned} y(-\pi) &= y(\pi) \\ y'(-\pi) &= y'(\pi). \end{aligned}$$

It is straightforward to calculate that the eigenfunctions for this system are 1, $\cos nx$, and $\sin nx$ for n a positive integer. The corresponding eigenvalues are n^2. Note that, for $n > 0$, there are two distinct eigenfunctions with the same eigenvalue n^2. But, for $n = 0$, there is only one eigenfunction. ∎

EXAMPLE 4.2.3 The Mathieu equation is

$$y'' + (\lambda + 16d\cos 2x)y = 0 , \quad 0 \le x \le \pi .$$

Notice that $p = 1$, $q = 1$, and $r = -16d\cos 2x$. Of course d is a constant and λ is a constant. We can impose the endpoint conditions $y(0) = y(\pi)$, $y'(0) = y'(\pi)$ or $y(0) = -y(\pi)$, $y'(0) = -y'(\pi)$. ∎

An important feature of the eigenfunctions of a Sturm–Liouville problem is their "orthogonality." This idea we now explain.

Proposition 4.2.4 *Eigenfunctions u and v having different eigenvalues for a Sturm–Liouville problem as in equation (4.1.1) are orthogonal with weight q in the sense that*

$$\int_a^b \big[u(x)v(x)\big]q(x)\, dx = 0 . \qquad (4.2.4.1)$$

Proof: Suppose that u is an eigenfunction corresponding to eigenvalue λ and v is an eigenfunction corresponding to eigenvalue μ. And assume that $\lambda \ne \mu$.
 Let us use the operator notation

$$L[y] = [p(x)y']' + r(x)y .$$

Observe that saying that u is an eigenfunction with eigenvalue λ is just the same as saying that

$$L[u] = -\lambda q u \qquad (4.2.4.2)$$

and likewise

$$L[v] = -\mu q v . \qquad (4.2.4.3)$$

 Then we know that

$$L[u] + \lambda q(x)u = L[v] + \mu q(x)v = 0 .$$

 Now observe, by direct calculation, that

$$uL[v] - vL[u] = \frac{d}{dx}\Big\{ p(x)[u(x)v'(x) - v(x)u'(x)] \Big\} .$$

We integrate this identity from a to b, using the equalities in (4.2.4.2) and (4.2.4.3). The result is

$$(\lambda - \mu)\int_a^b q(x)u(x)v(x)\, dx = p(x)[u(x)v'(x) - v(x)u'(x)]\Big|_{x=a}^{x=b} . \qquad (4.2.4.4)$$

If we now use our endpoint conditions $\alpha u(a) + \alpha' u'(a) = 0$ and $\beta u(b) + \beta' u'(b) = 0$, we may see that

$$p(a)[u(a)v'(a) - v(a)u'(a)] = [\alpha p(a)/\alpha'][u(a)v(a) - v(a)u(a)] = 0$$

provided that $\alpha' \neq 0$. If $\alpha \neq 0$ then the right-hand side of (4.2.4.4) similarly reduces at $x = a$ to

$$[\alpha' p(a)/\alpha][u'(a)v'(a) - v'(a)u'(a)] = 0.$$

As a result,

$$p(a)[u(a)v'(a) - v(a)u'(a)] = 0$$

unless $\alpha = \alpha' = 0$. Similar calculations apply to the endpoint $x = b$. Since we do not allow both $\alpha = \alpha' = 0$ and $\beta = \beta' = 0$, we find therefore that the right-hand side of (4.2.4.4) vanishes. Formula (4.2.4.1) now follows from formula (4.2.4.4) after dividing through by the nonzero factor $\lambda - \mu$. □

In a sense, Sturm–Liouville theory is a generalization of the Fourier theory that we shall learn in Chapter 6. One of the key ideas in the Fourier theory is that most any function can be expanded in a series of sine and cosine functions. And of course these two families of functions are the eigenfunctions for the problem $y'' + \lambda y = 0$. The eigenvalues (values of λ) that arise form an infinite increasing sequence, and they tend to infinity. And only one eigenfunction corresponds to each eigenvalue. We shall learn now that these same phenomena take place for a Sturm–Liouville problem.

Now suppose that, for a given Sturm–Liouville problem on an interval $[a, b]$, the eigenvalues are λ_j and the corresponding eigenfunctions are y_j. Our goal is to take a fairly arbitrary function f defined on the interval of definition $[a, b]$ and write it as

$$f(x) = \sum_{j=1}^{\infty} a_j y_j(x). \tag{4.2.5}$$

How can we determine the coefficients a_j?

Fix an index n. We multiply both sides of the equation (4.2.5) by $y_n(x)q(x)$ and integrate from a to b. The result is

$$\int_a^b f(x)y_n(x)q(x)\, dx = \int_a^b \left[\sum_{j=1}^{\infty} a_j y_j(x) \right] y_n(x)q(x)\, dx.$$

We switch the summation and integration operations on the right-hand side (a move that is justified by an advanced theorem from real analysis—see [KRA2]). So we now have

$$\int_a^b f(x)y_n(x)q(x)\, dx = \sum_{j=1}^{\infty} \int_a^b a_j y_j(x)y_n(x)q(x)\, dx.$$

Now the orthogonality of the eigenfunctions y_j tells us that each summand on the right vanishes *except* for the term with $j = n$. So we have

$$\int_a^b f(x)y_n(x)q(x)\,dx = \int_a^b a_n y_n(x)y_n(x)q(x)\,dx\,.$$

Thus we have learned that

$$a_n = \frac{\int_a^b f(x)y_n(x)q(x)\,dx}{\int_a^b a_n(y_n(x))^2 q(x)\,dx}\,. \tag{4.2.6}$$

The following theorem summarizes the key facts about the series expansion (4.2.5) and the formula (4.2.6) for calculating the coefficients.

Theorem 4.2.7 *Suppose that y_1, y_2, ... are the eigenfunctions of a regular Sturm–Liouville problem posed on the interval $[a, b]$. This simply means that the interval is finite and that p and q are positive and continuous on the entire interval. Suppose that f is a continuous function on the interval $[a, b]$. Then the series*

$$f(x) = \sum_{j=1}^{\infty} a_j y_j(x)\,,$$

with the coefficients a_j calculated according to the formula

$$a_j = \frac{\int_a^b f(x)y_j(x)q(x)\,dx}{\int_a^b (y_j(x))^2 q(x)\,dx}$$

converges at each point x of the open interval (a, b) to the value $f(x)$.

EXAMPLE 4.2.8 Consider the Sturm–Liouville problem

$$y'' + \lambda y = 0$$

on the interval $[0, L]$ with boundary conditions $y'(0) = y(L) = 0$. It has eigenfunctions

$$y_j(x) = \cos\frac{(2j-1)\pi x}{2L}$$

(we leave this calculation as an exercise for you). Notice that, for this Sturm–Liouville problem, $q(x) \equiv 1$ and $r(x) = 0$.

The corresponding eigenfunction series expansion for a continuous function f is

$$f(x) = \sum_{j=1}^{\infty} a_j \cos \frac{(2j-1)\pi x}{2L},$$

where

$$a_j = \frac{\int_0^L f(x) \cos[(2j-1)\pi x/(2L)]\, dx}{\int_0^L \cos^2[(2j-1)\pi x/(2L)]\, dx}.$$

But a simple calculation shows that

$$\int_0^L \cos^2[(2j-1)\pi x/(2L)]\, dx = \frac{L}{2}.$$

So we may write

$$a_j = \frac{2}{L} \int_0^L f(x) \cos[(2j-1)\pi x/(2L)]\, dx.$$

We may similarly analyze the Sturm–Liouville problem

$$y'' + \lambda y = 0$$

on the interval $[0, L]$ with boundary conditions $y(0) = y'(L) = 0$. For this setup, the eigenfunctions (this is an exercise for you) are

$$y_j(x) = \sin \frac{(2j-1)\pi x}{2L}$$

and the series expansion for a continuous function f on $[0, L]$ is

$$f(x) = \sum_{j=1}^{\infty} a_j \sin \frac{(2j-1)\pi x}{2L},$$

with

$$a_j = \frac{2}{L} \int_0^L f(x) \sin[(2j-1)\pi x/(2L)]\, dx. \qquad \blacksquare$$

Exercises

1. Explain how the equation

$$\frac{d^2 y}{dx^2} + \alpha(x)\frac{dy}{dx} + (\lambda\beta(x) - \gamma(x))y = 0$$

can be written in the form of a Sturm–Liouville equation by setting

$$p(x) = \exp\left(\int \alpha(x)dx\right).$$

Describe $q(x)$ and $r(x)$ in terms of $\alpha(x)$, $\beta(x)$, and $\gamma(x)$.

2. Find a way to write the generalized Legendre equation

$$(1 - x^2)\frac{d^2y}{dx^2} - 2x\frac{dy}{dx} + \left(n(n+1) - \frac{m^2}{1-p^2}\right)y = 0$$

in Sturm–Liouville form.

3. Consider the Sturm–Liouville problem $y'' + \lambda y = 0$ with endpoint conditions $y(0) = 0$ and $y(\pi) + y'(\pi) = 0$. Show that there is an infinite sequence of eigenfunctions with distinct eigenvalues. What are those eigenvalues?

4. Show that, for any eigenvalue λ, the Mathieu equation has either an odd eigenfunction or an even eigenfunction.

5. Consider the equation $y'' + \lambda y = 0$. Verify that the eigenvalues for the endpoint conditions

   (a) $y(0) = y(\pi) = 0$
   (b) $y(0) = y'(\pi) = 0$
   (c) $y'(0) = y(\pi) = 0$
   (d) $y'(0) = y'(\pi) = 0$

   are, respectively, $\{k^2\}$, $\{(k+1/2)^2\}$, $\{(k+1/2)^2\}$, $\{k^2\}$. Can you find the eigenfunctions?

6. Find all eigenvalues λ so that the differential equation $y'''' + \lambda y = 0$ has a nontrivial eigenfunction satisfying $y(0) = y'(0) = y(\pi) = y'(\pi) = 0$.

4.3   Applications of the Sturm–Liouville Theory

In the last section of this chapter, we explore applications of Sturm–Liouville theory to quantum mechanics. In the present section we look at more basic applications to classical physics.

Perhaps the most classical application to mechanics is the vibrating string. We treated the vibrating string in detail in Section 2.5, so we shall not repeat those ideas here.

A second application is to the study of the longitudinal vibrations of an elastic bar of local stiffness $p(x)$ and density $q(x)$. The mean longitudinal displacement $w(x, t)$ of the section of this bar from its equilibrium position x satisfies the wave equation (see Section 10.2)

$$q(x)\frac{\partial^2 w}{\partial t^2} = \frac{\partial}{\partial x}\left[p(x)\frac{\partial w}{\partial x}\right].$$

The simple harmonic vibrations given by the separation of variables

$$u = \varphi(x) \cos k(t - t_0)$$

are in fact the solutions of the Sturm–Liouville equation

$$\frac{d}{dx}\left[p(x)\frac{du}{dx}\right] + k^2 p(x)u = 0.$$

Note that this new equation is in fact the special instance of our original equation (4.1.1) with $r = 0$ and $\lambda = k^2$.

For a finite bar, represented mathematically by the interval $[a, b]$, there are various natural sets of endpoint conditions:

- $u(a) = u(b) = 0$ (rigidly fixed ends),

- $u'(a) = u'(b) = 0$ (free ends),

- $u'(a) + \alpha u(a) = u'(b) + \beta u(b) = 0$ (elastically held ends),

- $u(a) = u(b), \ u'(a) = u'(b)$ (periodic constraints).

Each one of these endpoint conditions on u implies a corresponding condition on the original function u. The natural frequencies of longitudinal vibration (musical fundamental tones and overtones) of a bar whose ends are held in one of the four manners described above are solutions of a Sturm–Liouville system.

Finally we briefly discuss a (two-dimensional) vibrating membrane. The partial differential equation describing this situation is (with x, y being spatial coordinates in the plane and t being time)

$$w_{tt} = c^2(w_{xx} + w_{yy}).$$

In polar coordinates this would be

$$w_{tt} = c^2(w_{rr} + r^{-1}w_r + r^{-2}w_{\theta\theta}).$$

A basis of standing wave solutions can be found with the separation of variables

$$
\begin{aligned}
w(r, \theta, t) &= u(r)\cos n\theta \cos \kappa(t - t_0) \\
w(r, \theta, t) &= u(r)\sin n\theta \cos \kappa(t - t_0).
\end{aligned}
$$

For w to satisfy the membrane equation with $\kappa = ck$, it is enough for u to be a solution of the Bessel equation (4.2.1.1).

If the membrane we are studying is a disc with radius a, such as a vibrating drumhead, then the physically natural boundary conditions for this problem are $u(a) = 0$ and $u(0)$ nonsingular. These conditions characterize the Bessel functions among other solutions to the Bessel equation (up to a constant factor).

EXAMPLE 4.3.1 Consider a metal bar of length L, density δ, and cross-sectional area A. Further assume that the bar has tensile modulus[1] E. The bar has total mass $M\delta AL$.

The end $x = 0$ of the bar is fixed, and a mass m is attached to the end $x = L$. We initially stretch the bar linearly by moving the mass m a distance $d = bL$ to the right. At time $t = 0$ the mass is released and the system is set in motion. We wish to determine the subsequent vibrations of the bar.

So we need to solve the boundary value problem

$$u_{tt} = a^2 u_{xx} \quad \text{for } 0 < x < L \text{ and } t > 0$$

with boundary conditions

$$
\begin{aligned}
u(0,t) &= 0 \\
m u_{tt}(L,t) &= -AE u_x(L,t) \\
u(x,0) &= bx \\
u_t(x,0) &= 0 \,.
\end{aligned}
$$

Note that the differential equation is the wave equation, which we study in detail in Section 10.2 below. The first, third, and fourth boundary conditions are physically obvious. The second one takes into account what the tensile modulus represents. Put in other words, what we are doing in this equation is equating $ma = m u_{tt}$ for the mass m and the force $F = -AE u_x$.

Following the method of separation of variables, we set

$$u(x,t) = X(x) \cdot T(t)$$

and substitute this expression into the partial differential equation. This leads, as usual, to the two ordinary differential equations

$$X'' + \lambda X = 0 \tag{4.3.1.1}$$

and

$$T'' + \lambda a^2 T = 0 \,. \tag{4.3.1.2}$$

We know that

$$0 = u(0,t) = X(0)T(t)$$

so that $X(0) = 0$. Since $u_{tt} = X(x)T''(t)$ and $u_x = X'(x)T(t)$, we can interpret the second boundary condition as

$$mX(L)T''(t) = -AEX'(L)T(t) \,. \tag{4.3.1.3}$$

---

[1]The tensile modulus or Young's modulus is a measure of the stiffness of an elastic material and is a quantity used to characterize materials. It is defined to be the ratio of the stress (force per unit area) along an axis to the strain (ratio of deformation over initial length) along that axis in the range of stress in which Hooke's law holds.

But equation (4.3.1.2) tells us that

$$T''(t) = -\lambda a^2 T(t)$$

and a little physical analysis shows that $a^2 = E/\delta$. So the last line gives

$$T''(t) = -\frac{\lambda E}{\delta} T(t) \,. \tag{4.3.1.4}$$

Now we can substitute equation (4.3.1.4) into (4.3.1.3) to obtain

$$mX(L)\frac{\lambda E}{\delta}T = AEX'(L)T \,.$$

Dividing through by ET/δ now finally yields

$$m\lambda X(L) = A\delta X'(L) \,.$$

Thus we conclude that the eigenvalue problem for $X(x)$ is

$$X'' + \lambda X = 0$$

with the boundary conditions

$$\begin{aligned} X(0) &= 0 \\ m\lambda X(L) &= A\delta X'(L) \,. \end{aligned}$$

This is *not* a standard Sturm–Liouville problem, just because of the presence of λ in the endpoint condition for L. It can be shown, however, that all the eigenvalues are positive. So we may write $\lambda = \beta^2$ and observe that $X(x) = \sin \beta x$ satisfies $X(0) = 0$.

Now our right endpoint condition yields

$$m\beta^2 \sin \beta L = A\delta\beta \cos \beta L \,.$$

Therefore

$$\tan \beta L = \frac{A\delta}{m\beta} = \frac{M/m}{\beta L} \,.$$

Here $M = A\delta L$. Now set $\gamma = \beta L$. Then the eigenvalues and eigenfunctions of our system are

$$\lambda_j = \frac{\gamma_j^2}{L^2} \quad \text{and} \quad X_j(x) = \sin \frac{\gamma_j x}{L}$$

for $j = 1, 2, \ldots$ and γ_j are the positive roots of the equation $\tan x = (M/m)/x$.

Now we perform our usual analysis on the equation

$$T_j'' + \frac{\gamma_j^2 a^2}{L^2} T_j = 0$$

with endpoint condition $T_j'(0) = 0$ to find that

$$T_j(t) = \cos\left(\frac{\gamma_j at}{L}\right)$$

up to a multiplicative constant.

What we have left to do now is to find coefficients a_j so that the series

$$u(x,t) = \sum_{j=1}^{\infty} a_j \cos\frac{\gamma_j at}{L} \sin\frac{\gamma_j x}{L} \tag{4.3.1.5}$$

satisfies the condition

$$u(x,0) = \sum_{j=1}^{\infty} a_j \sin\frac{\gamma_j x}{L} = f(x) = bx. \tag{4.3.1.6}$$

Since our problem is *not* a Sturm–Liouville problem, it turns out that the eigenfunctions are not orthogonal. So some extra care will be required.

In practice, from the physical point of view, we are not so much interested in the displacement function $u(x,t)$ itself. Rather, we care about how the bar's natural frequencies of vibration are affected by the mass m on the free end. Whatever the coefficients a_j in equation (4.3.1.6) may turn out to be, we find that equation (4.3.1.5) tells us that the jth circular frequency is

$$\omega_j = \frac{\gamma_j a}{L} = \frac{\gamma_j}{L} \cdot \sqrt{\frac{E}{\delta}},$$

where the γ_j are defined as above. This last may be rewritten as

$$\cot x = \frac{mx}{M}.$$

We conclude then that the natural frequencies of this system are determined by the ratio of the mass m to the total mass M of the bar.   ∎

Exercises

1. Consider the problem

    $$u_t = u_{xx}, \ 0 < x < L \ t > 0,$$

    with endpoint conditions

    $$u_x(0,t) = hu(L,t) + u_x(L,t) = 0$$

and
$$u(x,0) = f(x).$$

Find a series solution of this system.

[**Hint:** Your solutions should have the form

$$u(x,t) = \sum_{j=1}^{\infty} c_j \exp\left(-\frac{\beta_j^2 kt}{L^2}\right) \cos\frac{\beta_j x}{L}.$$

Here the β_j are the positive roots of the equations $\tan x = hL/x$.]

2. Consider the equation

$$u_{xx} + u_{yy} = 0 , \ 0 < x < L , \ 0 < y < L,$$

with the endpoint conditions

$$u(0,y) = hu(L,y) + u_x(L,y) = 0$$

and
$$u(x,L) = 0$$

and
$$u(x,0) = f(x).$$

Find a solution as in Exercise 1.

3. Consider the equation

$$u_{xx} = u_{yy} = 0 , \ 0 < x < L , \ y > 0,$$

with the endpoint conditions

$$u(0,y) = hu(L,y) + u_x(L,y) = 0$$

and
$$u(x,0) = f(x)$$

and in addition $u(x,y)$ is bounded as $y \to +\infty$. Find a solution as in Exercise 1.

4. Consider the heat equation

$$u_t = ku_{xx} , \ 0 < x < L , \ t > 0,$$

with the endpoint conditions

$$hu(0,t) - u_x(0,t) = 0$$

and
$$hu(L,t) + u_x(L,t) = 0$$

and
$$u(x,0) = f(x).$$

Find a solution as in Exercise 1.

5. Calculate the speed in miles per hour of the longitudinal sound wave in the following problem. The medium is water, $\delta = 1\,\mathrm{g/cm^2}$, and the bulk modulus is $K = 2.25 \times 10^{10}$ in cgs units.

6. A problem concerning the diffusion of a gas through a membrane leads to the boundary value problem

$$u_t = ku_{xx} , \ 0 < x < L , \ t > 0 ,$$

with endpoint conditions

$$u(0, t) = 0 ,$$
$$u_t(L, t) + hku_x(L, t) = 0 ,$$
$$u(x, 0) = 1 .$$

If β_j are the positive roots of the equation $x \tan x = hL$, then derive the solution

$$u(x, t) = \sum_{j=1}^{\infty} c_j \exp\left(-\frac{\beta_j^2 kt}{L^2}\right) \sin \frac{\beta_j x}{L} .$$

7. Show that the equation

$$\frac{d^2 y}{dx^2} + A(x)\frac{dy}{dx} + [\lambda B(x) - C(x)]y = 0$$

can be written in Sturm–Liouville form. Do so by defining $p(x) = e^{\int A(x)\, dx}$. What are q and r in terms of A, B, and C?

4.4   Singular Sturm–Liouville

In a regular or standard Sturm–Liouville problem we assume that p and q are nonvanishing on the entire closed interval $[a, b]$. In a singular Sturm–Liouville problem we allow vanishing at the endpoints, and this gives rise to many new phenomena. And many of these actually come up in physical situations. We consider some of them here.

EXAMPLE 4.4.1 Consider the differential equation

$$xy'' + y' + \lambda xy = 0 . \tag{4.4.1.1}$$

We can rewrite this as

$$-(xy')' = \lambda xy , \tag{4.4.1.2}$$

and we assume that $0 < x < 1$ and that $\lambda > 0$. This equation arises in the study of a disc-shaped elastic membrane.

If we introduce a new independent variable defined by $t = \sqrt{\lambda}x$, then we have

$$\frac{dy}{dx} = \sqrt{\lambda}\frac{dy}{dt} \quad \text{and} \quad \frac{d^2 y}{dx^2} = \lambda\frac{d^2 y}{dt^2} .$$

Thus equation (4.4.1.1) becomes

$$\frac{t}{\sqrt{\lambda}}\lambda\frac{d^2 y}{dt^2} + \sqrt{\lambda}\frac{dy}{dt} + \lambda\frac{t}{\sqrt{\lambda}}y = 0$$

or, simplifying,

$$t\frac{d^2y}{dt^2} + \frac{dy}{dt} + ty = 0. \tag{4.4.1.3}$$

This last equation is Bessel's equation of order 0. See Chapter 3 and the first part of Chapter 4.

The general solution of this last equation is

$$y = C_1 J_0(t) + C_2 Y_0(t),$$

where J_0 is the Bessel function of the first kind and Y_0 is the Bessel function of the second kind. Hence the general solution of (4.4.1.2) is

$$y = C_1 J_0(\sqrt{\lambda}x) + C_2 Y_0(\sqrt{\lambda}x). \tag{4.4.1.4}$$

It is known that

$$J_0(\sqrt{\lambda}x) = 1 + \sum_{j=1}^{\infty} \frac{(-1)^j \lambda^j x^{2j}}{2^{2j}(j!)^2}, \quad x > 0$$

and

$$Y_0(\sqrt{\lambda}x) = \frac{2}{\pi}\left[\left(\gamma + \ln\frac{\sqrt{\lambda}x}{2}\right)J_0(\sqrt{\lambda}x) + \sum_{j=1}^{\infty} \frac{(-1)^{j+1} H_j \lambda^j x^{2j}}{2^{2j}(j!)^2}\right], \quad x > 0.$$

Here $H_j = 1 + (1/2) + \cdots + (1/j)$ and $\gamma = \lim_{j\to\infty}(H_j - \ln j).$[2]

Now suppose that we seek a solution to (4.4.1.2) that satisfies the endpoint conditions

$$y(0) = 0 \tag{4.4.1.5}$$
$$y(1) = 0. \tag{4.4.1.6}$$

Because $J_0(0) = 1$ and $Y_0(x) \to -\infty$ as $x \to 0$, we see that $y(0) = 0$ can hold only if $C_1 = 0$ and $C_2 = 0$ in equation (4.4.1.4). So the boundary value problem given by (4.4.1.2), (4.4.1.5), (4.4.1.6) has only the trivial solution.

We may endeavor to understand this situation by hypothesizing that the endpoint condition (4.4.1.5) is too restrictive for the ordinary differential equation (4.4.1.2). This illustrates the idea that, at a singular point for the Sturm–Liouville problem, we need to consider a modified boundary condition. What we will do then is, instead of our usual endpoint conditions, we shall require that the solution (4.4.1.4) and its derivative remain bounded. That is to say, our boundary condition at $x = 0$ will now be

$$y, y' \text{ remain bounded as } x \to 0.$$

---

[2]These are standard constructions in the subject of special functions. The ideas are due to Euler.

This last condition can be achieved by choosing $C_2 = 0$ in Equation (4.4.1.4). For this eliminates the unbounded solution Y_0. The second boundary condition $y(1) = 0$ now gives us that

$$J_0(\sqrt{\lambda}) = 0. \tag{4.4.1.7}$$

It is possible to show that Equation (4.4.1.7) has an infinite set of positive roots,[3] and this gives the sequence

$$\lambda_1 < \lambda_2 < \cdots.$$

Corresponding eigenfunctions are

$$\phi_j(x) = J_0\left(\sqrt{\lambda_j}x\right).$$

Thus we see that the boundary value problem (4.4.1.2), (4.4.1.5), (4.4.1.6) is a *singular Sturm–Liouville problem*. We have learned that, if the boundary conditions for such a singular problem are relaxed in a suitable fashion, then we may find an infinite distinct sequence of eigenvalues and corresponding eigenfunctions—just as for a regular Sturm–Liouville problem. ∎

Now we give some other quick examples of singular Sturm–Liouville just to show how naturally they fit into the context that we have been studying.

EXAMPLE 4.4.2 The Legendre equation is given by

$$(1 - x^2)y'' - 2xy' + \ell(\ell+1)y = 0 , \quad x \in [-1, 1].$$

We can rewrite the equation as

$$-\frac{d}{dx}[(1 - x^2)y'] = \ell(\ell+1)y.$$

Here $p(x) = 1 - x^2$. Since $p(-1) = p(1) = 0$, we see that the problem is singular. ∎

EXAMPLE 4.4.3 Chebyshev's equation is given by

$$(1 - x^2)y'' - xy' + n^2 y = 0 , \quad x \in [-1, 1].$$

Dividing by $\sqrt{1 - x^2}$, we can convert the equation to Sturm–Liouville form:

$$\sqrt{1 - x^2}y'' - \frac{x}{\sqrt{1 - x^2}}y' + \frac{n^2}{\sqrt{1 - x^2}}y = 0$$

---

[3]The roots of J_0 are very well understood. The first three roots $\sqrt{\lambda}$ are 2.405, 5.520, and 8.654 (to four significant places). For j large, it is known that $\sqrt{\lambda_j} \approx (j - 1/4)\pi$.

or

$$-\frac{d}{dx}\left[\frac{1}{\sqrt{1-x^2}}y'\right] = \frac{n^2}{\sqrt{1-x^2}}y\,.$$

We see that $p(x) = (1-x^2)^{-1/2}$ is singular at ± 1, so this is a singular Sturm–Liouville problem. ∎

EXAMPLE 4.4.4 The *Hermite equation* is given by

$$y'' - 2xy' + 2\alpha y = 0\,, \quad x \in (-\infty,\infty)\,.$$

Multiplying through by e^{-x^2} we can recast the equation in Sturm–Liouville form:

$$-\frac{d}{dx}\left[e^{-x^2}y'\right] = 2\alpha e^{-x^2}y\,.$$

This is a singular Sturm–Liouville problem because the interval is infinite. ∎

EXAMPLE 4.4.5 *Lagrange's equation* is given by

$$xy'' + (1-x)y' + \alpha y = 0\,, \quad x \in [0,\infty)\,.$$

This can be converted to Sturm–Liouville form by multiplying through by e^{-x}. The result is

$$-\frac{d}{dx}\left[xe^{-x}y'\right] = \alpha e^{-x}y\,.$$

Note that $p(x) = xe^{-x}$. This is a singular Sturm–Liouville problem for two reasons: **(i)** $p(0) = 0$ and **(ii)** the interval is infinite. ∎

Singular Sturm–Liouville problems arise quite frequently in applications and it is worthwhile to study them further. Natural questions that one would like to have answered are

- What types of boundary conditions are allowable in a singular Sturm–Liouville problem?

- To what degree do the eigenvalues and eigenfunctions of a singular Sturm–Liouville problem behave like the eigenvalues and eigenfunctions of a regular Sturm–Liouville problem? In particular, are the eigenvalues all real? Do they form a discrete set? Do they tend to infinity? Are the eigenfunctions orthogonal? Can a given function be expanded in a convergent series of eigenfunctions?

The key to answering these questions is to study the equation

$$\int_0^1 [L(u)v - uL(v)]\, dx = 0.\tag{4.4.6}$$

To be specific, we shall study the situation in which the endpoint condition at 0 is singular but the endpoint condition at 1 is not. We will by default assume that the endpoint condition at 1 is

$$b_1 y(1) + b_2 y'(1) = 0.\tag{4.4.7}$$

We shall spend some time developing what will be an appropriate singular boundary condition at 0.

One thing we must allow for is the possibility that the integral in (4.4.6) is singular at 0. Thus it is appropriate to instead consider $\int_\epsilon^1$ for $\epsilon > 0$ small. At the end we let $\epsilon \to 0^+$. We assume that u, v each have two continuous derivatives. Thus we may integrate by parts in (4.4.6) and find that

$$\int_\epsilon^1 [L(u)v - uL(v)]\, dx = -p(x)[u'(x)v(x) - u(x)v'(x)]\Big|_\epsilon^1.\tag{4.4.8}$$

If both u and v satisfy the boundary condition (4.4.7), then the boundary term on the right-hand side of (4.4.8) at $x = 1$ is 0. Thus (4.4.8) becomes

$$\int_\epsilon^1 [L(u)v - uL(v)]\, dx = -p(\epsilon)[u'(\epsilon)v(\epsilon) - u(\epsilon)v'(\epsilon)].$$

Letting $\epsilon \to 0$ then gives

$$\int_0^1 [L(u)v - uL(v)]\, dx = \lim_{\epsilon \to 0} -p(\epsilon)[u'(\epsilon)v(\epsilon) - u(\epsilon)v'(\epsilon)].$$

Thus

$$\int_0^1 [L(u)v - uL(v)]\, dx = 0\tag{4.4.9}$$

if and only if, in addition to the other technical hypotheses enunciated above, we have

$$\lim_{\epsilon \to 0} -p(\epsilon)[u'(\epsilon)v(\epsilon) - u(\epsilon)v'(\epsilon)] = 0\tag{4.4.10}$$

for every u and v in the class functions we are considering. In conclusion, equation (4.4.10) is our criterion for determining what conditions are allowed in order for 0 to be a singular boundary point. A similar condition applies at the boundary point 1.

To sum up: A singular value problem for the Sturm–Liouville equation is said to be *self-adjoint* if condition (4.4.9) is satisfied—possibly as an improper integral—for each pair of functions u and v selected as follows:

(a) Each of u and v is twice continuously differentiable on $(0, 1)$.

(b) They satisfy a boundary condition of the form $b_1 y(1) + b_2 y'(1) = 0$ at each regular boundary point.

(c) They satisfy a boundary criterion like (4.4.10) at each singular boundary point.

Notice, for instance, in the example that we worked out above, that $p(x) = x$ vanishes at 0. So 0 is a singular boundary point. But it is perfectly clear that Equation (4.4.10) holds. So that singular Sturm–Liouville problem is certainly self-adjoint.

Exercises

1. Find a formal solution of the nonhomogeneous boundary value problem

$$-(xy')' = \mu x y + f(x),$$

where y and y' are bounded as $x \to 0$ and $y(1) = 0$. Also, f is a continuous function on $[0, 1]$ and μ is *not* an eigenvalue of the corresponding homogeneous problem.

2. The equation

$$(1 - x^2)y'' - xy' + \lambda y = 0$$

is *Chebychev's equation*.

(a) Show that this ODE can be written in the form

$$-[(1 - x^2)^{1/2}y']' = \lambda(1 - x^2)^{-1/2}y, \quad -1 < x < 1. \qquad (*)$$

(b) Consider the boundary conditions

$$y, \ y' \ \text{bounded as} \ x \to -1 \qquad (**)$$

and

$$y, \ y' \ \text{bounded as} \ x \to 1. \qquad (***)$$

Show that the boundary value problem given by $(*)$, $(**)$, $(***)$ is self-adjoint.

(c) It can be verified that the boundary value problem given by $(*)$, $(**)$, $(***)$ has the eigenvalues $\lambda_0 = 0$, $\lambda_1 = 1$, $\lambda_2 = 4$, and in general $\lambda_j = j^2$. The corresponding eigenfunctions are the Chebyshev polynomials $T_j(x)$, where $T_0(x) = 1$, $T_1(x) = x$, $T_2(x) = 1 - 2x^2$, and so forth. Prove that

$$\int_{-1}^{1} \frac{T_j(x)T_k(x)}{(1 - x^2)^{1/2}} \, dx = 0$$

for $j \neq k$. Technically speaking, this is a convergent improper integral.

3. Consider Legendre's equation

$$-[(1-x^2)y']' = \lambda y.$$

We subject this equation to the boundary conditions $y(0) = 0$ and y and y' are bounded as $x \to 1$. The eigenfunctions of this problem are the Legendre polynomials $\phi_0(x) = 1$, $\phi_1(x) = P_1(x) = x$, $\phi_2(x) = P_3(x) = (5x^3 - 3x)/2$, and in general $\phi_j(x) = P_{2j-1}(x)$. Here

$$P_n(x) = \frac{1}{2^n n!} \frac{d^n}{dx^n} \left[(1-x^2)^n \right].$$

The ϕ_j correspond to the eigenvalues $\lambda_1 = 2$, $\lambda_2 = 4 \cdot 3$, ..., $\lambda_j = 2j(2j-1)$.

(a) Show that

$$\int_0^1 \phi_j(x)\phi_k(x)\, dx = 0$$

when $j \neq k$.

(b) Find a formal solution of the nonhomogeneous problem

$$-[(1-x^2)y']' = \mu y + f(x),$$

where $y(0) = 0$, and y, y' are bounded as $x \to 1$. Also, f is a continuous function on $[0, 1]$, and μ is not an eigenvalue of the corresponding homogeneous problem.

4. Show that the eigenvalues of the singular Sturm–Liouville system defined by

$$\frac{d}{dx}\left[(1-x^2)^{a+1}\frac{du}{dx} \right] + \lambda(1-x^2)^a u = 0, \; a > -1,$$

with the additional condition that u be bounded on $(-1, 1)$, are given by $\lambda_j = j/(j + 2a)$.

5. Refer to Exercise 3 for terminology. Show that the Legendre polynomials (and their constant multiples) are the only solutions of the Legendre ODE that are bounded on $(-1, 1)$.

6. Show that the Sturm–Liouville system given by

$$[(x-a)(b-x)u']' + \lambda u = 0$$

for $a < b$, with $u(x)$ bounded on (a, b), has eigenvalues $\lambda = 4j(j+1)(b-a)^2$. Describe the eigenfunctions.

7. Consider the singular Sturm–Liouville system

$$(xe^{-x}u')' + \lambda e^{-x}u = 0$$

on the interval $(0, +\infty)$. We impose the endpoint conditions that $u(0^+)$ be bounded and further that $e^{-x}u(x) \to 0$ as $x \to +\infty$. Is it true that the values $\lambda = j$ give polynomial eigenfunctions?

4.5 Some Ideas from Quantum Mechanics

Sturm–Liouville problems arise in many parts of mathematical physics—both elementary and advanced. One of these is quantum mechanics. Here we see how the study of matter on a very small scale can be effected with some of the ideas that we have been learning. We derive our exposition from the lovely book [KBO].

We think of a system as being specified by a *state function* $\psi(\mathbf{r}, t)$. Here $\mathbf{r}$ represents a position vector for a point in space (of dimension one, two, or three), and t is time. We think of ψ as a probability distribution, so it is appropriate to assume that

$$1 = \langle \psi, \psi \rangle = \iint_{\mathbf{R}^3} |\psi(\mathbf{r}, t)|^2 \, d\mathbf{r} \, .$$

That is, for each fixed time t, ψ has total mass 1.

One of the basic tenets of quantum mechanics is that, in each system, there is a linear operator H such that

$$i\overline{h} \frac{\partial \psi}{\partial t} = H\psi \, .$$

Here $\overline{h}$ is a constant specified by the condition that

$$2\pi\overline{h} \approx 6.62 \cdot 10^{-16} \, J \cdot s \, .$$

Here $J \cdot s$ stands for *Joule seconds*. The actual value of the constant $\overline{h}$ is of no interest for the present discussion. The operator H has the nice (Hermitian) property that

$$\langle Hy_1, y_2 \rangle = \langle y_1, Hy_2 \rangle \, .$$

Finally—and this is one of the key ideas of von Neumann's model for quantum mechanics—to each observable property of the system there corresponds a linear, Hermitian operator A. Moreover, any measurement of the observable property gives rise to an eigenvalue of A. For example, as you learn in your physics course, the operator that corresponds to momentum is $-i\overline{h}\nabla$ and the operator that corresponds to energy is $i\overline{h}\partial/\partial t$.

Now that we have dispensed with the background, let us examine a specific system and see how a Sturm–Liouville problem comes into play. Consider a particle of mass m moving in a potential field $V(\mathbf{r}, t)$. Then, if p is the momentum of the particle, we have that

$$E = \frac{p^2}{2m} + V(\mathbf{r}, t) \, . \tag{4.5.1}$$

Observe that the first expression on the right is the kinetic energy and the second expression is the potential energy. Now we quantize this classical relation by substituting in appropriate operators for the potential and the energy. We then obtain

$$i\bar{h}\frac{\partial\psi}{\partial t} = -\frac{\bar{h}^2}{2m}\nabla^2\psi + V(\mathbf{r},t)\psi. \tag{4.5.2}$$

This important identity is known as *Schrödinger's equation*. It controls the evolution of the wave function.

To simplify matters, let us suppose that the potential V does not depend on time t. We may then seek to solve (4.5.2) by separation of variables. Let us write $\psi(\mathbf{r},t) = \alpha(\mathbf{r})T(t)$. Substituting into the differential equation, we find that

$$i\bar{h}\alpha(\mathbf{r})\frac{dT}{dt} = -\frac{\bar{h}^2}{2m}\nabla^2\alpha\cdot T + V(\mathbf{r})\alpha(\mathbf{r})T$$

or

$$i\bar{h}\frac{(dT/dt)(t)}{T(t)} = -\frac{\bar{h}^2}{2m\alpha(\mathbf{r})}\nabla^2\alpha(\mathbf{r}) + V(\mathbf{r}). \tag{4.5.3}$$

Observe that the left-hand side depends only on t, and the right-hand side only on $\mathbf{r}$. Thus both are equal to some constant μ.

In particular,

$$i\bar{h}\frac{dT}{dt} = \mu T.$$

So we may write

$$i\bar{h}\frac{\partial\psi/\partial t}{\psi} = i\bar{h}\frac{dT/dt}{T} = \mu$$

hence

$$i\bar{h}\frac{\partial\psi}{\partial t} = \mu\psi.$$

Thus we see that μ is the energy of the particle in our system.

Looking at the right-hand side of (4.5.3), we now see that

$$-\frac{\bar{h}^2}{2m}\nabla^2\alpha + (V(\mathbf{r}) - \mu)\alpha = 0. \tag{4.5.4}$$

This is the *time-independent Schrödinger equation*. It will turn out, contrary to the philosophy of classical physics, that the energy of this one-particle system must be one of the eigenvalues of the boundary value problem that we shall construct from equation (4.5.4). The energy is said to be *quantized*.

To consider a simple instance of the ideas we have been discussing, we examine a particle of mass m that is trapped in a region of zero potential by the infinite potentials at $x = 0$ and $x = a$. See Figure 4.1. Let us consider the possible energies that such a particle can have.

Thinking of ψ as a (continuous) probability distribution, we see that $\psi = 0$ outside the interval $(0, a)$, since the probability is zero that the particle will be found outside the interval. Thus the graph of ψ is as in Figure 4.2.

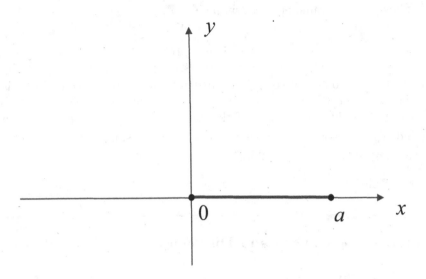

FIGURE 4.1
Particle trapped in a region of zero potential.

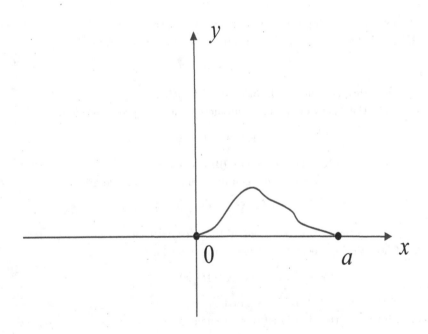

FIGURE 4.2
Graph of the continuous probability distribution ψ.

Thus our eigenvalue problem simplifies to

$$-\frac{\bar{h}^2}{2m}\frac{\partial^2 \alpha}{\partial x^2} = \mu\alpha,$$

subject to the boundary conditions $\alpha(0) = \alpha(a) = 0$. Of course this is a familiar problem. Observe that the role of μ in that example is now being played by $2m\mu/h^2$ and the role of the function y is now played by α. Thus we find that the allowed energies in our system are simply $\bar{h}^2/2m$ times the eigenvalues, or $\mu_n = \bar{h}^2 n^2\pi^2/[2ma^2]$.

Problems for Review and Discovery

A. Drill Exercises

1. Solve the equation

$$y'' - \left(2x + \frac{3}{x}\right)y' - 4y = 0$$

by finding a simple solution of the adjoint equation.

2. Find the eigenvalues and eigenfunctions for the problem

$$u'' + \lambda^2 u = 0 \quad, \quad 0 < x < a$$

with the endpoint conditions $u(0) = 0$, $u'(a) = 0$.

3. Find the eigenvalues and eigenfunctions for the problem

$$u'' + \lambda^2 u = 0 \quad, \quad 0 < x < a$$

with the endpoint conditions $u'(0) = 0$, $u(a) = 0$.

4. Find the eigenvalues and eigenfunctions for the problem

$$u'' + \lambda^2 u = 0 \quad, \quad 0 < x < a$$

with the endpoint conditions $u(0) - u'(0) = 0$, $u(a) + u'(a) = 0$.

5. Show that the eigenfunctions for the equation

$$u'' + \lambda^2(1 + x)u = 0$$

with $u(0) = 0$, $u'(a) = 0$ are orthogonal.

6. Show that the eigenfunctions for the equation

$$u'' + \frac{\lambda^2}{x^2}u = 0$$

with $u(1) = 0$, $u'(2) = 0$ are orthogonal.

B. Challenge Problems

1. The equation
$$x^2 y'' + xy' - n^2 y = 0$$
is Euler's equidimensional equation. Here n is a positive integer. Determine those values of n for which the equation is exact (meaning that the equation can be written in the form $\frac{d}{dx}[A(x)y' + B(x)y]$). For those values, find the general solution of the equation.

2. Show that a self-adjoint differential equation can be written in the form
$$[P(x)y']' + R(x)y = 0.$$
This is called the *standard form* for a self-adjoint equation.

3. Show that any equation
$$P(x)y'' + Q(x)y' + R(x)y = 0$$
can be made self-adjoint by multiplying through by the factor
$$\frac{1}{P} e^{\int (Q/P)\, dx}.$$

4. Legendre's equation
$$(1 - x^2)y'' - 2xy' + p(p + 1) = 0$$
is self-adjoint. Put it into standard form as described in Exercise 2.

5. Bessel's equation
$$x^2 y'' + xy' + (x^2 - p^2)y = 0$$
is *not* self-adjoint. Use the technique of Exercise 3 to make it self-adjoint, and then put the equation in standard form as described in Exercise 2.

6. Hermite's equation
$$y'' - 2xy' + (x^2 + 2py) = 0$$
is *not* self-adjoint. Use the technique of Exercise 3 to make it self-adjoint, and then put the equation in standard form as described in Exercise 2.

C. Problems for Discussion and Exploration

1. Consider the differential equation $y'' + \lambda y = 0$ with the endpoint conditions $y(0) = 0$ and $y(\pi) + y'(\pi) = 0$. Show that there is an infinite sequence of eigenfunctions with distinct eigenvalues. Identify the eigenvalues explicitly.

2. In what sense is the theory of Sturm–Liouville equations a generalization of the Fourier theory developed in Chapter 6? Answer this question in as much detail as you can. Where do the exponentials $e^{ij\theta}$ come from? From which Sturm–Liouville problem?

3. Read about the Fourier transform in the last section of Chapter 6. Now determine which functions f have the property that $\widehat{f} = f$. If, instead of using the $\widehat{}$ notation, we denote the Fourier transform by $\mathcal{F}$, then show that $\mathcal{F} \circ \mathcal{F} \circ \mathcal{F} \circ \mathcal{F}(f) = f$ for any function f. Conclude from this last calculation that there should be a function g with $\mathcal{F}(g) = -g$ and a function h with $\mathcal{F}(h) = ih$ and a function k with $\mathcal{F}(k) = -ik$.

5

Numerical Methods

- The idea of a numerical method

- Approximation

- Error terms

- Euler's method

- Improved Euler method

- Runge–Kutta method

The presentation in this book, or in any standard introductory text on differential equations, can be misleading. A casual reading might lead the student to think that "most" differential equations can be solved explicitly, with the solution given by a formula. Such is not the case. Although it can be proved abstractly that most any ordinary differential equation has a solution—at least locally—it is in general quite difficult to say in any explicit manner what the solution might be. It is sometimes possible to say something qualitative about solutions. And we have also seen that certain important equations that come from physics are fortuitously simple, and can be attacked effectively. But the bottom line is that many of the equations that we *must* solve for engineering or other applications simply do not have closed-form solutions. Just as an instance, the equations that govern the shape of an airplane wing cannot be solved. Yet we fly every day. How do we come to terms with the intractability of differential equations?

The advent of high-speed digital computers has made it both feasible and, indeed, easy to perform numerical approximation of solutions. The subject of the numerical solution of differential equations is a highly developed one, and is applied daily to problems in engineering, physics, biology, astronomy, and many other parts of science. Solutions may generally be obtained to any desired degree of accuracy, graphs drawn, and any desired analysis performed.

Not surprisingly—and like many of the other fundamental ideas related to calculus—the basic techniques for the numerical solution of differential equations go back to Newton and Euler. This is quite amazing, for these men had no notion of the computing equipment that we have available today. Their insights were quite prescient and powerful.

In the present chapter, we shall only introduce the most basic ideas in the subject of numerical analysis of differential equations. We refer the reader to [GER], [HIL], [ISK], [STA], [TOD] for further development of the subject.

5.1   Introductory Remarks

When we create a numerical or discrete model for a differential equation, we make several decisive replacements or substitutions. First, the derivatives in the equation are replaced by *differences* (as in replacing the derivative by a difference quotient). Second, the continuous variable x is replaced by a discrete variable. Third, the real number line is replaced by a discrete set of values. Any type of approximation argument involves some sort of loss of information; that is to say, there will always be an *error term*. It is also the case that these numerical approximation techniques can give rise to instability phenomena and other unpredictable behavior.

The practical significance of these remarks is that numerical methods should never be used in isolation. Whenever possible, the user should also employ qualitative techniques. Endeavor to determine whether the solution is bounded, periodic, or stable. What are its asymptotics at infinity? How do the different solutions interact with each other? In this way the scientist is not using the computing machine blindly, but is instead using the machine to aid and augment his/her understanding.

The *spirit* of the numerical method can be illustrated with a basic example. Consider the simple differential equation

$$y' = y, \quad y(0) = 1.$$

The initial condition tells us that the point $(0, 1)$ lies on the graph of the solution y. The equation itself tells us that, at that point, the slope of the solution is

$$y' = y = 1.$$

Thus the graph will proceed to the right, with slope 1. Let us assume that we shall do our numerical calculation with mesh 0.1. So we proceed to the right to the point $(0.1, 1.1)$. This is the second point on our "approximate solution graph."

Now we return to the differential equation to obtain the slope of the solution at this new point. It is

$$y' = y = 1.1.$$

Thus, when we proceed to sketch our approximate solution graph to the right of $(0.1, 1.1)$, we draw a line segment of slope 1.1 to the point $(0.2, 1.21)$. And so forth. See Figure 5.1.

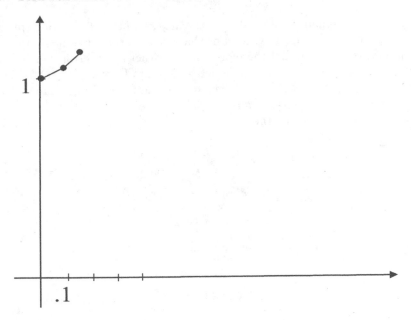

FIGURE 5.1
A simple approximation scheme.

Of course this is a very simple-minded example, and it is easy to imagine that the approximate solution is diverging rather drastically and unpredictably with each iteration of the method. In subsequent sections we shall learn techniques of Euler (which formalize the method just described) and Runge–Kutta (which give much better, and more reliable, results).

5.2   The Method of Euler

Consider an initial value problem of the form

$$y' = f(x, y), \quad y(x_0) = y_0.$$

We may integrate from x_0 to $x_1 = x_0 + h$ to obtain

$$y(x_1) - y(x_0) = \int_{x_0}^{x_1} y' \, dx = \int_{x_0}^{x_1} f(x, y) \, dx$$

or

$$y(x_1) = y(x_0) + \int_{x_0}^{x_1} f(x, y) \, dx.$$

Since the unknown function y occurs in the integrand on the right, we cannot proceed unless we have some method of approximating the integral.

The Euler method is obtained from the most simple technique for approximating the integral. Namely, we assume that the integrand does not vary much on the interval $[x_0, x_1]$, and therefore that a rather small error will result if we replace $f(x, y)$ by its value at the left endpoint. To wit, we put in place a partition $a = x_0 < x_1 < x_2 < \cdots < x_k = b$ of the interval $[a, b]$ under study. We assume that the x_j are equally spaced, with $|x_j - x_{j-1}| = h$ for every j. We set $y_0 = y(x_0)$. Now we take

$$
\begin{aligned}
y(x_1) &= y(x_0) + \int_{x_0}^{x_1} f(x, y)\, dx \\
&\approx y(x_0) + \int_{x_0}^{x_1} f(x_0, y_0)\, dx \\
&= y_0 + h \cdot f(x_0, y_0).
\end{aligned}
$$

Based on this calculation, we simply *define*

$$ y_1 = y_0 + h \cdot f(x_0, y_0). $$

Continuing in this fashion, we think of $x_k = x_{k-1} + h$ and define

$$ y_{k+1} = y_k + h \cdot f(x_k, y_k). $$

Then the points $(x_0, y_0), (x_1, y_1), \ldots, (x_k, y_k), \ldots$ are the points of our "approximate solution" to the differential equation. Figure 5.2 illustrates the exact solution, the approximate solution, and how they might deviate.

It is sometimes convenient to measure the *total relative error* $\overline{E}_n$ at the nth step; this quantity is defined to be

$$ \overline{E}_n = \frac{|y(x_n) - y_n|}{|y(x_n)|}. $$

We usually express this quantity as a percentage, and we obtain thereby a comfortable way of measuring how well the numerical technique under consideration is performing.

Now we are going to focus on a particular ordinary differential equation that will be the benchmark for all of our numerical techniques. Throughout this chapter, we are going to examine the initial value problem

$$ y' = x + y, \qquad y(0) = 1 $$

over and over again using different methods of numerical analysis. Our benchmark will be to calculate $y(1)$ numerically and compare it with the exact value of $y(1)$ that we may obtain by an explicit solution method.

EXAMPLE 5.2.1 Apply the Euler technique to the ordinary differential equation

$$ y' = x + y, \qquad y(0) = 1 \tag{5.2.1.1} $$

using increments of size $h = 0.2$ and $h = 0.1$.   ∎

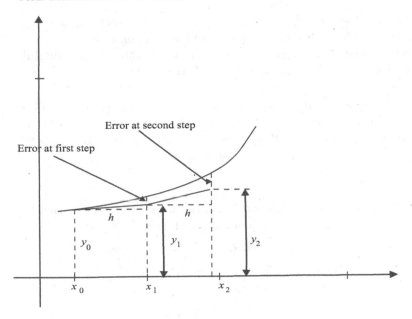

FIGURE 5.2
Euler's approximation scheme.

Table 5.2.1.2. Tabulated values for exact and numerical
solutions to equation (5.2.1.1) with $h = 0.2$

x_n	y_n	Exact	$\overline{E}_n$ (%)
0.0	1.00000	1.00000	0.0
0.2	1.20000	1.24281	3.4
0.4	1.48000	1.5836	6.5
0.6	1.85600	2.04424	9.2
0.8	2.720	2.65108	11.5
1.0	2.97664	3.43656	13.4

Solution: Of course we use the fact that an explicit solution of the differential equation (which is first-order linear) with initial condition is given by

$$y = -x - 1 + 2e^x .$$

We exhibit the calculations in Table 5.2.1.2. In the first line of this table, the initial condition $y(0) = 1$ determines the slope $y' = x + y = 1.00$. Since $h = 0.2$ and $y_1 = y_0 + h \cdot f(x_0, y_0)$, the next value is given by $1.00 + 0.2 \cdot (1.00) = 1.20$. This process is iterated in the following lines. As noted above, the expression $\overline{E}_n$ represents the percent of error. For instance, in the second line of the table it is calculated as

$$\overline{E}_1 = \frac{1.24281 - 1.2}{1.24281} \approx .034446\dots$$

and we represent the quantity in the table as a percent—so it is 3.4%. We shall retain five decimal places in this and succeeding tables.

For comparison purposes, we also record in Table 5.2.1.3 the tabulated values for $h = 0.1$. That is the solution of this example.   ∎

The displayed data makes clear that reducing the step size will increase accuracy. But the trade-off is that significantly more computation is required. In the next section we shall discuss errors, and in particular at what point there is no advantage to reducing the step size.

Table 5.2.1.3. Tabulated values for exact and numerical solutions to equation (5.2.1.1) with $h = 0.1$

x_n	y_n	Exact	$\overline{E}_n$ (%)
0.0	1.00000	1.00000	0.0
0.1	1.10000	1.110	0.9
0.2	1.22	1.24281	1.8
0.3	1.362	1.39972	2.7
0.4	1.52820	1.581	3.5
0.5	1.72102	1.79744	4.3
0.6	1.94312	2.04424	4.9
0.7	2.19743	2.32751	5.6
0.8	2.48718	2.65108	6.2
0.9	2.81590	3.01921	6.7
1.0	3.18748	3.41	7.2

Exercises

For each of Exercises 1–5, use the Euler method with $h = 0.1, 0.05$, and 0.01 to estimate the solution at $x = 1$. In each case, compare your results to the *exact* solution and discuss how well (or poorly) the Euler method has worked.

1. $y' = 2x + 2y, \quad y(0) = 1$
2. $y' = 1/y, \quad y(0) = 1$
3. $y' = e^y, \quad y(0) = 0$
4. $y' = y - \sin x, \quad y(0) = -1$
5. $y' = (x + y - 1)^2, \quad y(0) = 0$

6. Refer to Figure 5.2. Use geometric arguments to determine for what kind of exact solutions the Euler method would give accurate results. Do these results depend on h in any way? Construct two different examples to illustrate your point.

7. The ordinary differential equation

$$y' = y(1 - y^2)$$

possesses two equilibrium solutions: the solution $\phi_1 \equiv 0$, which is un-
stable, and the solution $\phi_2 \equiv 1$, which is stable. With the initial condition $y(0) = 0.1$, predict what *should* happen to the solution. Then, with
$h = 0.1$, use the Euler method to run the solution out to $x = 3$. What
happens to this numerical solution?

8. This exercise illustrates the danger in blindly applying numerical techniques. Apply the Euler method to the following initial value problem.

$$y' = \sec^2 x, \qquad y(0) = 0.$$

Use a step size of $h = 0.1$ and determine the numerical solution at $x = 1$.
Now explain why the initial value problem actually has *no solution* at
$x = 1$.

5.3   The Error Term

The notion of error is central to any numerical technique. Numerical methods
only give *approximate answers*. In order for the approximate answer to be
useful, we must know how close to the true answer it is. Since the whole
reason that we went after an approximate answer in the first place was that
we had no method for finding the exact answer, this whole discussion raises
tricky questions. How do we get our hands on the error, and how do we
estimate it? Any time decimal approximations are used, there is a rounding
off procedure involved. *Round-off error* is another critical phenomenon that
we must examine.

EXAMPLE 5.3.1 Examine the differential equation

$$y' = x + y, \qquad y(0) = 1 \qquad\qquad (5.3.1.1)$$

and consider what happens if the step size h is made too small.

Solution: Suppose that we are working with a computer having ordinary
precision—which is eight decimal places. This means that all numerical answers are rounded to eight places.

  Let $h = 10^{-10}$, a very small step size indeed (but one that could be
required for work in microtechnology). Let $f(x, y) = x + y$. Applying the
Euler method and computing the first step, we find that the computer yields

$$y_1 = y_0 + h \cdot f(x_0, y_0) = 1 + 10^{-10} = 1.$$

The last equality may seem rather odd—in fact it appears to be false. But this is how the computer will reason: It rounds to eight decimal places! The same phenomenon will occur with the calculation of y_2. In this situation, we see therefore that the Euler method will produce a constant solution—namely, $y \equiv 1$. And of course that is *not* a solution at all.   ∎

The last example is to be taken quite seriously. It describes what would actually happen if you had a canned piece of software to implement Euler's method, and you actually used it on a computer running in the most standard and familiar computing environment. If you are not aware of the dangers of round-off error, and why such errors occur, then you will be a very confused scientist indeed. One way to address the problem is with *double precision*, which gives 16-place decimal accuracy. Another way is to use a symbol manipulation program like `Mathematica` or `Maple` (in which one can pre-set any number of decimal places of accuracy).

In the present book, we cannot go very deeply into the subject of round-off error. What is most feasible for us is to acknowledge that round-off error must be dealt with in advance, and we shall assume that we have set up our problem so that round-off error is negligible. We shall instead concentrate our discussions on *discretization error*, which is a problem less contingent on artifacts of the computing environment and more central to the theory.

The local discretization error at the nth step is defined to be $\epsilon_n = y(x_n) - y_n$. Here $y(x_n)$ is the exact value at x_n of the solution of the differential equation, and y_n is the Euler approximation. In fact we may use Taylor's formula to obtain a useful estimate on this error term. To wit, we may write

$$y(x_0 + h) = y_0 + h \cdot y'(x_0) + \frac{h^2}{2} \cdot y''(\xi),$$

for some value of ξ between x_0 and $x_0 + h$. But we know, from the differential equation, that

$$y'(x_0) = f(x_0, y_0).$$

Thus

$$y(x_0 + h) = y_0 + h \cdot f(x_0, y_0) + \frac{h^2}{2} \cdot y''(\xi),$$

so that

$$y(x_1) = y(x_0 + h) = y_0 + h \cdot f(x_0, y_0) + \frac{h^2}{2} \cdot y''(\xi) = y_1 + \frac{h^2}{2} \cdot y''(\xi).$$

We may conclude that

$$\epsilon_1 = \frac{h^2}{2} \cdot y''(\xi).$$

Usually on the interval $[x_0, x_n]$ we may see on a priori grounds that $|y''|$ is bounded by some constant M. Thus our error estimate takes the form

$$|\epsilon_1| \leq \frac{Mh^2}{2}.$$

More generally, the same calculation shows that

$$|\epsilon_j| \le \frac{Mh^2}{2}.$$

Such an estimate shows us directly, for instance, that if we decrease the step size from h to $h/2$ then the accuracy is increased by a factor of 4.

Unfortunately, in practice, things are not as simple as the last paragraph might suggest. For an error is made at *each step* of the Euler method—or of any numerical method—so we must consider the *total discretization error*. This is just the aggregate of all the errors that occur at all steps of the approximation process.

To get a rough estimate of this quantity, we notice that our Euler scheme iterates in n steps, from x_0 to x_n, in increments of length h. So $h = [x_n - x_0]/n$ or $n = [x_n - x_0]/h$. If we assume that the errors accumulate without any cancellation, then the aggregate error is bounded by

$$|E_n| \le n \cdot \frac{Mh^2}{2} = (x_n - x_0) \cdot \frac{Mh}{2} \equiv C \cdot h.$$

Here $C = (x_n - x_0) \cdot M/2$, and $(x_n - x_0)$ is of course the length of the interval under study. Thus, for this problem, C is a universal constant. We see that, for Euler's method, the total discretization error is bounded by a constant times the step size.

EXAMPLE 5.3.2 Estimate the discretization error, for a step size of 0.2 and for a step size of 0.1, for the differential equation with initial data given by

$$y' = x + y, \qquad y(0) = 1. \qquad (5.3.2.1)$$

Solution: In order to get the maximum information about the error, we are going to proceed in a somewhat artificial fashion. Namely, we shall use the fact that *we can solve the initial value problem explicitly.* The solution is given by $y = 2e^x - x - 1$. Thus $y'' = 2e^x$. Thus, on the interval $[0,1]$,

$$|y''| \le 2e^1 = 2e.$$

Hence

$$|\epsilon_j| \le \frac{Mh^2}{2} \le \frac{2eh^2}{2} = eh^2$$

for each j. The total discretization error is then bounded (since we calculate this error by summing about $1/h$ terms) by

$$|E_n| \le eh. \qquad (5.3.2.2)$$

Referring to Table 5.2.1.2 in Section 5.2 for incrementing by $h = 0.2$, we see that the total discretization error at $x = 1$ is *actually equal to* 0.46 (rounded to two decimal places). (We calculate this error from the table by subtracting y_n from the exact solution.) The error bound given by (5.3.2.2) is $e \cdot (0.2) \approx 0.54$. Of course the actual error is less than this somewhat crude bound. With $h = 0.1$, the actual error from Table 5.2.1.2 is 0.25 while the error *bound* is $e \cdot (0.1) \approx 0.27$.   ∎

REMARK 5.3.3 In practice, we shall not be able to explicitly solve the differential equation being studied. That is, after all, why we are using numerical techniques and a computer. So how do we, in practice, determine when h is small enough to achieve the accuracy we desire? A rough-and-ready method, that is used commonly in the field, is this: Do the calculation for a given h, then for $h/2$, then for $h/4$, and so forth. When the distance between two successive calculations is within the desired tolerance for the problem, then it is quite likely that they both are also within the desired tolerance of the exact solution.

REMARK 5.3.4 How do we, in practice, check to see whether h is too small, and thus causing round-off error? One commonly used technique is to re-do the calculation in double precision (on a computer using one of the standard software packages, this would mean 16-place decimal accuracy instead of the usual 8-place accuracy). If the answer seems to change substantially, then some round-off error is probably present in the regular precision (8-place accuracy) calculation.

Exercises

In each of Exercises 1–5, use the exact solution, together with step sizes $h = 0.2$ and 0.1, to estimate the total discretization error that occurs with the Euler method at $x = 1$.

1. $y' = 2x + 2y, \quad y(0) = 1$
2. $y' = 1/y, \quad y(0) = 1$
3. $y' = e^y, \quad (0) = 0$
4. $y' = y - \sin x, \quad y(0) = -1$
5. $y' = (x + y - 1)^2, \quad y(0) = 0$

6. Consider the problem $y' = \sin 3\pi x$ with $y(0) = 0$. Determine the exact solution and sketch the graph on the interval $0 \le x \le 1$. Use the Euler method with $h = 0.2$ and $h = 0.1$ and sketch those results on the same set of axes. Compare and discuss. Now use the results of the present section of the text to determine a step size sufficient to guarantee a total error of

0.01 at $x = 1$. Apply the Euler method with *this* step size, and compare with the exact solution. Why is this step size necessarily so small?

5.4   An Improved Euler Method

We improve the Euler method by following the logical scheme that we employed when learning numerical methods of integration in calculus class. Namely, our first method of numerical integration was to approximate a desired integral by a sum of areas of rectangles. (This is analogous to the Euler method, where we approximate the integrand by the constant value at its left endpoint.) Next, in integration theory, we improved our calculations by approximating by a sum of areas of trapezoids. That amounts to averaging the values at the two endpoints. This is the philosophy that we now employ.

Recall that our old equation is

$$y_1 = y_0 + \int_{x_0}^{x_1} f(x, y)\, dx\,.$$

Our idea for Euler's method was to replace the integrand by $f(x_0, y_0)$. This generated the iterative scheme of the last section. Now we propose to instead replace the integrand with $[f(x_0, y_0) + f(x_1, y(x_1))]/2$. Thus we find that

$$y_1 = y_0 + \frac{h}{2}[f(x_0, y_0) + f(x_1, y(x_1))]\,. \tag{5.4.1}$$

The trouble with this proposed equation is that $y(x_1)$ is unknown—just because we do not know the exact solution y. What we can do instead is to replace $y(x_1)$ by its approximate value as found by the Euler method. Denote this new value by $z_1 = y_0 + h \cdot f(x_0, y_0)$. Then (5.4.1) becomes

$$y_1 = y_0 + \frac{h}{2} \cdot [f(x_0, y_0) + f(x_1, z_1)]\,.$$

The reader should pause to verify that each quantity on the right-hand side can be calculated from information that we have—*without* knowledge of the exact solution of the differential equation. More generally, our iterative scheme is

$$y_{j+1} = y_j + \frac{h}{2} \cdot [f(x_j, y_j) + f(x_{j+1}, z_{j+1})]\,,$$

where

$$z_{j+1} = y_j + h \cdot f(x_j, y_j)$$

and $j = 0, 1, 2, \ldots$.

This new method, usually called the *improved Euler method* or *Heun's*

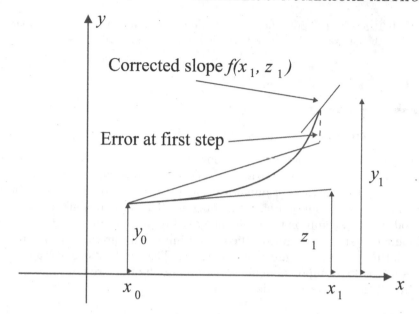

FIGURE 5.3
The improved Euler method.

method, first *predicts* and then *corrects* an estimate for y_j. It is an example of a class of numerical techniques called *predictor-corrector methods.* It is possible, using subtle Taylor series arguments, to show that the local discretization error is

$$\epsilon_j = -y'''(\xi) \cdot \frac{h^3}{12},$$

for some value of ξ between x_0 and x_n. Thus, in particular, the total discretization error is proportional to h^2 (instead of h, as before), so we expect more accuracy for the same step size. Figure 5.3 gives a way to visualize the improved Euler method. First, the point at (x_1, z_1) is predicted using the original Euler method, then this point is used to estimate the slope of the solution curve at x_1. This result is then averaged with the original slope estimate at (x_0, y_0) to make a better prediction of the solution—namely, (x_1, y_1).

We shall continue to examine our old friend

$$y' = x + y, \qquad y(0) = 1$$

and use the value $y(1)$ as a benchmark.

EXAMPLE 5.4.1 Apply the improved Euler method to the differential equation

$$y' = x + y, \qquad y(0) = 1 \qquad (5.4.1.1)$$

with step size 0.2 and gauge the improvement in accuracy over the ordinary

Euler method used in Examples 5.2.1 and 5.3.2.   ■

Table 5.4.1.2
Tabulated values for exact and numerical
solutions to (5.4.1.1) with $h = 0.2$
using the improved Euler method

x_n	y_n	Exact	$\overline{E}_n$ (%)
0.0	1.00000	1.00000	0.00
0.2	1.24000	1.24281	0.23
0.4	1.57680	1.581.12.25	0.43
0.6	2.03170	2.04424	0.61
0.8	2.63067	2.65108	0.77
1.0	3.40542	3.41.12.256	0.91

Solution: We see (remembering that $f(x, y) = x + y$) that

$$z_{k+1} = y_k + 0.2 \cdot f(x_k, y_k) = y_k + 0.2 \cdot (x_k + y_k)$$

and

$$y_{k+1} = y_k + 0.1 \cdot [(x_k + y_k) + (x_{k+1} + z_{k+1})] \, .$$

We begin the calculation by setting $k = 0$ and using the initial values $x_0 = 0.0000$, $y_0 = 1.0000$. Thus

$$z_1 = 1.0000 + 0.2 \cdot (0.0000 + 1.0000) = 1.2000$$

and

$$y_1 = 1.0000 + 0.1 \cdot [(0.0000 + 1.0000) + (0.2 + 1.2000)] = 1.2400 \, .$$

We continue this process and obtain the values shown in Table 5.4.1.2.

We see that the resulting approximate value for $y(1)$ is 3.40542. The aggregate error is about 1 percent, whereas with the former Euler method it was more than 13 percent. This is a substantial improvement.

Of course a smaller step size results in even more dramatic improvement in accuracy. Table 5.4.1.3 displays the results of applying the improved Euler method to our differential equation using a step size of $h = 0.1$. The relative error at $x = 1.00000$ is now about 0.2 percent, which is another order of magnitude of improvement in accuracy. We have predicted that halving the step size will decrease the aggregate error by a factor of 4. These results bear out that prediction.

Table 5.4.1.3
Tabulated values for exact and numerical
solutions to (5.4.1.1) with $h = 0.1$
using the improved Euler method

x_n	y_n	Exact	$\overline{E}_n$ (%)
0.0	1.00000	1.00000	0.0
0.1	1.11000	1.11034	0.0
0.2	1.24205	1.24281	0.1
0.3	1.39847	1.39972	0.1
0.4	1.58180	1.5836	0.1
0.5	1.79489	1.79744	0.1
0.6	2.04086	2.04424	0.2
0.7	2.32315	2.32751	0.2
0.8	2.64558	2.65108	0.2
0.9	3.0121.12.2	3.01921	0.2
1.0	3.42816	3.43656	0.2

In the next section we shall use a method of subdividing the intervals of our step sequence to obtain greater accuracy. This results in the Runge–Kutta method.

Math Nugget

Carl Runge (1856–1927) was professor of applied mathematics at Göttingen from 1904 to 1925. He is known for his work in complex variable theory, and for his discovery of a theorem that foreshadowed the famous Thue–Siegel–Roth theorem in diophantine equations. He also taught Hilbert to ski. M. W. Kutta (1867–1944), another German applied mathematician, is remembered for his contribution to the Kutta–Joukowski theory of airfoil lift in aerodynamics.

Runge's name is also remembered in connection with an incident involving the distinguished mathematician Gábor Szegő. While returning on a train from a conference, Szegő got into a fistfight with a young man who was sharing his compartment (it seems that the point at issue was whether the window should remain open or closed). Szegő had been a wrestler, and he did quite well in the fisticuffs. Seems that the young man came from a wealthy and influential family, one that was particularly important in Göttingen. So Szegő was brought up on charges. Now Runge's father-in-law was an attorney, and he defended Szegő—but to no avail. Szegő had to leave Göttingen, and ultimately ended up at Stanford.

Exercises

For each of Exercises 1–5, use the improved Euler method with $h = 0.1, 0.05$, and 0.01 to estimate the solution at $x = 1$. Compare your results to the *exact* solution and the results obtained with the original Euler method in Exercises 1–5 of Section 5.2.

  1. $y' = 2x + 2y, \quad y(0) = 1$
  2. $y' = 1/y, \quad y(0) = 1$
  3. $y' = e^y, \quad y(0) = 0$
  4. $y' = y - \sin x, \quad y(0) = -1$
  5. $y' = (x + y - 1)^2, \quad y(0) = 0$

5.5   The Runge–Kutta Method

Just as the trapezoid rule provides an improvement over the rectangular method for approximating integrals, so Simpson's rule gives an even better means for approximating integrals. With Simpson's rule we approximate not by rectangles or trapezoids but by parabolas.

Check your calculus book (for instance, [STE, p. 421] to review how Simpson's rule works. When we apply it to the integral of f, we find that

$$\int_{x_1}^{x_2} f(x,y)\, dx = \frac{1}{6}[f(x_0, y_0) + 4f(x_{1/2}, y(x_{1/2})) + f(x_1, y(x_1))]. \quad (5.5.1)$$

Here $x_{1/2} \equiv x_0 + h/2$, the midpoint of x_0 and x_1.

The Runge–Kutta method proceeds analogously. We cannot provide all the rigorous details of the derivation of the fourth-order Runge–Kutta method. We instead give an intuitive development.

Just as we did in obtaining our earlier numerical algorithms, we must now estimate both $y_{1/2}$ and y_1. The first estimate of $y_{1/2}$ comes from Euler's method. Thus

$$y_{1/2} = y_0 + \frac{m_1}{2}.$$

Here

$$m_1 = h \cdot f(x_0, y_0).$$

(The factor of $1/2$ here comes from the step size from x_0 to $x_{1/2}$.) To correct the estimate of $y_{1/2}$, we calculate it again in this manner:

$$y_{1/2} = y_0 + \frac{m_2}{2},$$

where

$$m_2 = h \cdot f(x_0 + h/2, y_0 + m_1/2).$$

Now, to predict y_1, we use the expression for $y_{1/2}$ and the Euler method:

$$y_1 = y_{1/2} + \frac{m_3}{2},$$

where $m_3 = h \cdot f(x_0 + h/2, y_0 + m_2/2)$.

Finally, let $m_4 = h \cdot f(x_0 + h, y_0 + m_3)$. The Runge–Kutta scheme is then obtained by substituting each of these estimates into (5.5.1) to yield

$$y_1 = y_0 + \frac{1}{6}(m_1 + 2m_2 + 2m_3 + m_4).$$

Just as in our earlier work, this algorithm can be applied to any number of mesh points in a natural way. At each step of the iteration, we first compute the four numbers m_1, m_2, m_3, m_4 given by

$$
\begin{aligned}
m_1 &= h \cdot f(x_k, y_k) \\
m_2 &= h \cdot f\left(x_k + \frac{h}{2}, y_k + \frac{m_1}{2}\right) \\
m_3 &= h \cdot f\left(x_k + \frac{h}{2}, y_k + \frac{m_2}{2}\right) \\
m_4 &= h \cdot f(x_k + h, y_k + m_3).
\end{aligned}
$$

Then y_{k+1} is given by

$$y_{k+1} = y_k + \frac{1}{6}(m_1 + 2m_2 + 2m_3 + m_4).$$

This new analytic paradigm, the Runge–Kutta technique, is capable of giving extremely accurate results without the need for taking very small values of h (thus making the work computationally expensive). The local truncation error is

$$\epsilon_k = -\frac{y^{(2)}(\xi) \cdot h^5}{180},$$

where ξ is a point between x_0 and x_n. The total truncation error is thus of the order of magnitude of h^4.

Now let us apply our new methodology—by far the best one yet—to our benchmark problem

$$y' = x + y, \qquad y(0) = 1.$$

As usual, we shall calculate $y(1)$ as a test.

EXAMPLE 5.5.2 Apply the Runge–Kutta method to the differential equation

$$y' = x + y, \qquad y(0) = 1. \tag{5.5.2.1}$$

Take $h = 1$, so that the process has only a single step.

Solution: We determine that

$$
\begin{aligned}
m_1 &= 1 \cdot (0+1) = 1 \\
m_2 &= 1 \cdot (0+0.5+1+0.5) = 2 \\
m_3 &= 1 \cdot (0+0.5+1+1) = 2.5 \\
m_4 &= 1 \cdot (0+1+1+2.5) = 4.5.
\end{aligned}
$$

Thus

$$y_1 = 1 + \frac{1}{6}(1 + 4 + 5 + 4.5) = 3.417.$$

Observe that this approximate solution is even better than that obtained with the improved Euler method for $h = 0.2$ (with five steps). And the amount of computation involved was absolutely minimal.

Table 5.5.2.2 shows the result of applying Runge–Kutta to our differential equation with $h = 0.2$. Notice that our approximate value for $y(1)$ is 3.436596, which agrees with the exact value to four decimal places. The relative error is less than 0.002 percent.

If we cut the step size in half—to 0.1, then the accuracy is increased dramatically—see Table 5.5.2.3. Now the relative error is less than 0.0002 percent.

Table 5.5.2.2

Tabulated values for exact and numerical
solutions to (5.5.2.1) with $h = 0.2$
using the Runge–Kutta method

x_n	y_n	Exact	$\bar{E}_n$ (%)
0.0	1.00000	1.00000	0.00000
0.2	1.24280	1.24281	0.00044
0.4	1.58364	1.58365	0.00085
0.6	2.04421	2.04424	0.00125
0.8	2.65104	2.65108	0.00152
1.0	3.436596	3.43656	0.00179

Table 5.5.2.3

Tabulated values for exact and numerical
solutions to (2) with $h = 0.1$
using the Runge–Kutta method

x_n	y_n	Exact	$\bar{E}_n$ (%)
0.0	1.00000	1.00000	0.0
0.1	1.1103417	1.1103418	0.00002
0.2	1.24281	1.24281	0.00003
0.3	1.39972	1.39972	0.00004
0.4	1.583652	1.58365	0.00006
0.5	1.79744	1.79744	0.00007
0.6	2.04424	2.04424	0.00008
0.7	2.32750	2.32751	0.00009
0.8	2.65108	2.65108	0.00010
0.9	3.01920	3.01921	0.00011
1.0	3.43656	3.43656	0.00012

Exercises

For each of Exercises 1–5, use the Runge–Kutta method with $h = 0.1, 0.05$, and
$h = 0.01$, to estimate the solution at $x = 1$. Compare your results to the *exact*
solution and the results obtained with both the Euler method (Exercises 1–5 of
Section 5.2) and the improved Euler method (Exercises 1–5 of Section 5.4).

1. $y' = 2x + 2y, \quad y(0) = 1$
2. $y' = 1/y, \quad y(0) = 1$
3. $y' = e^y, \quad y(0) = 0$
4. $y' = y - \sin x, \quad y(0) = -1$
5. $y' = (x + y - 1)^2, \quad y(0) = 0$

6. Use the Runge–Kutta method with $h = 0.2$ to find an approximate solution of the following initial value problem.

$$t^2 y'' - 3ty' + 3y = 1, \quad y(1) = 0, \quad y'(1) = 0.$$

Determine the exact solution and compare your results. Does the differential equation possess a solution at $t = 0$? How might the Runge–Kutta method be employed to compute the solution at 0?

7. Use your favorite scientific computing language—BASIC or Fortran or C or APL—to write a routine to implement the Runge–Kutta method. Apply the program to the initial value problem

$$y' = y + x, \ y(0) = 1.$$

Now solve the same initial value problem using your symbol manipulation software (such as Maple or Mathematica). You will probably find that the symbol manipulation software is faster and has a higher degree of accuracy. Can you speculate why this is so?

Now apply both methodologies to the initial value problem

$$y' = y - \sin y + xy, \ y(0) = 1.$$

Supply similar comparisons and commentary.

5.6 A Constant Perturbation Method for Linear, Second-Order Equations

The philosophy that we have employed in each of the numerical techniques of this chapter is a very simple one, and it parallels the philosophy that was used to develop numerical techniques of integration in calculus. Namely, we approximate the differential equation (more precisely, we approximate the *coefficients* of the differential equation) by polynomials. In the most rudimentary technique—Euler's method—we approximate by constant functions. In the improved Euler method we approximate by linear functions. And in the last, most sophisticated technique (the Runge–Kutta method) we approximate by quadratic functions (or parabolas).

This methodology just described, while straightforward and logical, has its limitations. First of all, it is not adapted to the particular differential equation that is being studied. It is a universal technique. Second, it only works on very small intervals where the approximation (by constant, linear, or quadratic functions) is good. Third, it is not very flexible.

In this section, we shall introduce a technique that is much more adaptable. It will actually be different in its implementation for each differential equation to which we apply it. Instead of approximating by a universal object like a linear function, we shall *approximate by a constant-coefficient differential equation*. The setup is as follows.

Suppose that we are given an initial value problem

$$y'' + a(x)y' + b(x)y = c(x), \quad x \in [a,b], \ y(a) = y_0, \ y'(a) = y_1.$$

Here $a(x), b(x), c(x)$ are given functions. Thus our differential equation has *variable coefficients*. As is customary and familiar, we introduce a partition

$$a = x_0 < x_1 < \cdots x_{k-1} < x_k = b.$$

Now, on each interval $I_j = [x_{j-1}, x_j]$, we approximate each of the coefficient functions by a constant:

$$a(x) \leftrightarrow \widetilde{a}_j, \ b(x) \leftrightarrow \widetilde{b}_j, \ c(x) \leftrightarrow \widetilde{c}_j .$$

A convenient way (but by no means the only way) of choosing these constants is

$$
\begin{aligned}
\widetilde{a}_j &= \frac{a(x_{j-1}) + a(x_j)}{2}, \\
\widetilde{b}_j &= \frac{b(x_{j-1}) + b(x_j)}{2}, \\
\widetilde{c}_j &= \frac{c(x_{j-1}) + c(x_j)}{2}.
\end{aligned}
$$

Thus, on each interval I_j, we solve the approximating differential equation

$$\widetilde{y}'' + \widetilde{a}_j \widetilde{y}' + \widetilde{b}_j \widetilde{y} = \widetilde{c}_j, \ x \in [x_{j-1}, x_j]. \tag{5.6.1}$$

This is, of course, an equation that can be solved by hand. Let us assume for convenience that $\widetilde{a}_j^2 \neq 4\widetilde{b}_j$ for each j. With this assumption, we let

$$\omega^+ = \frac{-\widetilde{a}_j + \sqrt{\widetilde{a}_j^2 - 4\widetilde{b}_j}}{2}$$

$$\omega^- = \frac{-\widetilde{a}_j - \sqrt{\widetilde{a}_j^2 - 4\widetilde{b}_j}}{2}.$$

Then it is easy to see that the general solution of the *associated homogeneous equation*

$$\widetilde{y}'' + \widetilde{a}_j \widetilde{y}' + \widetilde{b}_j \widetilde{y} = 0$$

is

$$\widetilde{y}(x) = A_j e^{\omega^+ \cdot (x - x_{j-1})} + B_j e^{\omega^- \cdot (x - x_{j-1})}.$$

(Note that the use of $(x - x_{j-1})$ instead of just x only contributes a multiplicative constant, which is harmless.) Here, as usual, A_j and B_j are arbitrary constants.

A particular solution of equation (5.6.1) is

$$\widetilde{y}_p = -\frac{\widetilde{c}_j}{\widetilde{b}_j} \equiv \widetilde{u}_j$$

provided that $\widetilde{b}_j \neq 0$. In fact, in what follows, we shall assume that $\widetilde{b}_j \neq 0$, that $\widetilde{a}_j \neq 0$, and further that $\widetilde{a}_j^2 \neq 4\widetilde{b}_j$. The interested reader may work out the details for the other cases, but the case that we treat is both typical and indicative.

Thus we find that the general solution of (5.6.1) is given by taking a linear combination of the particular solution we have found and the general solution to the associated homogeneous equation:

$$\widetilde{y}(x) = A_{j-1} e^{\omega^+ (x - x_{j-1})} + B_{j-1} e^{\omega^- (x - x_{j-1})} + \widetilde{u}_j.$$

The first derivative of this equation is

$$\widetilde{y}'(x) = A_{j-1} \omega^+ e^{\omega^+ (x - x_{j-1})} + B_{j-1} \omega^- e^{\omega^- (x - x_{j-1})}.$$

The values of A_{j-1} and B_{j-1} on the jth interval $[x_{j-1}, x_j]$ are then determined by suitable initial conditions $\widetilde{y}(x_{j-1}) = \widetilde{y}^{j-1}$, $\widetilde{y}'(x_{j-1}) = \widetilde{y}_1^{j-1}$. Thus we have

$$\widetilde{y}^{j-1} = A_{j-1} + B_{j-1} + u_j$$

and

$$\widetilde{y}'^{j-1} = \omega^+ A_{j-1} + \omega^- B_{j-1}.$$

It follows that

$$A_{j-1} = \frac{1}{\omega^+ - \omega^-} \left(-\omega^- \widetilde{y}^{j-1} + \omega^- u_j + \widetilde{y}'^{j-1} \right)$$

and

$$B_{j-1} = \frac{1}{\omega^+ - \omega^-} \left(\omega^+ \widetilde{y}^{j-1} - \omega^+ u_j - \widetilde{y}'^{j-1} \right).$$

Now we need to explain how the algorithm advances from step to step (as j increases, beginning at $j = 1$). In the first interval $I_1 = [x_0, x_1]$, we construct $\widetilde{a}_0, \widetilde{b}_0, \widetilde{c}_0$ and also ω^+, ω^-, u_0. The solution, in particular the values of A_0 and B_0, are determined by the initial conditions $\widetilde{y}(a) = y_0$, $\widetilde{y}'(a) = y_1$. The value of this unique solution, and the value of its first derivative, at the point x_1, are taken to be the initial conditions when we next perform our algorithm on the interval $I_2 = [x_1, x_2]$. This will determine a unique solution on the second interval I_2.

In general, we take the unique solution produced by our algorithm on the interval I_{j-1} and use the value of it and its first derivative to give initial conditions for the initial value problem on I_j.

The advantage of this new methodology is that the approximations come directly from the coefficients $a(x), b(x), c(x)$ of the original equation. The user can control the size of the deviations $a(x) - \tilde{a}_j$, $b(x) - \tilde{b}_j$, $c(x) - \tilde{c}_j$ by adjusting the size of the intervals in the partition. One upshot is that the range of values for h under which we get an acceptable approximation to the desired solution can, in principle, be quite large.

The reader may wish to practice with this new method by applying it to the initial value problem

$$y'' - 4xy' + (4x^2 + \alpha^2 - 2)y + \alpha^2 e^{x^2} = 0, \ y_0 = 1, y_1 = \beta$$

for $x \in [0, 5]$. See how the behavior of the approximation changes for different values of $\alpha \in [1, 25]$ and $\beta \in [0, 25]$.

Problems for Review and Discovery

A. Drill Exercises

1. For each of these exercises, use the Euler method with $h = 0.1, 0.05$, and 0.01 to estimate the solution at $x = 1$. In each case, compare your results to the *exact* solution and discuss how well (or poorly) the Euler method has worked.

   (a) $y' = x - 2y$, $\quad y(0) = 2$
   (b) $y' = 1/y^2$, $\quad y(0) = 1$
   (c) $y' = e^{-y}$, $\quad y(0) = 1$
   (d) $y' = y + \cos x$, $\quad y(0) = -2$
   (e) $y' = (x - y + 1)^2$, $\quad y(0) = 0$
   (f) $y' = \frac{1}{x+y}$, $\quad y(0) = 1$

2. In each of these exercises, use the exact solution, together with step sizes $h = 0.2$ and 0.1, to estimate the total discretization error that occurs with the Euler method at $x = 1$.

   (a) $y' = 2x + y$, $\quad y(0) = 0$
   (b) $y' = \frac{1}{x-2y}$, $\quad y(0) = 2$
   (c) $y' = e^{-y}$, $\quad y(0) = 0$
   (d) $y' = y + \cos y$, $\quad y(0) = -2$
   (e) $y' = (x - y + 1)^2$, $\quad y(0) = 0$

(f) $y' = x - 3y, \quad y(0) = 1$

3. In each of these exercises, use the improved Euler method with $h = 0.1, 0.05,$ and 0.01 to estimate the solution at $x = 1$. Compare your results to the *exact* solution and the results obtained with the original Euler method in Exercise 1 above.

(a) $y' = x - 2y, \quad y(0) = 2$
(b) $y' = 1/y^2, \quad y(0) = 1$
(c) $y' = e^{-y}, \quad y(0) = 1$
(d) $y' = y + \cos x, \quad y(0) = -2$
(e) $y' = (x - y + 1)^2, \quad y(0) = 0$
(f) $y' = \frac{1}{x+y}, \quad y(0) = 1$

4. For each of these exercises, use the Runge–Kutta method with $h = 0.1, 0.05,$ and $h = 0.01$ to estimate the solution at $x = 1$. Compare your results to the *exact* solution and the results obtained with both the Euler method (Exercise 1) and the improved Euler method (Exercise 3).

(a) $y' = x - 2y, \quad y(0) = 2$
(b) $y' = 1/y^2, \quad y(0) = 1$
(c) $y' = e^{-y}, \quad y(0) = 1$
(d) $y' = y + \cos x, \quad y(0) = -2$
(e) $y' = (x - y + 1)^2, \quad y(0) = 0$
(f) $y' = \frac{1}{x+y}, \quad y(0) = 1$

B. Challenge Problems

1. Consider the initial value problem

$$y' = -\frac{1}{2}y, \quad y(0) = 3.$$

Apply the Euler method at $x = 2$ with step size h and show that the resulting approximation is

$$\mathcal{A} \approx \left(1 - \frac{h}{2}\right)^{2/h}.$$

2. Apply the improved Euler method to the initial value problem

$$y' = y, \quad y(0) = 1.$$

Use step sizes $h = 1, 0.1, 0.01, 0.001, 0.0001$ to get better and better approximations to Euler's constant e. What number of decimal places of accuracy do you obtain?

3. Use the Runge–Kutta method with step size $h = 0.1$ to approximate the solution to

$$y' = \sin(4y) - 2x, y(0) = 0$$

at the points $0, 0.1, 0.2, \ldots, 1.9, 2.0$. Use this numerical data to make a rough sketch of the solution $y(x)$ on the interval $[0, 2]$.

4. Use the Runge–Kutta method with step size $h = 0.05$ to approximate the solution to

$$y' = 4\sin(y - 3x), \quad y(0) = 1$$

at the points $0, 0.05, 0.1, 0.15, 0.2, \ldots, 0.95, 1$. Use this numerical data to make a rough sketch of the solution $y(x)$ on the interval $[0, 1]$.

C. Problems for Discussion and Exploration

1. Devise an initial value problem that will enable you to get a numerical approximation to the value of the number π. Use this device to compute π to four decimal places of accuracy.

2. The logistic equation

$$\frac{dp}{dt} = ap - bp^2 \quad p(0) = p_0$$

is often used as a simple model of population growth. Take $a = 1, b = 2, p_0 = 50$, and step size 0.1. Use Euler's method to approximate the value of the solution at $x = 2$. Now use the improved Euler method. What increase in accuracy do you obtain? Conclude by applying the Runge–Kutta method. What increase in accuracy do you see now?

3. Replace the logistic equation in Exercise 2 with the more general equation

$$\frac{dp}{dt} = ap - bp^r, \quad p(0) = p_0$$

for some parameter $r > 1$. Take $a = 2, b = 1, p_0 = 1.5$, and explore the effect of varying the parameter r. Conduct this exploration using Euler's method with step size $h = 0.1$. Now use the improved Euler method and see how things change.

4. It is standard to model the velocity of a falling body with the initial value problem

$$m\frac{dv}{dt} = mg - kv, \quad v(0) = v_0, \tag{*}$$

where g is the acceleration due to gravity, $-kv$ is air resistance, and m is the mass of the body. Explain why this is a correct physical model, just using Newton's laws from elementary physics. Of course this equation may be solved explicitly.

In some settings it is appropriate to replace the air resistance terms with $-kv^r$ for some $r > 1$. Then the initial value problem becomes

$$m\frac{dv}{dt} = mg - kv^r, \quad v(0)0 = v_0. \tag{**}$$

Explore the effect of changing the parameter r by taking $m = 5, g = 9.81$, $k = 4$, and $v_0 = 0$. Use the improved Euler method with step size $h = 0.1$ on the interval $[0, 10]$. Now use the Runge–Kutta method and see whether you can learn more.

5. In the study of nonisothermal flow of a Newtonian fluid between parallel
    plates, one encounters the ordinary differential equation

    $$\frac{d^2 y}{dt^2} + x^2 e^y = 0, \quad x > 0.$$

    There is a sequence of changes of variable that will transform this equa-
    tion to

    $$\frac{dv}{du} = u\left(\frac{u}{2} + 1\right) v^3 + \left(u + \frac{5}{2}\right) v^2 .$$

    See whether you can discover the changes of variable that will effect this
    transformation.

    Now use the Runge–Kutta method to approximate $v(2)$ if $v(2.1) = 0.1$.

6

Fourier Series: Basic Concepts

- The idea of Fourier series

- Calculating a Fourier series

- Convergence of Fourier series

- Odd and even functions

- Fourier series on arbitrary intervals

- Orthogonality

6.1 Fourier Coefficients

Trigonometric and Fourier series constitute one of the oldest parts of analysis. They arose, for instance, in classical studies of the heat and wave equations. Today they play a central role in the study of sound, heat conduction, electromagnetic waves, mechanical vibrations, signal processing, and image analysis and compression. Whereas power series (see Chapter 3) can only be used to represent very special functions (most functions, even smooth ones, do *not* have convergent power series), Fourier series can be used to represent very broad classes of functions.

For us, a trigonometric series or Fourier series is one of the form

$$f(x) = \frac{1}{2}a_0 + \sum_{n=1}^{\infty} \left(a_n \cos nx + b_n \sin nx \right). \qquad (6.1.1)$$

We shall be concerned with three main questions:

1. Given a function f, how do we calculate the coefficients a_n, b_n?

2. Once the series for f has been calculated, can we determine that it converges, and that it converges to f?

3. How can we use Fourier series to solve a differential equation?

We begin our study with some classical calculations that were first per-formed by Euler (1707–1783). It is convenient to assume that our function f is defined on the interval $[-\pi, \pi] = \{x \in \mathbf{R} : -\pi \leq x \leq \pi\}$. We shall temporarily make the important assumption that the *trigonometric series (6.1.1) for f converges uniformly*. While this turns out to be true for a large class of functions (continuously differentiable functions, for example), for now this is merely a convenience so that our calculations are justified.

We apply the integral to both sides of (6.1.1). The result is

$$
\int_{-\pi}^{\pi} f(x)\, dx = \int_{-\pi}^{\pi} \left(\frac{1}{2}a_0 + \sum_{n=1}^{\infty} \left(a_n \cos nx + b_n \sin nx \right) \right) dx
$$

$$
= \int_{-\pi}^{\pi} \frac{1}{2}a_0\, dx + \sum_{n=1}^{\infty} \int_{-\pi}^{\pi} a_n \cos nx\, dx + \sum_{n=1}^{\infty} \int_{-\pi}^{\pi} b_n \sin nx\, dx .
$$

The change in order of summation and integration is justified by the uniform convergence of the series (see [KRA2, page 202, ff.]).

Now each of $\cos nx$ and $\sin nx$ integrates to 0. The result is that

$$
a_0 = \frac{1}{\pi} \int_{-\pi}^{\pi} f(x)\, dx .
$$

In effect, then, a_0 is (twice) the *average* of f over the interval $[-\pi, \pi]$.

To calculate a_j for $j > 1$, we multiply the formula (6.1.1) by $\cos jx$ and then integrate as before. The result is

$$
\int_{-\pi}^{\pi} f(x) \cos jx\, dx = \int_{-\pi}^{\pi} \left\{ \frac{1}{2}a_0 + \sum_{n=1}^{\infty} \left(a_n \cos nx + b_n \sin nx \right) \right\} \cos jx\, dx
$$

$$
= \int_{-\pi}^{\pi} \frac{1}{2}a_0 \cos jx\, dx + \sum_{n=1}^{\infty} \int_{-\pi}^{\pi} a_n \cos nx \cos jx\, dx
$$

$$
+ \sum_{n=1}^{\infty} \int_{-\pi}^{\pi} b_n \sin nx \cos jx\, dx . \qquad (6.1.2)
$$

Now the first integral on the right vanishes, as we have already noted. Further recall that

$$
\cos nx \cos jx = \frac{1}{2} \left(\cos(n+j)x + \cos(n-j)x \right)
$$

and

$$
\sin nx \cos jx = \frac{1}{2} \left(\sin(n+j)x + \sin(n-j)x \right) .
$$

It follows immediately that

$$
\int_{-\pi}^{\pi} \cos nx \cos jx\, dx = 0 \qquad \text{when } n \neq j
$$

and

$$\int_{-\pi}^{\pi} \sin nx \cos jx \, dx = 0 \qquad \text{for all } n, j.$$

Thus our formula (6.1.2) reduces to

$$\int_{-\pi}^{\pi} f(x) \cos jx \, dx = \int_{-\pi}^{\pi} a_j \cos jx \cos jx \, dx.$$

We may use our formula above for the product of cosines to integrate the right-hand side. The result is

$$\int_{-\pi}^{\pi} f(x) \cos jx \, dx = a_j \cdot \pi$$

or

$$a_j = \frac{1}{\pi} \int_{-\pi}^{\pi} f(x) \cos jx \, dx.$$

A similar calculation shows that

$$b_j = \frac{1}{\pi} \int_{-\pi}^{\pi} f(x) \sin jx \, dx.$$

In summary, we now have formulas for calculating all the a_j's and b_j's:

$$a_j = \frac{1}{\pi} \int_{-\pi}^{\pi} f(x) \cos jx \, dx, \quad j = 0, 1, \ldots$$

and

$$b_j = \frac{1}{\pi} \int_{-\pi}^{\pi} f(x) \sin jx \, dx, \quad j = 1, 2, \ldots.$$

EXAMPLE 6.1.3 Find the Fourier series of the function

$$f(x) = x, \qquad -\pi \le x \le \pi.$$

Solution:
Of course

$$a_0 = \frac{1}{\pi} \int_{-\pi}^{\pi} x \, dx = \frac{1}{\pi} \cdot \frac{x^2}{2} \Big|_{-\pi}^{\pi} = 0.$$

For $j \ge 1$, we calculate a_j as follows:

$$\begin{aligned}
a_j \quad &= \quad \frac{1}{\pi} \int_{-\pi}^{\pi} x \cos jx \, dx \\
&\overset{\text{(parts)}}{=} \quad \frac{1}{\pi} \left(x \frac{\sin jx}{j} \Big|_{-\pi}^{\pi} - \int_{-\pi}^{\pi} \frac{\sin jx}{j} \, dx \right) \\
&= \quad \frac{1}{\pi} \left\{ 0 - \left(-\frac{\cos jx}{j^2} \Big|_{-\pi}^{\pi} \right) \right\} \\
&= \quad 0.
\end{aligned}$$

Similarly, we calculate the b_j:

$$
\begin{aligned}
b_j &= \frac{1}{\pi}\int_{-\pi}^{\pi} x\sin jx\, dx \\
&\overset{\text{(parts)}}{=} \frac{1}{\pi}\left(x\cdot\frac{-\cos jx}{j}\Big|_{-\pi}^{\pi} - \int_{-\pi}^{\pi}\frac{-\cos jx}{j}\, dx \right) \\
&= \frac{1}{\pi}\left\{ -\frac{2\pi\cos j\pi}{j} - \left(-\frac{\sin jx}{j^2}\Big|_{-\pi}^{\pi} \right) \right\} \\
&= \frac{2\cdot(-1)^{j+1}}{j}.
\end{aligned}
$$

Now that all the coefficients have been calculated, we may summarize the result as

$$
x = f(x) = 2\left(\sin x - \frac{\sin 2x}{2} + \frac{\sin 3x}{3} - + \cdots \right).
$$

∎

It is sometimes convenient, in the study of Fourier series, to think of our functions as defined on the entire real line. We extend a function that is initially given on the interval $[-\pi, \pi]$ to the entire line using the idea of *periodicity*. The sine function and cosine function are periodic in the sense that $\sin(x + 2\pi) = \sin x$ and $\cos(x + 2\pi) = \cos x$. We say that sine and cosine are *periodic with period* 2π. Thus it is natural, if we are given a function f on $[-\pi, \pi)$, to define $f(x + 2\pi) = f(x)$, $f(x + 2\cdot 2\pi) = f(x)$, $f(x - 2\pi) = f(x)$, etc.[1]

Figure 6.1 exhibits the periodic extension of the function $f(x) = x$ on $[-\pi, \pi)$ to the real line.

Figure 6.2 shows the first four summands of the Fourier series for $f(x) = x$. The finest dashes show the curve $y = 2\sin x$, the next finest is $-\sin 2x$, the next is $(2/3)\sin 3x$, and the coarsest is $-(1/2)\sin 4x$.

Figure 6.3 shows the sum of the first four terms of the Fourier series and also of the first six terms, as compared to $f(x) = x$. Figure 6.4 shows the sum of the first eight terms of the Fourier series and also of the first ten terms, as compared to $f(x) = x$.

EXAMPLE 6.1.4 Calculate the Fourier series of the function

$$
g(x) = \begin{cases} 0 & \text{if} \quad -\pi \le x < 0 \\ \pi & \text{if} \quad 0 \le x \le \pi. \end{cases}
$$

Solution:

---

[1] Notice that we take the original function f to be defined on $[-\pi, \pi)$ rather than $[-\pi, \pi]$ to avoid any ambiguity at the endpoints.

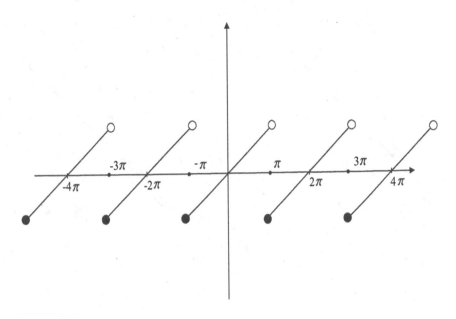

FIGURE 6.1
Periodic extension of $f(x) = x$.

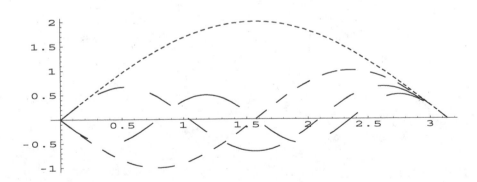

FIGURE 6.2
The first four summands for $f(x) = x$.

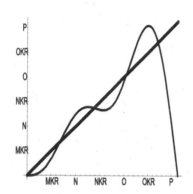 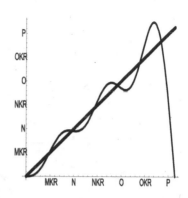

FIGURE 6.3
The sum of four terms and of six terms of the Fourier series of $f(x) = x$.

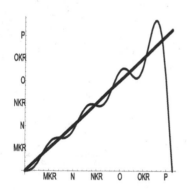 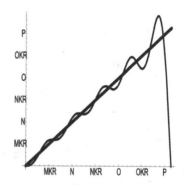

FIGURE 6.4
The sum of eight terms and of ten terms of the Fourier series of $f(x) = x$.

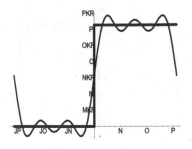 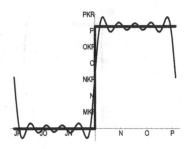

FIGURE 6.5

The sum of four and of six terms of the Fourier series of g.

Following our formulas, we calculate

$$a_0 = \frac{1}{\pi} \int_{-\pi}^{\pi} g(x)\, dx = \frac{1}{\pi} \int_{-\pi}^{0} 0\, dx + \frac{1}{\pi} \int_{0}^{\pi} \pi\, dx = \pi\,.$$

$$a_n = \frac{1}{\pi} \int_{0}^{\pi} \pi \cos nx\, dx = \left.\frac{\sin nx}{n}\right|_{0}^{\pi} = 0, \quad \text{all } n \geq 1\,.$$

$$b_n = \frac{1}{\pi} \int_{0}^{\pi} \pi \sin nx\, dx = \frac{1}{n}\left(1 - \cos n\pi\right) = \frac{1}{n}\left(1 - (-1)^n\right)\,.$$

Another way to write this last calculation is

$$b_{2j} = 0\,,\ b_{2j-1} = \frac{2}{2j-1}\quad j = 1, 2, \ldots\,.$$

In sum, the Fourier expansion for g is

$$g(x) = \frac{\pi}{2} + 2\left(\sin x + \frac{\sin 3x}{3} + \frac{\sin 5x}{5} + \cdots\right)\,.$$

Figure 6.5 shows the fourth and sixth partial sums, compared against the function $g(x)$. Figure 6.6 shows the eighth and tenth partial sums, compared against the function $g(x)$. ∎

EXAMPLE 6.1.5 Find the Fourier series of the function given by

$$h(x) = \begin{cases} -\frac{\pi}{2} & \text{if} \quad -\pi \leq x < 0 \\[2mm] \frac{\pi}{2} & \text{if} \quad 0 \leq x \leq \pi\,. \end{cases}$$

Solution:

This is the same function as in the last example, with $\pi/2$ subtracted. Thus

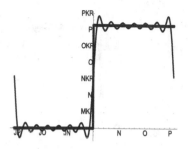 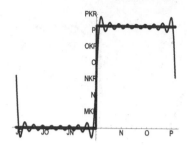

FIGURE 6.6
The sum of eight terms and of ten terms of the Fourier series of g.

the Fourier series may be obtained by subtracting $\pi/2$ from the Fourier series
that we obtained in that example. The result is

$$h(x) = 2\left(\sin x + \frac{\sin 3x}{3} + \frac{\sin 5x}{5} + \cdots\right).$$

The graph of this function, suitably periodized, is shown in Figure 6.7.   ∎

EXAMPLE 6.1.6 Calculate the Fourier series of the function

$$k(x) = \begin{cases} -\frac{\pi}{2} - \frac{x}{2} & \text{if} \quad -\pi \leq x < 0 \\ \frac{\pi}{2} - \frac{x}{2} & \text{if} \quad 0 \leq x \leq \pi. \end{cases}$$

Solution:
This function is simply the function h from Example 6.1.5 minus half the
function f from Example 6.1.3. In other words, $k(x) = h(x) - [1/2]f(x)$. Thus
we may obtain the requested Fourier series by subtracting half the series from
Example 6.1.3 from the series in Example 6.1.5. The result is

$$\begin{aligned} f(x) &= 2\left(\sin x + \frac{\sin 3x}{3} + \frac{\sin 5x}{5} + \cdots\right) \\ &\quad - \left(\sin x - \frac{\sin 2x}{2} + \frac{\sin 3x}{3} - + \cdots\right) \\ &= \sin x + \frac{\sin 2x}{2} + \frac{\sin 3x}{3} + \cdots \\ &= \sum_{n=1}^{\infty} \frac{\sin nx}{n}. \end{aligned}$$

The graph of this series is the sawtooth wave shown in Figure 6.8.   ∎

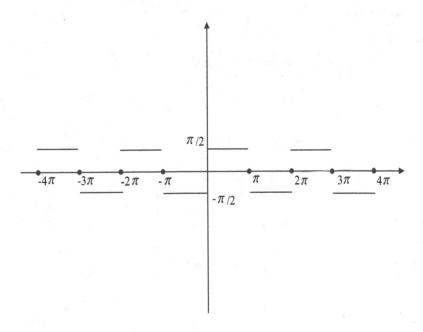

FIGURE 6.7
Graph of the function f in Example 6.1.5.

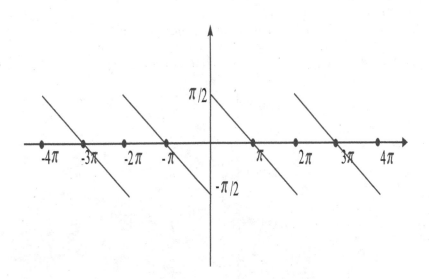

FIGURE 6.8
The sawtooth wave.

Exercises

1. Find the Fourier series of the function

$$f(x) = \begin{cases} \pi & \text{if} & -\pi \leq x \leq \dfrac{\pi}{2} \\ 0 & \text{if} & \dfrac{\pi}{2} < x \leq \pi. \end{cases}$$

2. Find the Fourier series for the function

$$f(x) = \begin{cases} 0 & \text{if} & -\pi \leq x < 0 \\ 1 & \text{if} & 0 \leq x \leq \frac{\pi}{2} \\ 0 & \text{if} & \dfrac{\pi}{2} < x \leq \pi. \end{cases}$$

3. Find the Fourier series of the function

$$f(x) = \begin{cases} 0 & \text{if} & -\pi \leq x < 0 \\ \sin x & \text{if} & 0 \leq x \leq \pi. \end{cases}$$

4. Solve Exercise 3 with $\sin x$ replaced by $\cos x$.

5. Find the Fourier series for each of these functions. Pay special attention to the reasoning used to establish your conclusions; consider alternative lines of thought.

    (a) $f(x) = \pi, \quad -\pi \leq x \leq \pi$

    (b) $f(x) = \sin x, \quad -\pi \leq x \leq \pi$

    (c) $f(x) = \cos x, \quad -\pi \leq x \leq \pi$

    (d) $f(x) = \pi + \sin x + \cos x, \quad -\pi \leq x \leq \pi$

    Solve Exercises 6 and 7 by using the methods of Examples 6.1.5 and 6.1.6, without actually calculating the Fourier coefficients.

6. Find the Fourier series for the function given by

    (a)

    $$f(x) = \begin{cases} -a & \text{if} & -\pi \leq x < 0 \\ a & \text{if} & 0 \leq x \leq \pi \end{cases}$$

    for a a positive real number.

    (b)

    $$f(x) = \begin{cases} -1 & \text{if} & -\pi \leq x < 0 \\ 1 & \text{if} & 0 \leq x \leq \pi \end{cases}$$

    (c)

    $$f(x) = \begin{cases} -\frac{\pi}{4} & \text{if} & -\pi \leq x < 0 \\ \frac{\pi}{4} & \text{if} & 0 \leq x \leq \pi \end{cases}$$

    (d)

    $$f(x) = \begin{cases} -1 & \text{if} & -\pi \leq x < 0 \\ 2 & \text{if} & 0 \leq x \leq \pi \end{cases}$$

    (e)

    $$f(x) = \begin{cases} 1 & \text{if} & -\pi \leq x < 0 \\ 2 & \text{if} & 0 \leq x \leq \pi \end{cases}$$

7. Without using the theory of Fourier series at all, show graphically that the sawtooth wave of Figure 6.1 can be represented as the sum of a sawtooth wave of period π and a square wave of period π.

6.2 Some Remarks about Convergence

The study of convergence of Fourier series is both deep and subtle. It would take us far afield to consider this matter in any detail. In the present section we shall very briefly describe a few of the basic results, but we shall not prove them. See [KRA3] for a more thoroughgoing discussion of these matters.

Our basic pointwise convergence result for Fourier series, which finds its genesis in work of Dirichlet (1805–1859), is this:

Definition 6.2.1 Let f be a function on $[-\pi, \pi]$. We say that f is *piecewise smooth* if the graph of f consists of finitely many continuously differentiable curves, and furthermore that the one-sided derivatives exist at each of the endpoints $\{p_1, \ldots, p_k\}$ of the definition of the curves, in the sense that

$$\lim_{h \to 0^+} \frac{f(p_j + h) - f(p_j)}{h} \quad \text{and} \quad \lim_{h \to 0^-} \frac{f(p_j + h) - f(p_j)}{h}$$

exist. Further, we require that f' extend continuously to $[p_j, p_{j+1}]$ for each $j = 1, \ldots, k - 1$. See Figure 6.9.

Theorem 6.2.2 *Let f be a function on $[-\pi, \pi]$ which is piecewise smooth and overall continuous. Then the Fourier series of f converges at each point c of $[-\pi, \pi]$ to $f(c)$.*

Let f be a function on the interval $[-\pi, \pi]$. We say that f has a *simple discontinuity* (or a *discontinuity of the first kind*) at the point $c \in (-\pi, \pi)$ if the limits $\lim_{x \to c^-} f(x)$ and $\lim_{x \to c^+} f(x)$ exist and

$$\lim_{x \to c^-} f(x) \neq \lim_{x \to c^+} f(x).$$

The reader should understand that a simple discontinuity is in contradistinction to the other kind of discontinuity. That is to say, f has a *discontinuity of the second kind* at c if either $\lim_{x \to c^-} f(x)$ or $\lim_{x \to c^+} f(x)$ does not exist.

EXAMPLE 6.2.3 The function

$$f(x) = \begin{cases} 1 & \text{if} \quad -\pi \leq x \leq 1 \\ 2 & \text{if} \quad 1 < x \leq \pi \end{cases}$$

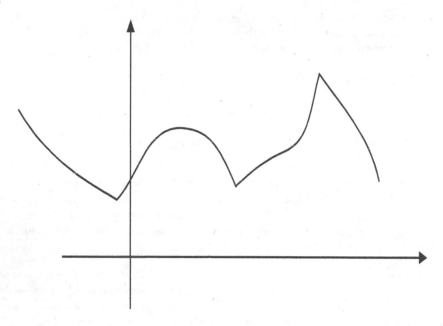

FIGURE 6.9
A piecewise smooth function.

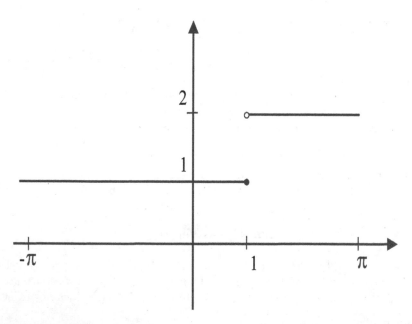

FIGURE 6.10
A simple discontinuity.

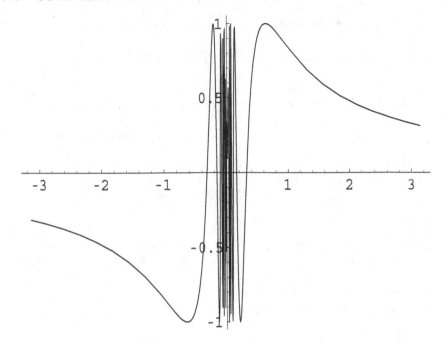

FIGURE 6.11
A discontinuity of the second kind.

has a simple discontinuity at $x = 1$. It is continuous at all other points of the interval $[-\pi, \pi]$. See Figure 6.10.

The function

$$g(x) = \begin{cases} \sin \frac{1}{x} & \text{if} \quad x \neq 0 \\ 0 & \text{if} \quad x = 0 \end{cases}$$

has a discontinuity of the second kind at the origin. See Figure 6.11. ■

Our next result about convergence is a bit more technical to state, but it is important in practice, and has historically been very influential. It is due to L. Fejér.

Definition 6.2.4 Let f be a function and let

$$\frac{1}{2}a_0 + \sum_{n=1}^{\infty} \left(a_n \cos nx + b_n \sin nx \right)$$

be its Fourier series. The Nth *partial sum* of this series is

$$S_N(f)(x) = \frac{1}{2}a_0 + \sum_{n=1}^{N} \left(a_n \cos nx + b_n \sin nx \right).$$

The *Cesàro mean* of the series is

$$\sigma_N(f)(x) = \frac{1}{N+1} \sum_{j=0}^{N} S_j(f)(x).$$

In other words, the Cesàro means are simply the averages of the partial sums.

Theorem 6.2.5 *Let f be a continuous function on the interval $[-\pi, \pi]$. Then the Cesàro means $\sigma_N(f)$ of the Fourier series for f converge uniformly to f.*

It is worth noting explicitly that if the Fourier series of a function f converges at a point x_0, then the Cesàro means of the series also converge at x_0—and to the very same limit.

A useful companion result is this:

Theorem 6.2.6 (Fejér) *Let f be a piecewise continuous function on $[-\pi, \pi]$—meaning that the graph of f consists of finitely many continuous curves. Let p be the endpoint of one of those curves, and assume that $\lim_{x \to p-} f(x) \equiv f(p^-)$ and $\lim_{x \to p+} f(x) \equiv f(p^+)$ both exist (and are possibly unequal). Then the Cesàro means of the Fourier series of f at p converge to $[f(p^-) + f(p^+)]/2$.*

In fact, with a few more hypotheses, we may make the result even sharper. Recall that a function f is *monotone increasing* if $x_1 \leq x_2$ implies $f(x_1) \leq f(x_2)$. The function is *monotone decreasing* if $x_1 \leq x_2$ implies $f(x_1) \geq f(x_2)$. If the function is either monotone increasing or monotone decreasing then we just call it *monotone*. Now we have this result of Dirichlet:

Theorem 6.2.7 (Dirichlet) *Let f be a function on $[-\pi, \pi]$ which is piecewise continuous. Assume that each piece of f is monotone. Then the Fourier series of f converges at each point of continuity c of f in $[-\pi, \pi]$ to $f(c)$. At other points x it converges to $[f(x^-) + f(x^+)]/2$.*

The hypotheses in this theorem are commonly referred to as the *Dirichlet conditions*.

By linearity, we may extend this last result to functions that are piece-wise the difference of two monotone functions. Such functions are said to be of *bounded variation*, and exceed the scope of the present book. See [KRA2] for a detailed discussion. The book [TIT] discusses convergence of the Fourier series of such functions.

Exercises

1. In Exercises 1, 2, 3, 4 of the last section, sketch the graph of the partial sum S_4 of each Fourier series on the interval $-\pi \leq x \leq \pi$. Also, in each case, sketch the graph of the full sum of the Fourier series. Of course use the theorems in this section to aid you in your work.

2. Find the Fourier series for the periodic function defined by

$$f(x) = \begin{cases} -\pi & \text{if} \quad -\pi \leq x < 0 \\ x & \text{if} \quad 0 \leq x < \pi \end{cases}$$

Sketch the graph of the sum of this series on the interval $-5\pi \leq x \leq 5\pi$ and find what numerical sums are implied by the convergence behavior at the points of discontinuity $x = 0$ and $x = \pi$, etc.

3. **(a)** Show that the Fourier series for the periodic function

$$f(x) = \begin{cases} 0 & \text{if} \quad -\pi \leq x < 0 \\ x^2 & \text{if} \quad 0 \leq x < \pi \end{cases}$$

is

$$f(x) \quad = \quad \frac{\pi^2}{6} + 2\sum_{j=1}^{\infty}(-1)^j \frac{\cos jx}{j^2}$$

$$+ \pi \sum_{j=1}^{\infty}(-1)^{j+1}\frac{\sin jx}{j} - \frac{4}{\pi}\sum_{j=1}^{\infty}\frac{\sin(2j-1)x}{(2j-1)^3} .$$

(b) Sketch the graph of the sum of this series on the interval $-5\pi \leq x \leq 5\pi$.

(c) Use the series in part **(a)** with $x = 0$ and $x = \pi$ to obtain the two sums

$$1 - \frac{1}{2^2} + \frac{1}{3^2} - \frac{1}{4^2} + - \cdots = \frac{\pi^2}{12}$$

and

$$1 + \frac{1}{2^2} + \frac{1}{3^2} + \frac{1}{4^2} + \cdots = \frac{\pi^2}{6} .$$

(d) Derive the second sum in **(c)** from the first. [**Hint:** Add $2\sum_j (1/[2j])^2$ to both sides.]

4. What can you say about the convergence of the Fourier series of the function

$$f(x) = \begin{cases} -1 & \text{if} & x < 0 \\ 0 & \text{if} & x = 0 \\ 1 & \text{if} & x > 0 \end{cases}$$

at the origin?

5. **(a)** Find the Fourier series for the periodic function defined by $f(x) = e^x$, $-\pi \le x \le \pi$. [**Hint:** Recall that $\cosh x = (e^x + e^{-x})/2$.]

    (b) Sketch the graph of the sum of this series on the interval $-5\pi \le x \le 5\pi$.

    (c) Use the series in **(a)** to establish the sums

$$\sum_{j=1}^{\infty} \frac{1}{j^2 + 1} = \frac{1}{2}\left(\frac{\pi}{\tanh \pi} - 1\right)$$

and

$$\sum_{j=1}^{\infty} \frac{(-1)^j}{j^2 + 1} = \frac{1}{2}\left(\frac{\pi}{\sinh \pi} - 1\right).$$

6. It is usually most convenient to study classes of functions that form linear spaces, that is, that are closed under the operations of addition and scalar multiplication. Unfortunately, this linearity condition does not hold for the class of functions defined on the interval $[-\pi, \pi]$ by the Dirichlet conditions. Verify this statement by examining the functions

$$f(x) = \begin{cases} x^2 \sin \frac{1}{x} + 2x & \text{if} & x \ne 0 \\ 0 & \text{if} & x = 0 \end{cases}$$

and

$$g(x) = -2x.$$

7. If f is defined on the interval $[-\pi, \pi]$ and satisfies the Dirichlet conditions there, then prove that $f(x^-) \equiv \lim_{\substack{t \to x \\ t < x}} f(t)$ and $f(x^+) \equiv \lim_{\substack{t \to x \\ t > x}} f(t)$ exist at every interior point, and also that $f(x^+)$ exists at the left endpoint and $f(x^-)$ exists at the right endpoint. [**Hint:** Each interior point of discontinuity is isolated from other such points, in the sense that the function is continuous at all nearby points. Also, on each side of such a point and near enough to it, the function does not oscillate; it is therefore increasing or decreasing.]

6.3   Even and Odd Functions: Cosine and Sine Series

A function f is said to be *even* if $f(-x) = f(x)$. A function g is said to be *odd* if $g(-x) = -g(x)$.

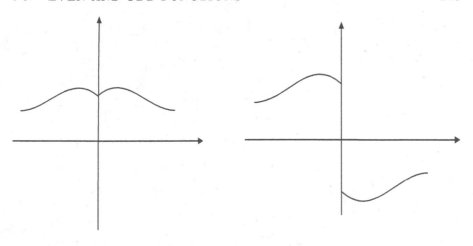

FIGURE 6.12
An even and an odd function.

EXAMPLE 6.3.1 The function $f(x) = \cos x$ is even because $\cos(-x) = \cos x$. The function $g(x) = \sin x$ is odd because $\sin(-x) = -\sin x$.   ■

The graph of an even function is symmetric about the y-axis. The graph of an odd function is skew-symmetric about the y-axis. Refer to Figure 6.12.
    If f is even on the interval $[-a, a]$ then

$$\int_{-a}^{a} f(x) \, dx = 2 \int_{0}^{a} f(x) \, dx \qquad (6.3.1)$$

and if f is odd on the interval $[-a, a]$ then

$$\int_{-a}^{a} f(x) \, dx = 0 . \qquad (6.3.2)$$

Finally, we have the following parity relations

$$(\text{even}) \cdot (\text{even}) = (\text{even}) \qquad (\text{even}) \cdot (\text{odd}) = (\text{odd})$$

$$(\text{odd}) \cdot (\text{odd}) = (\text{even}) .$$

Now suppose that f is an even function on the interval $[-\pi, \pi]$. Then $f(x) \cdot \sin nx$ is odd, and therefore

$$b_n = \frac{1}{\pi} \int_{-\pi}^{\pi} f(x) \sin nx \, dx = 0 .$$

For the cosine coefficients, we have

$$a_n = \frac{1}{\pi} \int_{-\pi}^{\pi} f(x) \cos nx \, dx = \frac{2}{\pi} \int_{0}^{\pi} f(x) \cos nx \, dx .$$

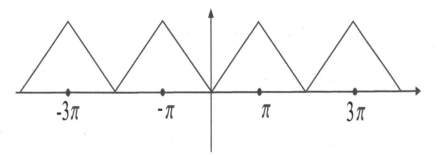

FIGURE 6.13
Periodic extension of $f(x) = |x|$.

Thus the Fourier series for an even function contains only cosine terms.

By the same token, suppose now that f is an odd function on the interval $[-\pi, \pi]$. Then $f(x) \cdot \cos nx$ is an odd function, and therefore

$$a_n = \frac{1}{\pi} \int_{-\pi}^{\pi} f(x) \cos nx \, dx = 0 \, .$$

For the sine coefficients, we have

$$b_n = \frac{1}{\pi} \int_{-\pi}^{\pi} f(x) \sin nx \, dx = \frac{2}{\pi} \int_{0}^{\pi} f(x) \sin nx \, dx \, .$$

Thus the Fourier series for an odd function contains only sine terms.

EXAMPLE 6.3.2 Examine the Fourier series of the function $f(x) = x$ from the point of view of even/odd.

Solution:
The function is odd, so the Fourier series must be a sine series. We calculated in Example 6.1.1 that the Fourier series is in fact

$$x = f(x) = 2\left(\sin x - \frac{\sin 2x}{2} + \frac{\sin 3x}{3} - + \cdots\right). \qquad (6.3.2.1)$$

The expansion is valid on $(-\pi, \pi)$, but not at the endpoints (since the series of course sums to 0 at $-\pi$ and π). ∎

EXAMPLE 6.3.3 Examine the Fourier series of the function $f(x) = |x|$ from the point of view of even/odd.

Solution: The function is even, so the Fourier series must be a cosine series. In fact we see that

$$a_0 = \frac{1}{\pi} \int_{-\pi}^{\pi} |x| \, dx = \frac{2}{\pi} \int_{0}^{\pi} x \, dx = \pi \, .$$

Also, for $n \geq 1$,

$$a_n = \frac{2}{\pi} \int_0^\pi |x| \cos nx \, dx = \frac{2}{\pi} \int_0^\pi x \cos nx \, dx.$$

An integration by parts gives that

$$a_n = \frac{2}{\pi n^2}(\cos n\pi - 1) = \frac{2}{\pi n^2}[(-1)^n - 1].$$

As a result,

$$a_{2j} = 0 \quad \text{and} \quad a_{2j-1} = -\frac{4}{\pi(2j-1)^2} \;, \quad j = 1, 2, \dots.$$

In conclusion,

$$|x| = \frac{\pi}{2} - \frac{4}{\pi}\left(\cos x + \frac{\cos 3x}{3^2} + \frac{\cos 5x}{5^2} + \cdots\right). \tag{6.3.3.1}$$

The periodic extension of the original function $f(x) = |x|$ on $[-\pi, \pi]$ is depicted in Figure 6.13. By Theorem 6.2.7 (see also Theorem 6.2.2), the series converges to f at every point of $[-\pi, \pi]$. ■

It is worth noting that $x = |x|$ on $[0, \pi]$. Thus the expansions (6.3.2.1) and (6.3.3.1) represent the same function on that interval. Of course (6.3.2.1) is the *Fourier sine series* for x on $[0, \pi]$ while (6.3.3.1) is the *Fourier cosine series* for x on $[0, \pi]$. More generally, if g is *any* integrable function on $[0, \pi]$, we may take its odd extension $\tilde{g}$ to $[-\pi, \pi]$ and calculate the Fourier series. The result will be the Fourier sine series expansion for g on $[0, \pi]$. Instead we could take the even extension $\tilde{\tilde{g}}$ to $[-\pi, \pi]$ and calculate the Fourier series. The result will be the Fourier cosine series expansion for g on $[0, \pi]$.

EXAMPLE 6.3.4 Find the Fourier sine series and the Fourier cosine series expansions for the function $f(x) = \cos x$ on the interval $[0, \pi]$.

Solution:
Of course the Fourier series expansion of the odd extension $\tilde{f}$ contains only sine terms. Its coefficients will be

$$b_n = \frac{2}{\pi} \int_0^\pi \cos x \sin nx \, dx = \begin{cases} 0 & \text{if} \quad n = 1 \\ \frac{2n}{\pi}\left(\frac{1+(-1)^n}{n^2-1}\right) & \text{if} \quad n > 1. \end{cases}$$

As a result,

$$b_{2j-1} = 0 \quad \text{and} \quad b_{2j} = \frac{8j}{\pi(4j^2 - 1)} \;, \quad j = 1, 2, \dots.$$

The sine series for f is therefore

$$\cos x = f(x) = \frac{8}{\pi} \sum_{j=1}^{\infty} \frac{j \sin 2jx}{4j^2 - 1} \; , \quad 0 < x < \pi .$$

To obtain the cosine series for f, we consider the even extension $\widetilde{\widetilde{f}}$. Of course all the b_n will vanish. Also

$$a_n = \frac{2}{\pi} \int_0^{\pi} \cos x \cos nx \, dx = \begin{cases} 1 & \text{if} \quad n = 1 \\ 0 & \text{if} \quad n \neq 1 . \end{cases}$$

We therefore see, not surprisingly, that the Fourier cosine series for cosine on $[0, \pi]$ is the single summand $\cos x$.   ∎

Exercises

1. Determine whether each of the following functions is even, odd, or neither.

$$x^5 \sin x, \; x^2 \sin 2x, \; e^x, \; (\sin x)^3, \; \sin x^2, \; \cos(x + x^3),$$

$$x + x^2 + x^3, \; \ln \frac{1+x}{1-x} .$$

2. Show that any function f defined on a symmetrically placed interval can be written as the sum of an even function and an odd function. [**Hint:** $f(x) = \frac{1}{2}[f(x) + f(-x)] + \frac{1}{2}[f(x) - f(-x)]$.]

3. Prove properties (6.3.1) and (6.3.2) analytically, by dividing the integral and making a suitable change of variables.

4. Show that the sine series of the constant function $f(x) = \pi/4$ is

$$\frac{\pi}{4} = \sin x + \frac{\sin 3x}{3} + \frac{\sin 5x}{5} + \cdots$$

for $0 < x < \pi$. What sum is obtained by setting $x = \pi/2$? What is the cosine series of this function?

5. Find the Fourier series for the function of period 2π defined by $f(x) = \cos x/2$, $-\pi \leq x \leq \pi$. Sketch the graph of the sum of this series on the interval $-5\pi \leq x \leq 5\pi$.

6. Find the sine and the cosine series for $f(x) = \sin x$.

7. Find the Fourier series for the 2π-periodic function defined on its fundamental period $[-\pi, \pi]$ by

$$f(x) = \begin{cases} x + \frac{\pi}{2} & \text{if} \quad -\pi \leq x < 0 \\ -x + \frac{\pi}{2} & \text{if} \quad 0 \leq x \leq \pi \end{cases}$$

(a) by computing the Fourier coefficients directly;

(b) using the formula

$$|x| = \frac{\pi}{2} - \frac{4}{\pi}\left(\cos x + \frac{\cos 3x}{3^2} + \frac{\cos 5x}{5^2} + \cdots\right)$$

from the text.

Sketch the graph of the sum of this series (a triangular wave) on the interval $-5\pi \le x \le 5\pi$.

8. For the function $f(x) = \pi - x$, find

   (a) its Fourier series on the interval $-\pi < x < \pi$;

   (b) its cosine series on the interval $0 \le x \le \pi$;

   (c) its sine series on the interval $0 < x \le \pi$.

   Sketch the graph of the sum of each of these series on the interval $[-5\pi, 5\pi]$.

9. Let

$$f(x) = \begin{cases} x & \text{if} & 0 \le x \le \pi/2 \\ \pi - x & \text{if} & \pi/2 < x \le \pi. \end{cases}$$

   Show that the cosine series for this function is

$$f(x) = \frac{\pi}{4} - \frac{2}{\pi}\sum_{j=1}^{\infty} \frac{\cos 2(2j-1)x}{(2j-1)^12}.$$

   Sketch the graph of the sum of this series on the interval $[-5\pi, 5\pi]$.

10. (a) Show that the cosine series for x^2 is

$$x^2 = \frac{\pi^2}{3} + 4\sum_{j=1}^{\infty}(-1)^j\frac{\cos jx}{j^2}, \qquad -\pi \le x \le \pi.$$

   (b) Find the sine series for x^2 and use this expansion together with the formula (6.3.2.1) to obtain the sum

$$1 - \frac{1}{3^3} + \frac{1}{5^3} - \frac{1}{7^3} + - \cdots = \frac{\pi^3}{32}.$$

   (c) Denote by s the sum of the reciprocals of the cubes of the odd positive integers:

$$s = \frac{1}{1^3} + \frac{1}{3^3} + \frac{1}{5^3} + \frac{1}{7^3} + \cdots,$$

   and show that then

$$\sum_{j=1}^{\infty}\frac{1}{j^3} = \frac{1}{1^3} + \frac{1}{2^3} + \frac{1}{3^3} + \frac{1}{4^3} + \cdots = \frac{8}{7}\cdot s.$$

   The exact numerical value of this last sum has been a matter of great interest since Euler first raised the question in 1736. It is closely related to the Riemann hypothesis. Roger Apéry proved, by an extremely ingenious argument in 1978, that s is irrational.[2]

---

[2] The Riemann hypothesis is perhaps the most celebrated open problem in modern mathematics. Originally formulated as a question about the zero set of a complex analytic function, this question has profound implications for number theory and other branches of mathematics. The recent books [DER], [SAB] discuss the history and substance of the problem.

11. (a) Show that the cosine series for x^3 is

$$x^3 = \frac{\pi^3}{4} + 6\pi \sum_{j=1}^{\infty}(-1)^j \frac{\cos jx}{j^2} + \frac{24}{\pi} \sum_{j=1}^{\infty} \frac{\cos(2j-1)x}{(2j-1)^4}, \quad 0 \le x \le \pi.$$

(b) Use the series in **(a)** to obtain

$$\textbf{(i)} \ \sum_{j=1}^{\infty} \frac{1}{(2j-1)^4} = \frac{\pi^4}{96} \quad \text{and} \quad \textbf{(ii)} \ \sum_{j=1}^{\infty} \frac{1}{j^4} = \frac{\pi^4}{90}.$$

12. (a) Show that the cosine series for x^4 is

$$x^4 = \frac{\pi^4}{5} + 8 \sum_{j=1}^{\infty}(-1)^j \frac{\pi^2 j^2 - 6}{j^4} \cos jx, \quad -\pi \le x \le \pi.$$

(b) Use the series in **(a)** to find a new derivation of the second sum in Exercise 11(b).

13. The functions $\sin^2 x$ and $\cos^2 x$ are both even. Show, without using any calculations, that the identities

$$\sin^2 x = \frac{1}{2}(1 - \cos 2x) = \frac{1}{2} - \frac{1}{2}\cos 2x$$

and

$$\cos^2 x = \frac{1}{2}(1 + \cos 2x) = \frac{1}{2} + \frac{1}{2}\cos 2x$$

are actually the Fourier series expansions of these functions.

14. Find the sine series of the functions in Exercise 13, and verify that these expansions satisfy the identity $\sin^2 x + \cos^2 x = 1$.

15. Prove the trigonometric identities

$$\sin^3 x = \frac{3}{4}\sin x - \frac{1}{4}\sin 3x \quad \text{and} \quad \cos^3 x = \frac{3}{4}x + \frac{1}{4}\cos 3x$$

and show briefly, without calculation, that these are the Fourier series expansions of the functions $\sin^3 x$ and $\cos^3 x$.

6.4   Fourier Series on Arbitrary Intervals

We have developed Fourier analysis on the interval $[-\pi, \pi]$ (resp. the interval $[0, \pi]$) just because it is notationally convenient. In particular,

$$\int_{-\pi}^{\pi} \cos jx \cos kx \, dx = 0 \quad \text{for } j \ne k,$$

$$\int_{-\pi}^{\pi} \sin jx \sin kx \, dx = 0 \quad \text{for } j \ne k,$$

$$\int_{-\pi}^{\pi} \sin jx \cos kx \, dx = 0 \quad \text{for all } j, k \,,$$

and so forth. This fact is special to the interval of length 2π. But many physical problems take place on an interval of some other length. We must therefore be able to adapt our analysis to intervals of any length. This amounts to a straightforward change of scale on the horizontal axis. We treat the matter in the present section.

Now we concentrate our attention on an interval of the form $[-L, L]$. As x runs from $-L$ to L, we shall have a corresponding variable t that runs from $-\pi$ to π. We mediate between these two variables using the formulas

$$t = \frac{\pi x}{L} \qquad \text{and} \qquad x = \frac{Lt}{\pi} \,.$$

Thus the function $f(x)$ on $[-L, L]$ is transformed to a new function $\tilde{f}(t) \equiv f(Lt/\pi)$ on $[-\pi, \pi]$.

If f satisfies the conditions for convergence of the Fourier series, then so will $\tilde{f}$, and vice versa. Thus we may consider the Fourier expansion

$$\tilde{f}(t) = \frac{1}{2}a_0 + \sum_{n=1}^{\infty} \left(a_n \cos nt + b_n \sin nt \right).$$

Here, of course,

$$a_n = \frac{1}{\pi} \int_{-\pi}^{\pi} \tilde{f}(t) \cos nt \, dt \quad \text{and} \quad b_n = \frac{1}{\pi} \int_{-\pi}^{\pi} \tilde{f}(t) \sin nt \, dt \,.$$

Now let us write out these last two formulas and perform the change of variables $x = Lt/\pi$. We find that

$$
\begin{aligned}
a_n &= \frac{1}{\pi} \int_{-\pi}^{\pi} f(Lt/\pi) \cos nt \, dt \\
&= \frac{1}{\pi} \int_{-L}^{L} f(x) \cos \frac{n\pi x}{L} \cdot \frac{\pi}{L} \, dx \\
&= \frac{1}{L} \int_{-L}^{L} f(x) \cos \frac{n\pi x}{L} \, dx \,.
\end{aligned}
$$

Likewise,

$$b_n = \frac{1}{L} \int_{-L}^{L} f(x) \sin \frac{n\pi x}{L} \, dx \,.$$

EXAMPLE 6.4.1 Calculate the Fourier series on the interval $[-2, 2]$ of the function

$$f(x) = \begin{cases} 0 & \text{if} \quad -2 \leq x < 0 \\ 1 & \text{if} \quad 0 \leq x \leq 2. \end{cases}$$

Solution:

Of course $L = 2$ so we calculate that

$$a_n = \frac{1}{2} \int_0^2 \cos \frac{n\pi x}{2} \, dx = \begin{cases} 1 & \text{if} & n = 0 \\ 0 & \text{if} & n \geq 1. \end{cases}$$

Also

$$b_n = \frac{1}{2} \int_0^2 \sin \frac{n\pi x}{2} \, dx = \frac{1}{n\pi}[(-1)^n - 1].$$

This may be rewritten as

$$b_{2j} = 0 \quad \text{and} \quad b_{2j-1} = \frac{-2}{(2j-1)\pi} \;,\quad j = 1, 2, \dots.$$

In conclusion,

$$f(x) = \widetilde{f}(t) = \frac{1}{2}a_0 + \sum_{n=1}^{\infty}\left(a_n \cos nt + b_n \sin nt\right)$$

$$= \frac{1}{2} + \sum_{j=1}^{\infty} \frac{-2}{(2j-1)\pi} \sin\left((2j-1)\cdot\frac{\pi x}{2}\right). \qquad \blacksquare$$

EXAMPLE 6.4.2 Calculate the Fourier series of the function $f(x) = \cos x$ on the interval $[-\pi/2, \pi/2]$.

Solution:

We calculate that

$$a_0 = \frac{2}{\pi} \int_{-\pi/2}^{\pi/2} \cos x \, dx = \frac{4}{\pi}.$$

Also, for $n \geq 1$,

$$\begin{aligned} a_n &= \frac{2}{\pi} \int_{-\pi/2}^{\pi/2} \cos x \cos(2nx)\, dx \\ &= \frac{2}{\pi} \int_{-\pi/2}^{\pi/2} \frac{1}{2}\left(\cos(2n+1)x + \cos(2n-1)x\right) dx \\ &= \frac{1}{\pi}\left(\frac{\sin(2n+1)x}{2n+1} + \frac{\sin(2n-1)x}{2n-1}\right)\Bigg|_{-\pi/2}^{\pi/2} \\ &= \begin{cases} \dfrac{2}{\pi}\left(\dfrac{-1}{2n+1} + \dfrac{1}{2n-1}\right) = \dfrac{4}{\pi(4n^2-1)} & \text{if } n \text{ is odd}, \\[3mm] \dfrac{2}{\pi}\left(\dfrac{1}{2n+1} + \dfrac{-1}{2n-1}\right) = \dfrac{-4}{\pi(4n^2-1)} & \text{if } n \text{ is even}. \end{cases} \end{aligned}$$

A similar calculation shows that

$$
\begin{aligned}
b_n &= \frac{1}{\pi} \int_{-\pi/2}^{\pi/2} \cos x \sin 2nx \, dx \\
&= \frac{2}{\pi} \int_{-\pi/2}^{\pi/2} \frac{1}{2}\left(\sin(2n+1)x + \sin(2n-1)x\right) dx \\
&= \frac{1}{\pi}\left(\frac{-\cos(2n+1)x}{2n+1} + \frac{-\cos(2n-1)x}{2n-1}\right)\Big|_{-\pi/2}^{\pi/2} \\
&= 0.
\end{aligned}
$$

This last comes as no surprise since the cosine function is even.

As a result, the Fourier series expansion for $\cos x$ on the interval $[-\pi/2, \pi/2]$ is

$$
\begin{aligned}
\cos x &= f(x) \\
&= \frac{2}{\pi} + \sum_{m=1}^{\infty} \frac{-4}{\pi(4(2m)^2 - 1)} \cos \frac{2m\pi x}{\pi/2} \\
&\quad + \sum_{k=1}^{\infty} \frac{4}{\pi(4(2k-1)^2 - 1)} \cos \frac{2(2k-1)\pi x}{\pi/2}. \quad \blacksquare
\end{aligned}
$$

Exercises

1. Calculate the Fourier series for the given function on the given interval.
   - (a) $f(x) = x$, $[-1, 1]$
   - (b) $g(x) = \sin x$, $[-2, 2]$
   - (c) $h(x) = e^x$, $[-3, 3]$
   - (d) $f(x) = x^2$, $[-1, 1]$
   - (e) $g(x) = \cos 2x$, $[-\pi/3, \pi/3]$
   - (f) $h(x) = \sin(2x - \pi/3)$, $[-1, 1]$

2. For the functions
   $$f(x) \equiv -3, \quad -2 \leq x < 0$$
   and
   $$g(x) \equiv 3, \quad 0 \leq x < 2,$$
   write down the Fourier expansion directly from Example 6.4.1 in the text—without any calculation.

3. Find the Fourier series for these functions.

(a)
$$f(x) = \begin{cases} 1+x & \text{if} & -1 \le x < 0 \\ 1-x & \text{if} & 0 \le x \le 1. \end{cases}$$

(b)
$$f(x) = |x|, \qquad -2 \le x \le 2.$$

4. Show that
$$\frac{L}{2} - x = \frac{L}{\pi} \sum_{j=1}^{\infty} \frac{1}{j} \sin \frac{2j\pi x}{L}, \qquad 0 < x < L.$$

5. Find the cosine series for the function defined on the interval $0 \le x \le 1$ by $f(x) = x^2 - x + 1/6$. This is a special instance of the Bernoulli polynomials.

6. Find the cosine series for the function defined by
$$f(x) = \begin{cases} 2 & \text{if} & 0 \le x \le 1 \\ 0 & \text{if} & 1 < x \le 2. \end{cases}$$

7. Expand $f(x) = \cos \pi x$ in a Fourier series on the interval $-1 \le x \le 1$.

8. Find the cosine series for the function defined by
$$f(x) = \begin{cases} \frac{1}{4} - x & \text{if} & 0 \le x < \frac{1}{2} \\ x - \frac{3}{4} & \text{if} & \frac{1}{2} \le x \le 1. \end{cases}$$

6.5   Orthogonal Functions

In the classical Euclidean geometry of 3-space, just as we learn in multivariable calculus class, one of the key ideas is that of orthogonality. Let us briefly review it now.

If $\mathbf{v} = \langle v_1, v_2, v_3 \rangle$ and $\mathbf{w} = \langle w_1, w_2, w_3 \rangle$ are vectors in $\mathbb{R}^3$ then we define their *dot product*, or *inner product*, or *scalar product* to be

$$\mathbf{v} \cdot \mathbf{w} = v_1 w_1 + v_2 w_2 + v_3 w_3.$$

What is the interest of the inner product? There are three answers:

- Two vectors are perpendicular or *orthogonal*, written $\mathbf{v} \perp \mathbf{w}$, if and only if $\mathbf{v} \cdot \mathbf{w} = 0$.

- The *length* of a vector is given by

$$\|\mathbf{v}\| = \sqrt{\mathbf{v} \cdot \mathbf{v}}.$$

- The *angle* θ between two vectors $\mathbf{v}$ and $\mathbf{w}$ is given by

$$\cos \theta = \frac{\mathbf{v} \cdot \mathbf{w}}{\|\mathbf{v}\| \|\mathbf{w}\|}.$$

In fact all of the geometry of 3-space is built on these three facts.

One of the great ideas of twentieth-century mathematics is that many other spaces—sometimes abstract spaces, and sometimes infinite-dimensional spaces—can be equipped with an inner product that endows that space with a useful geometry. That is the idea that we shall explore in this section.

Let X be a vector space. This means that X is equipped with (i) a notion of addition and (ii) a notion of scalar multiplication. These two operations are hypothesized to satisfy the expected properties: addition is commutative and associative, scalar multiplication is commutative, associative, and distributive, and so forth. We say that X is equipped with an *inner product* (which we now denote by $\langle \ , \ \rangle$) if there is a binary operation

$$\langle \, \bullet \, , \, \bullet \, \rangle : X \times X \to \mathbb{R}$$

satisfying the following properties for $\mathbf{u}, \mathbf{v}, \mathbf{w} \in X$ and $c \in \mathbb{R}$:

(a) $\langle \mathbf{v}, \mathbf{v} \rangle \geq 0$;

(b) $\langle \mathbf{v}, \mathbf{w} \rangle = \langle \mathbf{w}, \mathbf{v} \rangle$

(c) $\langle \mathbf{v}, \mathbf{v} \rangle = 0$ if and only if $\mathbf{v} = 0$;

(d) $\langle \alpha \mathbf{v} + \beta \mathbf{w}, \mathbf{u} \rangle = \alpha \langle \mathbf{v}, \mathbf{u} \rangle + \beta \langle \mathbf{w}, \mathbf{u} \rangle$ for any vectors $\mathbf{u}, \mathbf{v}, \mathbf{w} \in V$ and scalars α, β.

We shall give some interesting examples of inner products below. Before we do, let us note that an inner product as just defined gives rise to a notion of length, or a *norm*. Namely, we define

$$\|\mathbf{v}\| = \sqrt{\langle \mathbf{v}, \mathbf{v} \rangle}.$$

By Properties (a) and (c), we see that $\|\mathbf{v}\| \geq 0$ and $\|\mathbf{v}\| = 0$ if and only if $\mathbf{v} = 0$.

In fact the two key properties of the inner product and the norm are enunciated in the following proposition:

Proposition 6.5.1 *Let X be a vector space and $\langle\,,\,\rangle$ an inner product on that space. Let $\|\ \ \|$ be the induced norm. Then*

(1) The Cauchy–Schwarz–Buniakovski Inequality: *If $u, v \in X$ then*
$$|u \cdot v| \le \|u\| \cdot \|v\|.$$

(2) The Triangle Inequality: *If $u, v \in X$ then*
$$\|u + v\| \le \|u\| + \|v\|.$$

In fact, just as an exercise, we shall derive the Triangle Inequality from the Cauchy–Schwarz–Buniakovski Inequality. We have

$$
\begin{aligned}
\|u + v\|^2 &= \langle (u + v), (u + v) \rangle \\
&= \langle u, u \rangle + \langle u, v \rangle + \langle v, u \rangle + \langle v, v \rangle \\
&= \|u\|^2 + \|v\|^2 + 2\langle u, v \rangle \\
&\le \|u\|^2 + \|v\|^2 + 2\|u\| \cdot \|v\| \\
&= (\|u\| + \|v\|)^2.
\end{aligned}
$$

Now taking the square root of both sides completes the argument. We shall explore the proof of the Cauchy–Schwarz–Buniakovski Inequality in Exercise 5.

EXAMPLE 6.5.2 Let $X = C[0, 1]$, the continuous functions on the interval $[0, 1]$. This is certainly a vector space with the usual notions of addition of functions and scalar multiplication of functions. We define an inner product by

$$\langle f, g \rangle = \int_0^1 f(x) g(x)\, dx$$

for any $f, g \in X$.

Then it is straightforward to verify that this definition of inner product satisfies all our axioms. Thus we may *define* two functions to be orthogonal if

$$\langle f, g \rangle = 0.$$

We say that the angle θ between two functions is given by

$$\cos \theta = \frac{\langle f, g \rangle}{\|f\|\|g\|}.$$

The *length* or *norm* of an element $f \in X$ is given by

$$\|f\| = \sqrt{\langle f, f \rangle} = \left(\int_0^1 f(x)^2\, dx \right)^{1/2}. \qquad \blacksquare$$

EXAMPLE 6.5.3 Let X be the space of all real sequences $\{a_j\}_{j=1}^{\infty}$ with the property that $\sum_{j=1}^{\infty} |a_j|^2 < \infty$. This is a vector space with the obvious notions of addition and scalar multiplication:

$$\{a_j\} + \{b_j\} = \{a_j + b_j\}$$

and

$$c\{a_j\} = \{ca_j\}.$$

Define an inner product by

$$\langle \{a_j\}, \{b_j\} \rangle = \sum_{j=1}^{\infty} a_j b_j.$$

Then this inner product satisfies all our axioms. ∎

For the purposes of studying Fourier series, the most important inner product space is that which we call $L^2[-\pi, \pi]$. This is the space of real functions f on the interval $[-\pi, \pi]$ with the property that

$$\int_{-\pi}^{\pi} f(x)^2 \, dx < \infty.$$

The inner product on this space is

$$\langle f, g \rangle = \int_{-\pi}^{\pi} f(x)g(x) \, dx.$$

One must note here that, by a variant of the Cauchy–Schwarz–Buniakovski inequality, it holds that if $f, g \in L^2$ then the integral $\int f \cdot g \, dx$ exists and is finite. So our inner product makes sense.

Exercises

1. Verify that each pair of functions f, g is orthogonal on the given interval $[a, b]$ using the inner product

$$\langle f, g \rangle = \int_a^b f(x)g(x) \, dx.$$

(a) $f(x) = \sin 2x$, $g(x) = \cos 3x$, $[-\pi, \pi]$
(b) $f(x) = \sin 2x$, $g(x) = \sin 4x$, $[0, \pi]$

(c) $f(x) = x^2$, $g(x) = x^3$, $[-1, 1]$

(d) $f(x) = x$, $g(x) = \cos 2x$, $[-2, 2]$

2. Prove the so-called *parallelogram law* in the space L^2:

$$2\|f\|^2 + 2\|g\|^2 = \|f + g\|^2 + \|f - g\|^2.$$

[**Hint:** Expand the right-hand side.]

3. Prove the *Pythagorean theorem* and its converse in L^2: The function f is orthogonal to the function g if and only if

$$\|f - g\|^2 = \|f\|^2 + \|g\|^2.$$

4. In the space L^2, show that if f is continuous and $\|f\| = 0$ then $f(x) = 0$ for all x.

5. Prove the Cauchy–Schwarz–Buniakovski Inequality in L^2: If $f, g \in L^2$ then

$$|\langle f, g \rangle| \le \|f\| \cdot \|g\|.$$

Do this by considering the auxiliary function

$$\varphi(\lambda) = \|f + \lambda g\|^2$$

and calculating the value of λ for which it is a minimum.

6. *Bessel's inequality* states that, if f is any square-integrable function on $[-\pi, \pi]$ (i.e., $f \in L^2$), then its Fourier coefficients a_j and b_j satisfy

$$\frac{1}{2}a_0^2 + \sum_{j=1}^{\infty}(a_j^2 + b_j^2) \le \frac{1}{\pi}\int_{-\pi}^{\pi} |f(x)|^2\, dx.$$

This inequality is fundamental in the theory of Fourier series.

(a) For any $n \ge 1$, define

$$s_n(x) = \frac{1}{2}a_0 + \sum_{j=1}^{n}(a_j \cos jx + b_j \sin jx)$$

and show that

$$\frac{1}{\pi}\int_{-\pi}^{\pi} f(x)s_n(x)\, dx = \frac{1}{2}a_0^2 + \sum_{j=1}^{\infty}(a_j^2 + b_j^2).$$

(b) By considering all possible products in the multiplication of $s_n(x)$ by itself, show that

$$\frac{1}{\pi}\int_{-\pi}^{\pi} |s_n(x)|^2\, dx = \frac{1}{2}a_0^2 + \sum_{k=1}^{n}(a_j^2 + b_j^2).$$

(c) By writing

$$\frac{1}{\pi}\int_{-\pi}^{\pi} |f(x) - s_n(x)|^2\, dx$$

$$= \frac{1}{\pi}\int_{-\pi}^{\pi} |f(x)|^2\, dx - \frac{2}{\pi}\int_{-\pi}^{\pi} f(x)s_n(x)\, dx + \frac{1}{\pi}\int_{-\pi}^{\pi} |s_n(x)|^2\, dx$$

$$= \frac{1}{\pi}\int_{-\pi}^{\pi} |f(x)|^2\, dx - \frac{1}{2}a_0^2 - \sum_{j=1}^{n}(a_j^2 + b_j^2),$$

conclude that

$$\frac{1}{2}a_0^2 + \sum_{j=1}^{n}(a_j^2 + b_j^2) \leq \frac{1}{\pi}\int_{-\pi}^{\pi}|f(x)|^2\,dx\,.$$

(d) Now complete the proof of Bessel's inequality.

7. Use your symbol manipulation software, such as `Maple` or `Mathematica`, to implement the Gram–Schmidt procedure to orthonormalize a given finite family of functions on the unit interval. Apply your software routine to the family $1, x, x^2, \cdots, x^{10}$.

Historical Note
Riemann

Bernhard Riemann (1826–1866) was the son of a poor country minister in northern Germany. He studied the works of Euler and Legendre while still in high school; indeed, it is said that he mastered Legendre's treatise on number theory in less than a week. Riemann was shy and modest, with little awareness of his own extraordinary powers; thus, at age nineteen, he went to the University of Göttingen with the aim of pleasing his father by studying theology. Riemann soon tired of this curriculum, and with his father's acquiescence he turned to mathematics.

Of course the great Carl Friedrich Gauss was the senior mathematician in Göttingen at the time. Unfortunately, Gauss's austere manner offered little for an apprentice mathematician like Riemann, so he soon moved to Berlin. There he fell in with Dirichlet and Jacobi, and he learned a great deal from both. Two years later he returned to Göttingen and earned his doctorate. During the next eight years Riemann suffered debilitating poverty and also produced his greatest scientific work. Unfortunately his health was broken. Even after Gauss's death, when Dirichlet took the helm of the Göttingen math institute and did everything in his power to help and advance Riemann, the young man's spirits and health were well in decline. At the age of 39 he died of tuberculosis in Italy, where he had traveled several times to escape the cold and wet of northern Germany.

Riemann made profound contributions to the theory of complex variables. The Cauchy–Riemann equations, the Riemann mapping theorem, Riemann surfaces, the Riemann–Roch theorem, and the Riemann hypothesis all bear his name. Incidentally, these areas are all studied intensely today.

Riemann's theory of the integral, and his accompanying ideas on Fourier series, have made an indelible impression on calculus and real analysis.

At one point in his career, Riemann was required to present a probationary lecture before the great Gauss. In this offering, Riemann developed a theory of geometry that unified and far generalized all existing geometric theories. This is of course the theory of Riemannian manifolds, perhaps the most important idea in modern geometry. Certainly Riemannian curvature plays a major role in mathematical physics, in partial differential equations, and in many other parts of the subject.

Riemann's single paper on number theory (published in 1859), just ten pages, is about the prime number theorem. In it, he develops the so-called Riemann zeta function and formulates a number of statements about that deep and important artifact. All of these statements, save one, have by now been proved. The one exception is the celebrated *Riemann hypothesis*, now thought to be perhaps the most central, the most profound, and the most difficult problem in all of mathematics. The question concerns the location of the zeros of the zeta function, and it harbors profound implications for the distribution of primes and for number theory as a whole.

In a fragmentary note found among his posthumous papers, Riemann wrote that he had a proof of the Riemann hypothesis, and that it followed from a formula for the Riemann zeta function which he had not simplified enough to publish. To this day, nobody has determined what that formula might be, and so Riemann has left a mathematical legacy that has baffled the greatest minds of our time.

6.6   Introduction to the Fourier Transform

Many problems of mathematics and mathematical physics are set on all of Euclidean space—not on an interval. Thus it is appropriate to have analytical tools designed for that setting. The Fourier transform is one of the most important of these devices. In this section we explore the basic ideas behind the Fourier transform. We shall present the concepts in Euclidean space of any dimension. Throughout, we shall use the standard notation $f \in L^1$ or $f \in L^1(\mathbb{R}^n)$ to mean that f is integrable. We define a norm on L^1 by

$$\|f\|_{L^1} \equiv \int_{\mathbb{R}^n} |f(t)| \, dt \, .$$

If $t, \xi \in \mathbb{R}^n$ then we let

$$t \cdot \xi \equiv t_1 \xi_1 + \cdots + t_N \xi_N.$$

We define the *Fourier transform* of a function $f \in L^1(\mathbb{R}^n)$ by

$$\widehat{f}(\xi) = \int_{\mathbb{R}^n} f(t) e^{it \cdot \xi} \, dt \, .$$

Here dt denotes N-dimensional volume: $dt = dt_1 dt_2 \cdots dt_n$. We sometimes write the Fourier transform as $\mathcal{F}(f)$ or simply $\mathcal{F}f$. It is also occasionally useful, as we shall see below, to write the Fourier transform as

$$(f)^{\widehat{}}.$$

Many references will insert a factor of 2π in the exponential or in the measure. Others will insert a minus sign in the exponent. There is no agreement on this matter. We have opted for this particular definition because of its simplicity.

Proposition   *If $f \in L^1(\mathbb{R}^N)$, then*

$$\sup_{\xi} |\widehat{f}(\xi)| \le \|f\|_{L^1(\mathbb{R}^N)}.$$

Proof: Observe that, for any $\xi \in \mathbb{R}^N$,

$$|\widehat{f}(\xi)| \le \int_{\mathbb{R}^n} |f(t)e^{it\cdot\xi}|\, dt = \int |f(t)|\, dt = \|f\|_{L^1}. \qquad\qquad \square$$

In our development of the ideas concerning the Fourier transform, it is frequently useful to restrict attention to certain "testing functions." We define them now. Let us say that $f \in C_c^k$ if f is k-times continuously differentiable and f is identically zero outside of some ball. Figure 6.14 exhibits such a function.

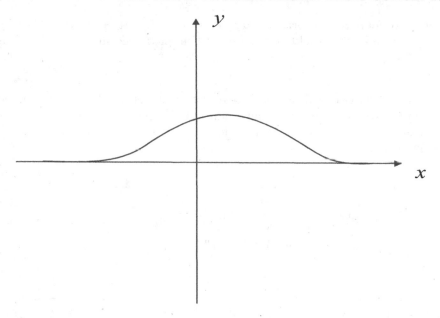

FIGURE 6.14
A C_c^k function.

Proposition   If $f \in L^1(\mathbb{R}^N)$, f is differentiable, and $\partial f / \partial x_j \in L^1(\mathbb{R}^N)$, then

$$\left(\frac{\partial f}{\partial x_j} \right)^{\widehat{}}(\xi) = -i\xi_j \widehat{f}(\xi).$$

Proof: Integrate by parts: if $f \in C_c^1$, then

$$
\begin{aligned}
\left(\frac{\partial f}{\partial x_j} \right)^{\widehat{}}(\xi) &= \int \frac{\partial f}{\partial t_j} e^{it \cdot \xi} \, dt \\
&= \int \cdots \int \left(\int \frac{\partial f}{\partial t_j} e^{it \cdot \xi} \, dt_j \right) dt_1 \ldots dt_{j-1} dt_{j+1} \ldots dt_N \\
&= -\int \cdots \int f(t) \left(\frac{\partial}{\partial t_j} e^{it \cdot \xi} \right) dt_j dt_1 \ldots dt_{j-1} dt_{j+1} \ldots dt_N \\
&= -i\xi_j \int \cdots \int f(t) e^{it \cdot \xi} \, dt \\
&= -i\xi_j \widehat{f}(\xi).
\end{aligned}
$$

(Of course the "boundary terms" in the integration by parts vanish since $f \in C_c^1$.) The general case follows from a limiting argument. $\qquad \square$

Proposition *If $f \in L^1(\mathbb{R}^N)$ and $ix_j f \in L^1(\mathbb{R}^N)$, then*

$$(ix_j f)\widehat{} = \frac{\partial}{\partial \xi_j} \widehat{f}.$$

Proof: Differentiate under the integral sign. $\qquad \square$

Proposition (The Riemann–Lebesgue Lemma) *If $f \in L^1(\mathbb{R}^N)$, then*

$$\lim_{\xi \to \pm\infty} |\widehat{f}(\xi)| = 0.$$

Proof: First assume that $g \in C_c^2(\mathbb{R}^N)$. We know that

$$\|\widehat{g}\|_{L^\infty} \leq \|g\|_{L^1} \leq C$$

and, for each j,

$$\sup_\xi |\xi_j^2 \widehat{g}(\xi)| = \sup \left| \left(\left(\frac{\partial^2}{\partial x_j^2} \right) g \right)^{\widehat{}} \right| \leq \int \left| \left(\frac{\partial^2}{\partial x_j^2} \right) g(x) \right| dx \equiv C_j.$$

Then $(1 + |\xi|^2)\widehat{g}$ is bounded. Therefore

$$|\widehat{g}(\xi)| \leq \frac{C''}{1 + |\xi|^2} \overset{|\xi|\to\infty}{\longrightarrow} 0.$$

This proves the result for $g \in C_c^2$.

Now let $f \in L^1$ be arbitrary. It is easy to see that there is a C_c^2 function ψ such that $\int |f - \psi| \, dx < \epsilon/2$.

Choose M so large that when $|\xi| > M$ then $|\widehat{\psi}(\xi)| < \epsilon/2$. Then, for

$|\xi| > M$, we have

$$
\begin{aligned}
|\widehat{f}(\xi)| &= |(f - \psi)\widehat{}(\xi) + \widehat{\psi}(\xi)| \\
&\leq |(f - \psi)\widehat{}(\xi)| + |\widehat{\psi}(\xi)| \\
&\leq \|f - \psi\|_{L^1} + \frac{\epsilon}{2} \\
&< \frac{\epsilon}{2} + \frac{\epsilon}{2} = \epsilon.
\end{aligned}
$$

This proves the result.   □

REMARK 6.6.1 The Riemann–Lebesgue lemma is intuitively clear when viewed in the following way. Fix an L^1 function f. An L^1 function is well-approximated by a continuous function, so we may as well suppose that f is continuous. But a continuous function is well-approximated by a smooth function, so we may as well suppose that f is smooth. On a small interval I—say of length $1/M$ for M large—a smooth function is nearly constant. So, if we let $|\xi| \gg 2\pi M^2$, then the character $e^{i\xi \cdot x}$ will oscillate at least M times on I, and will therefore integrate against a constant to a value that is very nearly zero. As M becomes larger, this statement becomes more and more accurate. That is the Riemann–Lebesgue lemma.

The three Euclidean groups that act naturally on $\mathbb{R}^N$ are

- rotations

- dilations

- translations

Certainly a large part of the utility of the Fourier transform is that it has natural invariance properties under the actions of these three groups. We shall now explicitly describe those properties.

We begin with the orthogonal group $O(N)$; an $N \times N$ matrix is *orthogonal* if it has real entries and its rows form an orthonormal system of vectors. A *rotation* is an orthogonal matrix with determinant 1 (also called a *special orthogonal* matrix).

Proposition   Let ρ be a rotation of $\mathbb{R}^N$. We define $\rho f(x) = f(\rho(x))$. Then we have the formula

$$
\widehat{\rho f} = \rho \widehat{f}.
$$

Proof: Remembering that ρ is orthogonal and has determinant 1, we calculate that

$$\widehat{\rho f}(\xi) = \int (\rho f)(t) e^{it \cdot \xi} \, dt = \int f(\rho(t)) e^{it \cdot \xi} \, dt$$

$$\stackrel{(s = \rho(t))}{=} \int f(s) e^{i \rho^{-1}(s) \cdot \xi} \, ds = \int f(s) e^{is \cdot \rho(\xi)} \, ds$$

$$= \widehat{f}(\rho \xi) = \rho \widehat{f}(\xi).$$

Here we have used the fact that $\rho^{-1} = {}^t\rho$ for an orthogonal matrix. The proof is complete. □

Definition   For $\delta > 0$ and $f \in L^1(\mathbb{R}^N)$ we set $\alpha_\delta f(x) = f(\delta x)$ and $\alpha^\delta f(x) = \delta^{-N} f(x/\delta)$. These are the dual *dilation operators* of Euclidean analysis.

Proposition   *The dilation operators interact with the Fourier transform as follows:*

$$(\alpha_\delta f)\widehat{\,} = \alpha^\delta \left(\widehat{f} \right)$$

$$\widehat{\alpha^\delta f} = \alpha_\delta \left(\widehat{f} \right).$$

Proof: We calculate that

$$(\alpha_\delta f)\widehat{\,}(\xi) = \int (\alpha_\delta f)(t) e^{it \cdot \xi} \, dt$$

$$= \int f(\delta t) e^{it \cdot \xi} \, dt$$

$$\stackrel{(s = \delta t)}{=} \int f(s) e^{i(s/\delta) \cdot \xi} \delta^{-N} \, ds$$

$$= \delta^{-N} \widehat{f}(\xi/\delta)$$

$$= \left(\alpha^\delta (\widehat{f}) \right)(\xi).$$

That proves the first assertion. The proof of the second is similar. □

For any function f on $\mathbb{R}^N$ and $a \in \mathbb{R}^N$ we define $\tau_a f(x) = f(x - a)$. Clearly τ_a is a *translation operator*.

Proposition   If $f \in L^1(\mathbb{R}^N)$ then

$$\widehat{\tau_a f}(\xi) = e^{ia \cdot \xi} \widehat{f}(\xi).$$

and

$$\left(\tau_a\{\widehat{f}\}\right)(\xi) = \left[e^{-ia \cdot t} f(t)\right]\widehat{\,}(\xi).$$

Proof: For the first equality, we calculate that

$$
\begin{aligned}
\widehat{\tau_a f}(\xi) \quad &= \quad \int_{\mathbb{R}^N} e^{ix \cdot \xi} (\tau_a f)(x)\, dx \\
&= \quad \int_{\mathbb{R}^N} e^{ix \cdot \xi} f(x-a)\, dx \\
\overset{(x-a)=t}{=} \quad &\quad \int_{\mathbb{R}^N} e^{i(t+a)\cdot \xi} f(t)\, dt \\
&= \quad e^{ia \cdot \xi} \int_{\mathbb{R}^N} e^{it \cdot \xi} f(t)\, dt \\
&= \quad e^{ia \cdot \xi} \widehat{f}(\xi).
\end{aligned}
$$

The second identity is proved similarly. $\square$

Much of the theory of classical harmonic analysis—especially in this century—concentrates on translation-invariant operators. An operator T on functions is called *translation-invariant*[3] if

$$T(\tau_a f)(x) = (\tau_a T f)(x)$$

for every x. It is a basic fact that any translation-invariant integral T operator is given by convolution with a kernel k:

$$T f(x) = \int_{\mathbb{R}^n} f(t)k(x-t)\, dt.$$

See the next subsection for more on this topic.

---

[3]It is perhaps more accurate to say that such an operator *commutes with translations*. However, the terminology "translation-invariant" is standard.

Proposition For $f \in L^1(\mathbb{R}^N)$ we let $\tilde{f}(x) = f(-x)$. Then
$$\widehat{\tilde{f}} = \tilde{\hat{f}}.$$

Proof: We calculate that

$$\widehat{\tilde{f}}(\xi) = \int \tilde{f}(t)e^{it\cdot\xi}\, dt = \int f(-t)e^{it\cdot\xi}\, dt$$

$$= \int f(t)e^{-it\cdot\xi}\, dt = \hat{f}(-\xi) = \tilde{\hat{f}}(\xi). \qquad \square$$

Proposition We have

$$\widehat{\overline{f}} = \overline{\tilde{\hat{f}}}.$$

Proof: We calculate that

$$\widehat{\overline{f}}(\xi) = \int \overline{f}(t)e^{it\cdot\xi}\, dt = \overline{\int f(t)e^{-it\cdot\xi}\, dt} = \overline{\hat{f}(-\xi)} = \overline{\tilde{\hat{f}}}(\xi). \qquad \square$$

Proposition If $f, g \in L^1$, then

$$\int \hat{f}(\xi)g(\xi)\, d\xi = \int f(\xi)\hat{g}(\xi)\, d\xi.$$

Proof: This is a straightforward change in the order of integration:

$$\int \hat{f}(\xi)g(\xi)\, d\xi = \int\int f(t)e^{it\cdot\xi}\, dt\, g(\xi)\, d\xi$$

$$= \int\int g(\xi)e^{it\cdot\xi}\, d\xi\, f(t)\, dt$$

$$= \int \hat{g}(t)f(t)\, dt. \qquad \square$$

6.6.1 Convolution and Fourier Inversion

If f and g are integrable functions then we define their *convolution* to be

$$f * g(x) = \int f(x-t)g(t)\,dt = \int f(t)g(x-t)\,dt.$$

Note that a simple change of variable confirms the second equality.

Proposition   If $f, g \in L^1$, then

$$\widehat{f * g} = \widehat{f} \cdot \widehat{g}.$$

Proof: We calculate that

$$\widehat{f * g}(\xi) = \int (f * g)(t)e^{it\cdot\xi}\,dt = \int\int f(t-s)g(s)\,ds\,e^{it\cdot\xi}\,dt$$

$$= \int\int f(t-s)e^{i(t-s)\cdot\xi}\,dt\,g(s)e^{is\cdot\xi}\,ds$$

$$= \int f(t)e^{it\cdot\xi}\,dt \int g(s)e^{is\cdot\xi}\,ds$$

$$= \widehat{f}(\xi) \cdot \widehat{g}(\xi).\qquad\qquad\qquad\square$$

6.6.2 The Inverse Fourier Transform

Our goal is to be able to recover f from $\widehat{f}$. This program entails several technical difficulties. First, we need to know that the Fourier transform is one-to-one in order to have any hope of success. Secondly, we would like to say that

$$f(t) = c \cdot \int \widehat{f}(\xi)e^{-it\cdot\xi}\,d\xi. \tag{6.6.1}$$

But in general the Fourier transform $\widehat{f}$ of an L^1 function f is not integrable (just calculate the Fourier transform of $\chi_{[0,1]}$—the characteristic function of the unit interval)—so the expression on the right of (6.6.1) does not necessarily make any sense.

Theorem *If $f, \widehat{f} \in L^1$ (and both are continuous), then*

$$f(0) = (2\pi)^{-N} \int \widehat{f}(\xi) \, d\xi. \qquad (6.6.2)$$

Of course there is nothing special about the point $0 \in \mathbb{R}^N$. We now exploit the compatibility of the Fourier transform with translations to obtain a more general formula. We apply formula (6.6.2) in our theorem to $\tau_{-h}f$: The result is

$$(\tau_{-h}f)(0) = (2\pi)^{-N} \int (\tau_{-h}f)\widehat{\,}(\xi) \, d\xi \qquad (6.6.3)$$

or

Theorem (The Fourier Inversion Formula) *If $f, \widehat{f} \in L^1$ (and if both $f, \widehat{f}$ are continuous), then for any $y \in \mathbb{R}^N$ we have*

$$f(y) = (2\pi)^{-N} \int \widehat{f}(\xi) e^{-iy \cdot \xi} \, d\xi. \qquad (6.6.4)$$

This result follows from (6.6.3)—just write out the integral. See [KRA3] for details.

Plancherel's Formula

We now give a treatment of the quadratic Fourier theory.

Proposition (Plancherel)   If $f \in C_c^\infty(\mathbb{R}^N)$, then

$$(2\pi)^{-N} \int |\widehat{f}(\xi)|^2 \, d\xi = \int |f(x)|^2 \, dx. \qquad (6.6.5)$$

Proof: Define $g = f * \widetilde{\overline{f}} \in C_c^\infty(\mathbb{R}^N)$. Then

$$\widehat{g} = \widehat{f} \cdot \widehat{\widetilde{\overline{f}}} = \widehat{f} \cdot \overline{\widehat{f}} = \widehat{f} \cdot \overline{\widehat{f}} = \widehat{f} \cdot \overline{\widehat{f}} = \widehat{f} \cdot \overline{\widehat{f}} = |\widehat{f}|^2. \qquad (6.6.6)$$

Now

$$g(0) = f * \widetilde{\overline{f}}\,(0) = \int f(-t)\overline{f}(-t) \, dt = \int f(t)\overline{f}(t) \, dt = \int |f(t)|^2 \, dt.$$

By Fourier inversion and formula (6.6.6) we may now conclude that

$$\int |f(t)|^2 \, dt = g(0) = (2\pi)^{-N} \int \widehat{g}(\xi) \, d\xi = (2\pi)^{-N} \int |\widehat{f}(\xi)|^2 \, d\xi.$$

That is the desired formula.   □

Definition   For any square integrable function f, the Fourier transform of f can be defined in the following fashion: Let $f_j \in C_c^\infty$ satisfy $f_j \to f$ in the L^2 topology. It follows from the Proposition that $\{\widehat{f_j}\}$ is Cauchy in L^2. Let g be the L^2 limit of this latter sequence. We set $\widehat{f} = g$.

It is easy to check that this definition of $\widehat{f}$ is independent of the choice of sequence $f_j \in C_c^\infty$ and that

$$(2\pi)^{-N} \int |\widehat{f}(\xi)|^2 \, d\xi = \int |f(x)|^2 \, dx.$$

Problems for Review and Discovery

A. Drill Exercises

1. Find the Fourier series for each of these functions.
    (a) $f(x) = x^2$, $-\pi \le x \le \pi$
    (b) $g(x) = x - |x|$, $-\pi \le x \le \pi$

(c) $h(x) = x + |x|$, $\quad -\pi \le x \le \pi$

(d) $f(x) = |x|$, $\quad -\pi \le x \le \pi$

(e) $g(x) = \begin{cases} -x^2 & \text{if} \quad -\pi \le x \le 0 \\ x^2 & \text{if} \quad 0 < x \le \pi \end{cases}$

(f) $h(x) = \begin{cases} |x| & \text{if} \quad -\pi \le x \le 1/2 \\ 1/2 & \text{if} \quad 1/2 < x \le \pi \end{cases}$

2. Calculate the Fourier series for each of these functions.

(a) $f(x) = \begin{cases} \sin x & \text{if} \quad -\pi \le x \le \pi/2 \\ 0 & \text{if} \quad \pi/2 < x \le \pi \end{cases}$

(b) $g(x) = \begin{cases} \cos x & \text{if} \quad -\pi \le x \le -\pi/2 \\ 0 & \text{if} \quad -\pi/2 < x \le \pi \end{cases}$

(c) $h(x) = \begin{cases} \cos x & \text{if} \quad -\pi \le x \le 0 \\ 1 & \text{if} \quad 0 < x \le \pi \end{cases}$

(d) $f(x) = \begin{cases} 1 & \text{if} \quad -\pi \le x \le 0 \\ \sin x & \text{if} \quad 0 < x \le \pi \end{cases}$

3. Sketch the graphs of the first three partial sums of the Fourier series for each of the functions in Exercise 2.

4. Calculate the sine series of each of these functions.

(a) $f(x) = \cos 2x$, $\quad 0 \le x \le \pi$

(b) $g(x) = x^2$, $\quad 0 \le x \le \pi$

(c) $h(x) = x - |x - 1/2|/2$, $\quad 0 \le x \le \pi$

(d) $f(x) = x^2 + |x + 1/4|$, $\quad 0 \le x \le \pi$

5. Calculate the cosine series of each of these functions.

(a) $f(x) = \sin 3x$, $\quad 0 \le x \le \pi$

(b) $g(x) = x^2$, $\quad 0 \le x \le \pi$

(c) $h(x) = x - |x - 1/2|/2$, $\quad 0 \le x \le \pi$

(d) $f(x) = x^2 + |x + 1/4|$, $\quad 0 \le x \le \pi$

6. Find the Fourier series expansion for the given function on the given interval.

(a) $f(x) = x^2 - x$, $\quad -1 \le x \le 1$

(b) $g(x) = \sin x$, $\quad -2 \le x \le 2$

(c) $h(x) = \cos x$, $\quad -3 \le x \le 3$

(d) $f(x) = |x|$, $\quad -1 \le x \le 1$

(e) $g(x) = |x - 1/2|$, $\quad -2 \le x \le 2$

(f) $h(x) = |x + 1/2|/2$, $\quad -3 \le x \le 3$

B. Challenge Problems

1. In Section 4.4 we learned about the Legendre polynomials. The first three Legendre polynomials are

$$P_0(x) \equiv 1 , \quad P_1(x) = x , \quad P_2(x) = \frac{3}{2}x^2 - \frac{1}{2} .$$

Verify that P_0, P_1, P_2 are mutually orthogonal on the interval $[-1, 1]$. Let

$$f(x) = \begin{cases} -1 & \text{if} & -1 \leq x \leq 0 \\ 2 & \text{if} & 0 < x \leq 1. \end{cases}$$

Find the first three coefficients in the expansion

$$f(x) = a_0 P_0(x) + a_1 P_1(x) + a_2 P_2(x) + \cdots .$$

2. Repeat Exercise 1 for the function $f(x) = x + |x|$.

3. The first three Hermite polynomials are

$$H_0(x) \equiv 1 , \quad H_1(x) = 2x , \quad H_2(x) = 4x^2 - 2 .$$

Verify that these functions are mutually orthogonal on the interval $(-\infty, \infty)$ with respect to the weight e^{-x^2} (this means that

$$\int_{-\infty}^{\infty} H_j(x) H_k(x) e^{-x^2} \, dx = 0$$

if $j \neq k$). Calculate the first three coefficients in the expansion

$$f(x) = b_0 H_0(x) + b_1 H_1(x) + b_2 H_2(x) + \cdots$$

for the function $f(x) = x - |x|$.

4. The first three Chebyshev polynomials are

$$T_0(x) \equiv 1 , \quad T_1(x) = x , \quad T_2(x) = 2x^2 - 1 .$$

Verify that these functions are mutually orthogonal on the interval $[-1, 1]$ with respect to the weight $(1-x^2)^{-1/2}$ (refer to Exercise 3 for the meaning of this concept). Calculate the first three coefficients in the expansion

$$f(x) = c_0 T_0(x) + c_1 T_1(x) + c_2 T_2(x) + \cdots$$

for the function $f(x) = x$.

C. Problems for Discussion and Exploration

1. Refer to Fejér's Theorem 6.2.6 about convergence of the Cesàro means. Confirm this result by direct calculation for these functions.

(a) $f(x) = \begin{cases} -1 & \text{if} & -\pi \leq x \leq 0 \\ 1 & \text{if} & 0 < x \leq \pi \end{cases}$

(b) $g(x) = \begin{cases} 0 & \text{if} & -\pi \leq x \leq 0 \\ \cos x & \text{if} & 0 < x \leq \pi \end{cases}$

(c) $h(x) = \begin{cases} \sin x & \text{if} & -\pi \leq x \leq 0 \\ 2 & \text{if} & 0 < x \leq \pi \end{cases}$

(d) $f(x) = \begin{cases} |x + 1|/2 & \text{if} & -\pi \leq x \leq 0 \\ 0 & \text{if} & 0 < x \leq \pi \end{cases}$

2. A celebrated result of classical Fourier analysis states that, if f is continuously differentiable on $[-\pi, \pi]$, then its Fourier series converges absolutely. Confirm this assertion (at all points except the endpoints of the interval) in the following specific examples.

(a) $f(x) = x$

(b) $g(x) = x^2$

(c) $h(x) = e^x$

3. In other expositions, it is convenient to define Fourier series using the language of complex numbers. Specifically, instead of expanding in terms of $\cos jx$ and $\sin jx$, we instead expand in terms of e^{ijx}. Specifically, we work with a function f on $[-\pi, \pi]$ and set

$$c_j = \frac{1}{2\pi} \int_{-\pi}^{\pi} f(t)e^{-ijt}\, dt\,.$$

We define the formal Fourier expansion of f to be

$$Sf \sim \sum_{j=-\infty}^{\infty} c_j e^{ijt}\,.$$

Explain why this new formulation of Fourier series is equivalent to that presented in Section 6.1 (i.e., explain how to pass back and forth from one language to the other).

What are the advantages and disadvantages of this new, complex form of the Fourier series?

7

Laplace Transforms

- The idea of the Laplace transform
- The Laplace transform and differential equations
- Derivatives and the Laplace transform
- Integrals and the Laplace transform
- Convolutions
- Step and impulse functions
- Discontinuous input

7.1 Introduction

The idea of the Laplace transform has had a profound influence on the development of mathematical analysis. It also plays a significant role in mathematical applications. More generally, the overall theory of transforms has become an important part of modern mathematics.

The concept of a *transform* is that it turns a given function into another function. We are already acquainted with several transforms:

I. The derivative D takes a differentiable function f (defined on some interval (a, b)) and assigns to it a new function $Df = f'$.

II. The integral I takes a continuous function f (defined on some interval $[a, b]$ and assigns to it a new function

$$I f(x) = \int_a^x f(t)\, dt \,.$$

III. The multiplication operator M_φ, which multiplies any given function f on the interval $[a, b]$ by a fixed function φ on $[a, b]$, is a transform:

$$M_\varphi f(x) = \varphi(x) \cdot f(x) \,.$$

273

We are particularly interested in transforms that are linear. A transform T is *linear* if

$$T[\alpha f + \beta g] = \alpha T(f) + \beta T(g)$$

for any real constants α, β. In particular (taking $\alpha = \beta = 1$),

$$T[f + g] = T(f) + T(g)$$

and (taking $\beta = 0$)

$$T(\alpha f) = \alpha T(f).$$

We would like to understand linear transforms that are given by integration. Let f be a function with domain $[0, \infty)$. The *Laplace transform* of f is defined by

$$L[f](s) = F(s) = \int_0^\infty e^{-sx} f(x)\, dx \qquad \text{for } s > 0.$$

Notice that we begin with a function f of x, and the Laplace transform L produces a new function $L[f]$ of s. We sometimes write the Laplace transform as $F(s)$. Notice that the Laplace transform is an improper integral; it exists precisely when

$$\int_0^\infty e^{-sx} f(x)\, dx = \lim_{N \to \infty} \int_0^N e^{-sx} f(x)\, dx$$

exists and is finite. Because of the presence of the factor e^{-sx}, the Laplace transform exists and is well defined for a large class of functions f.

Let us now calculate some Laplace transforms:

Function f	Laplace transform F
$f(x) \equiv 1$	$F(s) = \int_0^\infty e^{-sx}\, dx = \frac{1}{s}$
$f(x) = x$	$F(s) = \int_0^\infty e^{-sx} x\, dx = \frac{1}{s^2}$
$f(x) = x^n$	$F(s) = \int_0^\infty e^{-sx} x^n\, dx = \frac{n!}{s^{n+1}}$
$f(x) = e^{ax}$	$F(s) = \int_0^\infty e^{-sx} e^{ax}\, dx = \frac{1}{s-a}, \quad s > a$
$f(x) = \sin ax$	$F(s) = \int_0^\infty e^{-sx} \sin ax\, dx = \frac{a}{s^2+a^2}$
$f(x) = \cos ax$	$F(s) = \int_0^\infty e^{-sx} \cos ax\, dx = \frac{s}{s^2+a^2}$
$f(x) = \sinh ax$	$F(s) = \int_0^\infty e^{-sx} \sinh ax\, dx = \frac{a}{s^2-a^2}, \quad s > a$
$f(x) = \cosh ax$	$F(s) = \int_0^\infty e^{-sx} \cosh ax\, dx = \frac{s}{s^2-a^2}, \quad s > a$

We shall not actually perform all these integrations. We content ourselves with the third one, just to illustrate the idea. The student should definitely perform the others, just to get the feel of Laplace transform calculations.

Now

$$L[x^n](s) = \int_0^\infty e^{-sx} x^n \, dx$$

$$= -\frac{x^n e^{-sx}}{s}\Big|_0^\infty + \frac{n}{s} \int_0^\infty e^{-sx} x^{n-1} \, dx$$

$$= (0-0) + \frac{n}{s} L[x^{n-1}]$$

$$= \frac{n}{s}\left(\frac{n-1}{s}\right) L[x^{n-2}]$$

$$= \cdots$$

$$= \frac{n!}{s^n} L[1]$$

$$= \frac{n!}{s^n} \int_0^\infty e^{-sx} \cdot 1 \, dx$$

$$= \frac{n!}{s^{n+1}}.$$

The reader will find, as we just have, that integration by parts is eminently useful in the calculation of Laplace transforms.

It may be noted that the Laplace transform is a linear operator. Thus Laplace transforms of some compound functions may be readily calculated from the table just given:

$$L(5x^3 - 2e^x)(s) = \frac{5 \cdot 3!}{s^4} - \frac{2}{s-1}$$

and

$$L(4\sin 2x + 6x)(s) = \frac{4 \cdot 2}{s^2 + 2^2} + \frac{6}{s^2}.$$

Exercises

1. Evaluate all the Laplace transform integrals for the table in this section.
2. Without actually integrating, show that
    (a) $L[\sinh ax] = \dfrac{a}{s^2 - a^2}$
    (b) $L[\cosh ax] = \dfrac{s}{s^2 - a^2}$

3. Find $L[\sin^2 ax]$ and $L[\cos^2 ax]$ without integrating. How are these two transforms related to one another?

4. Use the formulas given in the text to find the Laplace transform of each of the following functions.

    (a) 10 (d) $4\sin x \cos x + 2e^{-x}$
    (b) $x^5 + \cos 2x$ (e) $x^6 \sin^2 3x + x^6 \cos^2 3x$
    (c) $2e^{3x} - \sin 5x$

5. Find a function f whose Laplace transform is:

    (a) $\dfrac{30}{s^4}$ (d) $\dfrac{1}{s^2 + s}$

    (b) $\dfrac{2}{s+3}$ (e) $\dfrac{1}{s^4 + s^2}$

    (c) $\dfrac{4}{s^3} + \dfrac{6}{s^2 + 4}$

    [**Hint:** The method of partial fractions will prove useful.]

6. Give a plausible definition of $\frac{1}{2}!$ (i.e., the factorial of the number $1/2$).

7. Use your symbol manipulation software, such as `Maple` or `Mathematica`, to calculate the Laplace transforms of each of these functions.

    (a) $f(x) = \sin(e^x)$
    (b) $g(x) = \ln(1 + \sin^2 x)$
    (c) $h(x) = \sin[\ln x]$
    (d) $k(x) = e^{\cos x}$

7.2   Applications to Differential Equations

The key to our use of Laplace transform theory in the subject of differential equations is the way that L treats derivatives. Let us calculate

$$
\begin{aligned}
L[y'](s) &= \int_0^\infty e^{-sx} y'(x)\, dx \\
&= y(x)e^{-sx}\Big|_0^\infty + s \int_0^\infty e^{-sx} y(x)\, dx \\
&= -y(0) + s \cdot L[y](s).
\end{aligned}
$$

In the second equality we of course used integration by parts.
     In summary,
$$
L[y'](s) = s \cdot L[y](s) - y(0).
$$

Likewise,

$$
\begin{aligned}
L[y''](s) &= L[(y')'] = s \cdot L[y'](s) - y'(0) \\
&= s\{s \cdot L[y](s) - y(0)\} - y'(0) \\
&= s^2 \cdot L[y](s) - sy(0) - y'(0).
\end{aligned}
$$

Now let us examine the differential equation

$$y'' + ay' + by = f(x), \tag{7.2.1}$$

with the initial conditions $y(0) = y_0$ and $y'(0) = y_1$. Here a and b are real constants. We apply the Laplace transform L to both sides of (7.2.1), of course using the linearity of L. The result is

$$L[y''](s) + aL[y'](s) + bL[y](s) = L[f](s).$$

For brevity, we often omit explicit mention of the new independent variable s for the Laplace transform function. Writing out what each term is, we find that

$$\{s^2 \cdot L[y] - sy(0) - y'(0)\} + a\{s \cdot L[y] - y(0)\} + bL[y] = L[f].$$

Now we can plug in what $y(0)$ and $y'(0)$ are. We may also gather like terms together. The result is

$$\{s^2 + as + b\}L[y] = (s + a)y_0 + y_1 + L[f]$$

or

$$L[y] = \frac{(s + a)y_0 + y_1 + L[f]}{s^2 + as + b}. \tag{7.2.2}$$

What we see here is a remarkable thing: The Laplace transform changes solving a differential equation from a rather complicated calculus problem to a simple algebra problem. The only thing that remains, in order to find an explicit solution to the original differential equation (7.2.1) with initial conditions, is to find the inverse Laplace transform of the right-hand side of (7.2.2). In practice we shall find that we can often perform this operation in a straightforward fashion. The following examples will illustrate the idea.

EXAMPLE 7.2.3 Use the Laplace transform to solve the differential equation

$$y'' + 4y = 4x \tag{7.2.3.1}$$

with initial conditions $y(0) = 1$ and $y'(0) = 5$.

Solution:
We proceed mechanically, by applying the Laplace transform to both sides of (7.2.3.1). Thus

$$L[y''] + L[4y] = L[4x].$$

We can use our various Laplace transform formulas to write this out more explicitly:

$$\{s^2 L[y] - sy(0) - y'(0)\} + 4L[y] = \frac{4}{s^2}$$

or

$$s^2 L[y] - s \cdot 1 - 5 + 4L[y] = \frac{4}{s^2}$$

or

$$(s^2 + 4)L[y] = s + 5 + \frac{4}{s^2} \,.$$

It is convenient to write this as

$$L[y] = \frac{s}{s^2 + 4} + \frac{5}{s^2 + 4} + \frac{4}{s^2 \cdot (s^2 + 4)} = \frac{s}{s^2 + 4} + \frac{5}{s^2 + 4} + \frac{1}{s^2} - \frac{1}{s^2 + 4},$$

where we have used a partial fractions decomposition in the last step. Simplifying, we have

$$L[y] = \frac{s}{s^2 + 4} + \frac{4}{s^2 + 4} + \frac{1}{s^2} \,.$$

Referring to our table of Laplace transforms, we may now deduce what y must be:

$$L[y] = L[\cos 2x] + L[2 \sin 2x] + L[x] = L[\cos 2x + 2 \sin 2x + x] \,.$$

We deduce then that

$$y = \cos 2x + 2 \sin 2x + x \,,$$

and this is the solution of our initial value problem.   ■

REMARK 7.2.4 It is useful to note that our formulas for the Laplace transform of the first and second derivative incorporated automatically the values $y(0)$ and $y'(0)$. Thus our initial conditions got built in during the course of our solution process.

A useful general property of the Laplace transform concerns its interaction with translations. Indeed, we have

$$\boxed{L[e^{ax} f(x)] = F(s - a) \,. \qquad\qquad (7.2.5)}$$

To see this, we calculate

$$
\begin{aligned}
L[e^{ax} f(x)] \;&=\; \int_0^\infty e^{-sx} e^{ax} f(x) \, dx \\
&=\; \int_0^\infty e^{-(s-a)x} f(x) \, dx \\
&=\; F(s - a) \,.
\end{aligned}
$$

We frequently find it useful to use the notation L^{-1} to denote the inverse operation to the Laplace transform.[1] For example, since

$$L[x^2] = \frac{2!}{s^3},$$

we may write

$$L^{-1}\left(\frac{2!}{s^3}\right) = x^2.$$

Since

$$L[\sin x - e^{2x}] = \frac{1}{s^2+1} - \frac{1}{s-2},$$

we may write

$$L^{-1}\left(\frac{1}{s^2+1} - \frac{1}{s-2}\right) = \sin x - e^{2x}.$$

EXAMPLE 7.2.6 Since

$$L[\sin bx] = \frac{b}{s^2+b^2},$$

we conclude that

$$L[e^{ax}\sin bx] = \frac{b}{(s-a)^2+b^2}.$$

Since

$$L^{-1}\left(\frac{1}{s^2}\right) = x,$$

we thus have

$$L^{-1}\left(\frac{1}{(s-a)^2}\right) = e^{ax}x. \qquad \blacksquare$$

EXAMPLE 7.2.7 Use the Laplace transform to solve the differential equation

$$y'' + 2y' + 5y = 3e^{-x}\sin x \qquad (7.2.7.1)$$

with initial conditions $y(0) = 0$ and $y'(0) = 3$.

Solution:
We calculate the Laplace transform of both sides, using our new formula (7.2.5) on the right-hand side, to obtain

$$\{s^2L[y] - sy(0) - y'(0)\} + 2\{sL[y] - y(0)\} + 5L[y] = 3 \cdot \frac{1}{(s+1)^2+1}.$$

---

[1] We tacitly use here the fact that the Laplace transform L is one-to-one: if $L[f] = L[g]$ then $f = g$. Thus L is invertible on its image. We are able to verify this assertion empirically through our calculations; the general result is proved rigorously in a more advanced treatment.

Plugging in the initial conditions, and organizing like terms, we find that

$$(s^2 + 2s + 5)L[y] = 3 + \frac{3}{(s+1)^2 + 1}$$

or

$$
\begin{aligned}
L[y] &= \frac{3}{s^2 + 2s + 5} + \frac{3}{(s^2 + 2s + 2)(s^2 + 2s + 5)} \\
&= \frac{3}{s^2 + 2s + 5} + \frac{1}{s^2 + 2s + 2} - \frac{1}{s^2 + 2s + 5} \\
&= \frac{2}{(s+1)^2 + 4} + \frac{1}{(s+1)^2 + 1} .
\end{aligned}
$$

Of course we have used partial fractions in the second equality.
We see therefore that

$$y = e^{-x} \sin 2x + e^{-x} \sin x .$$

This is the solution of our initial value problem.   ■

Exercises

1. Find the Laplace transforms of

   (a) $x^5 e^{-2x}$ (e) $e^{3x} \cos 2x$

   (b) $(1 - x^2)e^{-x}$ (f) xe^x

   (c) $e^{-x} \sin x$ (g) $x^2 \cos x$

   (d) $x \sin 3x$ (h) $\sin x \cos x$

2. Find the inverse Laplace transform of

   (a) $\dfrac{6}{(s+2)^2 + 9}$ (d) $\dfrac{12}{(s+3)^4}$

   (b) $\dfrac{s+3}{s^2 + 2s + 5}$ (e) $\dfrac{1}{s^4 + s^2 + 1}$

   (c) $\dfrac{s}{4s^2 + 1}$ (f) $\dfrac{6}{(s-1)^3}$

3. Solve each of the following differential equations with initial values using the Laplace transform.

   (a) $y' + y = e^{2x}$, $y(0) = 0$

   (b) $y'' - 4y' + 4y = 0$, $y(0) = 0$ and $y'(0) = 3$

(c) $y'' + 2y' + 2y = 2$, $y(0) = 0$ and $y'(0) = 1$

(d) $y'' + y' = 3x^2$, $y(0) = 0$ and $y'(0) = 1$

(e) $y'' + 2y' + 5y = 3e^{-x} \sin x$, $y(0) = 0$ and $y'(0) = 3$

4. Find the solution of $y'' - 2ay' + a^2 y = 0$ in which the initial conditions $y(0) = y_0$ and $y'(0) = y_0'$ are left unrestricted. (This provides an additional derivation of our earlier solution for the case in which the auxiliary equation has a double root.)

5. Apply the formula $L[y'] = sL[y] - y(0)$ to establish the formula for the Laplace transform of an integral:

$$L\left(\int_0^x f(t)\,dt\right) = \frac{F(s)}{s}\,.$$

Do so by finding

$$L^{-1}\left(\frac{1}{s(s+1)}\right)$$

in two different ways.

6. Solve the equation

$$y' + 4y + 5\int_0^x y\,dx = e^{-x}\,, \qquad y(0) = 0\,.$$

7.3   Derivatives and Integrals of Laplace Transforms

In some contexts it is useful to calculate the derivative of the Laplace transform of a function (when the corresponding integral makes sense). For instance, consider

$$F(s) = \int_0^\infty e^{-sx} f(x)\,dx\,.$$

Then

$$
\begin{aligned}
\frac{d}{ds}F(s) &= \frac{d}{ds}\int_0^\infty e^{-sx} f(x)\,dx \\
&= \int_0^\infty \frac{d}{ds}\left[e^{-sx} f(x)\right]\,dx \\
&= \int_0^\infty e^{-sx}\{-xf(x)\}\,dx = L[-xf(x)](s)\,.
\end{aligned}
$$

We see that the derivative[2] of $F(s)$ is the Laplace transform of $-xf(x)$. More generally, the same calculation shows us that

$$\frac{d^2}{ds^2}F(s) = L[x^2 f(x)](s)$$

---

[2] The passage of the derivative under the integral sign in this calculation requires advanced ideas from real analysis which we cannot treat here—see [KRA2].

and

$$\frac{d^j}{ds^j} F(s) = L[(-1)^j x^j f(x)](s).$$

EXAMPLE 7.3.1 Calculate

$$L[x \sin ax].$$

Solution:
We have

$$L[x \sin ax] = -L[-x \sin ax] = -\frac{d}{ds} L[\sin ax] = -\frac{d}{ds} \frac{a}{s^2 + a^2} = \frac{2as}{(s^2 + a^2)^2}.$$

■

EXAMPLE 7.3.2 Calculate the Laplace transform of $\sqrt{x}$.

Solution:
This calculation actually involves some tricky integrations. We first note that

$$L[\sqrt{x}] = L[x^{1/2}] = -L[-x \cdot x^{-1/2}] = -\frac{d}{ds} L[x^{-1/2}]. \tag{7.3.2.1}$$

Thus we must find the Laplace transform of $x^{-1/2}$.
　　Now

$$L[x^{-1/2}] = \int_0^\infty e^{-sx} x^{-1/2} \, dx.$$

The change of variables $sx = t$ yields

$$= s^{-1/2} \int_0^\infty e^{-t} t^{-1/2} \, dt.$$

The further change of variables $t = p^2$ gives the integral

$$L[x^{-1/2}] = 2s^{-1/2} \int_0^\infty e^{-p^2} \, dp. \tag{7.3.2.2}$$

　　Now we must evaluate the integral $I = \int_0^\infty e^{-p^2} \, dp$. Observe, introducing the dummy variable u, that

$$I \cdot I = \int_0^\infty e^{-p^2} \, dp \cdot \int_0^\infty e^{-u^2} \, du = \int_0^\infty \int_0^{\pi/2} e^{-r^2} \cdot r \, d\theta dr.$$

Here we have introduced polar coordinates in the standard way.
　　Now the last integral is easily evaluated and we find that

$$I^2 = \frac{\pi}{4}$$

hence $I = \sqrt{\pi}/2$. Thus $L[x^{-1/2}](s) = 2s^{-1/2}\{\sqrt{\pi}/2\} = \sqrt{\pi/s}$. Finally,

$$L[\sqrt{x}] = -\frac{d}{ds}\sqrt{\frac{\pi}{s}} = \frac{\sqrt{\pi}}{2s^{3/2}}.$$ ∎

We now derive some additional formulas that will be useful in solving differential equations. We let $y = f(x)$ be our function and $Y = L[f]$ be its Laplace transform. Then

$$L[xy] = -\frac{d}{ds}L[y] = -\frac{dY}{ds}.$$ (7.3.3)

Also

$$L[xy'] = -\frac{d}{ds}L[y'] = -\frac{d}{ds}[sY - y(0)] = -\frac{d}{ds}[sY]$$ (7.3.4)

and

$$L[xy''] = -\frac{d}{ds}L[y'']$$
$$= -\frac{d}{ds}[s^2Y - sy(0) - y'(0)]$$
$$= -\frac{d}{ds}[s^2Y] + y(0).$$ (7.3.5)

EXAMPLE 7.3.6 Use the Laplace transform to analyze Bessel's equation

$$xy'' + y' + xy = 0$$

with the single initial condition $y(0) = 1$.

Solution:
Apply the Laplace transform to both sides of the equation. Thus

$$L[xy''] + L[y'] + L[xy] = L[0] = 0.$$

We can apply our new formulas (7.3.5) and (7.3.3) to the first and third terms on the left. And of course we apply the usual formula for the Laplace transform of the derivative to the second term on the left. The result is

$$\left[-\frac{d}{ds}(s^2 Y) + 1 \right] + [sY - 1] + \left[-\frac{dY}{ds} \right] = 0$$

We may simplify this equation to

$$(s^2 + 1)\frac{dY}{ds} = -sY .$$

This is a *new* differential equation, and we may solve it by separation of variables. Now

$$\frac{dY}{Y} = -\frac{s \, ds}{s^2 + 1}$$

so

$$\ln Y = -\frac{1}{2}\ln(s^2 + 1) + C .$$

Exponentiating both sides gives

$$Y = D \cdot \frac{1}{\sqrt{s^2 + 1}} .$$

It is useful (with a view to calculating the inverse Laplace transform) to write this solution as

$$Y = \frac{D}{s} \cdot \left(1 + \frac{1}{s^2} \right)^{-1/2} . \tag{7.3.6.1}$$

Recall the binomial expansion:

$$(1 + z)^a = 1 + az + \frac{a(a - 1)}{2!}z^2 + \frac{a(a - 1)(a - 2)}{3!}z^3$$
$$+ \cdots + \frac{a(a - 1)\cdots(a - n + 1)}{n!}z^n + \cdots .$$

We apply this formula to the second term on the right of (7.3.6.1) with the role of z played by $1/s^2$. Thus

$$Y = \frac{D}{s} \cdot \left(1 - \frac{1}{2} \cdot \frac{1}{s^2} + \frac{1}{2!} \cdot \frac{1}{2} \cdot \frac{3}{2} \cdot \frac{1}{s^4} - \frac{1}{3!} \cdot \frac{1}{2} \cdot \frac{3}{2} \cdot \frac{5}{2} \cdot \frac{1}{s^6} \right.$$
$$\left. + - \cdots - + \frac{1 \cdot 3 \cdot 5 \cdots (2n - 1)}{2^n n!}\frac{(-1)^n}{s^{2n}} + \cdots \right)$$
$$= D \cdot \sum_{j=0}^{\infty} \frac{(2j)!}{2^{2j}(j!)^2} \cdot \frac{(-1)^j}{s^{2j+1}} .$$

The good news is that we can now calculate L^{-1} of Y (thus obtaining y) by just calculating the inverse Laplace transform of each term of this series.

The result is

$$y(x) \;=\; D \cdot \sum_{j=0}^{\infty} \frac{(-1)^j}{2^{2j}(j!)^2} \cdot x^{2j}$$

$$=\; D \cdot \left(1 - \frac{x^2}{2^2} + \frac{x^4}{2^2 \cdot 4^2} - \frac{x^6}{2^2 \cdot 4^2 \cdot 6^2} + \cdots \right).$$

Since $y(0) = 1$ (the initial condition), we see that $D = 1$ and

$$y(x) = 1 - \frac{x^2}{2^2} + \frac{x^4}{2^2 \cdot 4^2} - \frac{x^6}{2^2 \cdot 4^2 \cdot 6^2} + \cdots.$$

It should be noted that the origin $x_0 = 0$ is a regular singular point for this differential equation, and the Frobenius values of m are $m = 0, 0$. That explains why we do *not* have two undetermined constants in our solution.

The series we have just derived defines the celebrated and important *Bessel function* J_0. We have learned that the Laplace transform of J_0 is $1/\sqrt{s^2 + 1}$. ∎

It is also a matter of some interest to integrate the Laplace transform. We can anticipate how this will go by running the differentiation formulas in reverse. Our main result is

$$L\left(\frac{f(x)}{x} \right) = \int_s^{\infty} F(s)\, ds. \qquad (7.3.7)$$

In fact

$$\int_s^{\infty} F(t)\, dt \;=\; \int_s^{\infty} \left(\int_0^{\infty} e^{-tx} f(x)\, dx \right) dt$$

$$=\; \int_0^{\infty} f(x) \int_s^{\infty} e^{-tx}\, dt dx$$

$$=\; \int_0^{\infty} f(x) \left(\frac{e^{-tx}}{-x} \right)_s^{\infty} dt$$

$$=\; \int_0^{\infty} f(x) \cdot \frac{e^{-sx}}{x}\, dx$$

$$=\; \int_0^{\infty} \left(\frac{f(x)}{x} \right) \cdot e^{-sx}\, dx$$

$$=\; L\left[\frac{f(x)}{x} \right](s).$$

EXAMPLE 7.3.8 Use the fact that $L[\sin x] = 1/(s^2 + 1)$ to calculate $\int_0^\infty (\sin x)/x \, dx$.

Solution:

By formula (7.3.7) (with $f(x) = \sin x$),

$$\int_0^\infty \frac{\sin x}{x} \, dx = L[\sin x/x](0) = \int_0^\infty \frac{ds}{s^2 + 1} = \arctan s \Big|_0^\infty = \frac{\pi}{2} . \qquad \blacksquare$$

We conclude this section by summarizing the chief properties of the Laplace transform in a table. As usual, we let $F(s)$ denote the Laplace transform of $f(x)$ and $G(s)$ denote the Laplace transform of $g(x)$. The last property listed in this table concerns convolution, and we shall treat that topic in the next section.

Properties of the Laplace Transform

$$L[\alpha f(x) + \beta g(x)] = \alpha F(s) + \beta G(s)$$

$$L[e^{ax} f(x)] = F(s - a)$$

$$L[f'(x)] = sF(s) - f(0)$$

$$L[f''(x)] = s^2 F(s) - sf(0) - f'(0)$$

$$L\left(\int_0^x f(t) \, dt\right) = \frac{f(s)}{s}$$

$$L[-xf(x)] = F'(s)$$

$$L[(-1)^n x^n f(x)] = F^{(n)}(s)$$

$$L\left(\frac{f(x)}{x}\right) = \int_s^\infty F(s) \, ds$$

$$L\left(\int_0^x f(x - t)g(t) \, dt\right) = F(s)G(s)$$

Exercises

1. Verify that

$$L[x \cos ax] = \frac{s^2 - a^2}{(s^2 + a^2)^2} .$$

Use this result to find

$$L^{-1}\left(\frac{1}{(s^2+a^2)^2}\right).$$

2. Calculate each of the following Laplace transforms.
    (a) $L[x^2 \sin ax]$
    (b) $L[x^{3/2}]$
    (c) $L[x \cos ax]$
    (d) $L[xe^x]$

3. Solve each of the following differential equations.
    (a) $xy'' + (3x-1)y' - (4x+9)y = 0$
    (b) $xy'' + (2x+3)y' + (x+3)y = 3e^{-x}$

4. If a and b are positive constants, then evaluate the following integrals.
    (a) $\displaystyle\int_0^\infty \frac{e^{-ax}-e^{-bx}}{x}\, dx$
    (b) $\displaystyle\int_0^\infty \frac{e^{-ax}\sin bx}{x}\, dx$

5. Without worrying about convergence issues, verify that
    (a) $\displaystyle\int_0^\infty J_0(x)\, dx = 1$
    (b) $J_0(x) = \dfrac{1}{\pi}\displaystyle\int_0^\pi \cos(x\cos t)\, dt$

6. Without worrying about convergence issues, and assuming $x > 0$, show that
    (a) $f(x) = \displaystyle\int_0^\infty \frac{\sin xt}{t}\, dt \equiv \frac{\pi}{2}$
    (b) $f(x) = \displaystyle\int_0^\infty \frac{\cos xt}{1+t^2}\, dt = \frac{\pi}{2}e^{-x}$

7. (a) If f is periodic with period a, so that $f(x+a) = f(x)$, then show that
    $$F(s) = \frac{1}{1-e^{-as}}\int_0^a e^{-sx} f(x)\, dx.$$
    (b) Find $F(s)$ if $f(x) = 1$ in the intervals $[0,1]$, $[2,3]$, $[4,5]$, etc., and $f \equiv 0$ in the remaining intervals.

8. If y satisfies the differential equation
    $$y'' + x^2 y = 0,$$
    where $y(0) = y_0$ and $y'(0) = y_0'$, then show that its Laplace transform $Y(s)$ satisfies the equation
    $$Y'' + s^2 Y = sy_0 + y_0'.$$

    Observe that the new equation is of the same type as the original equation, so that no real progress has been made. The method of Example 7.2.6 is effective only when the coefficients are first-degree polynomials.

7.4   Convolutions

An interesting question, that occurs frequently with the use of the Laplace transform, is this: Let f and g be functions and F and G their Laplace transforms; what is $L^{-1}[F \cdot G]$? To discover the answer, we write

$$F(s) \cdot G(s) = \left(\int_0^\infty e^{-st} f(t)\, dt \right) \cdot \left(\int_0^\infty e^{-su} f(u)\, du \right)$$

$$= \int_0^\infty \int_0^\infty e^{-s(t+u)} f(t) g(u)\, dt du$$

$$= \int_0^\infty \left(\int_0^\infty e^{-s(t+u)} f(t)\, dt \right) g(u)\, du.$$

Now we perform the change of variable $x = t + u$ in the inner integral. The result is

$$F(s) \cdot G(s) = \int_0^\infty \left(\int_t^\infty e^{-sx} f(x-u)\, dx \right) g(u)\, du$$

$$= \int_0^\infty \int_t^\infty e^{-sx} f(x-u) g(u)\, dx du.$$

Reversing the order of integration, we may finally write

$$F(s) \cdot G(s) = \int_0^\infty \left(\int_0^x e^{-sx} f(x-u) g(u)\, du \right) dx$$

$$= \int_0^\infty e^{-sx} \left(\int_0^x f(x-u) g(u)\, du \right) dx$$

$$= L \left[\int_0^x f(x-u) g(u)\, du \right].$$

We call the expression $\int_0^x f(x-u) g(u)\, du$ the *convolution* of f and g. Many texts write

$$f * g(x) = \int_0^x f(x-u) g(u)\, du. \qquad (7.4.1)$$

Our calculation shows that

$$L[f * g](s) = F \cdot G = L[f] \cdot L[g].$$

The convolution formula is particularly useful in calculating inverse Laplace transforms.

EXAMPLE 7.4.2 Calculate

$$L^{-1}\left(\frac{1}{s^2(s^2+1)}\right).$$

Solution: We write

$$L^{-1}\left(\frac{1}{s^2(s^2+1)}\right) = L^{-1}\left(\frac{1}{s^2}\cdot\frac{1}{s^2+1}\right)$$

$$= \int_0^x (x-t)\cdot\sin t\, dt.$$

Notice that we have recognized that $1/s^2$ is the Laplace transform of x and $1/(s^2+1)$ is the Laplace transform of $\sin x$, and then applied the convolution result.

Now the last integral is easily evaluated (just integrate by parts) and seen to equal

$$x - \sin x.$$

We have thus discovered, rather painlessly, that

$$L^{-1}\left(\frac{1}{s^2(s^2+1)}\right) = x - \sin x. \qquad \qquad \blacksquare$$

The reader may note that this last example could also be done by using partial fractions.

An entire area of mathematics is devoted to the study of integral equations of the form

$$f(x) = y(x) + \int_0^x k(x-t)y(t)\, dt. \qquad (7.4.3)$$

Here f is a given forcing function, and k is a given function known as the *kernel*. Usually k is a mathematical model for the physical process being studied. The object is to solve for y. As you can see, the integral equation involves a convolution. And, not surprisingly, the Laplace transform comes to our aid in unraveling the equation.

In fact we apply the Laplace transform to both sides of (7.4.3). The result is

$$L[f] = L[y] + L[k]\cdot L[y]$$

hence

$$L[y] = \frac{L[f]}{1 + L[k]}.$$

Let us look at an example in which this paradigm occurs.

EXAMPLE 7.4.4 Use the Laplace transform to solve the integral equation

$$y(x) = x^3 + \int_0^x \sin(x-t)y(t)\, dt.$$

Solution: We apply the Laplace transform to both sides (using the convolution formula):

$$L[y] = L[x^3] + L[\sin x] \cdot L[y].$$

Solving for $L[y]$, we see that

$$L[y] = \frac{L[x^3]}{1 - L[\sin x]} = \frac{3!/s^4}{1 - 1/(s^2 + 1)}.$$

We may simplify the right-hand side to obtain

$$L[y] = \frac{3!}{s^4} + \frac{3!}{s^6}.$$

Of course it is easy to determine the inverse Laplace transform of the right-hand side. The result is

$$y(x) = x^3 + \frac{x^5}{20}. \qquad \blacksquare$$

7.4.1   Abel's Mechanics Problem

We now study an old problem from mechanics that goes back to Niels Henrik Abel (1802–1829). Imagine a wire bent into a smooth curve (Figure 7.1). The curve terminates at the origin. Imagine a bead sliding from the top of the wire, without friction, down to the origin. The only force acting on the bead is gravity, depending only on the weight of the bead. Say that the wire is the graph of a function $y = y(x)$. Then the total time for the descent of the bead is some number $T(y)$ that depends on the shape of the wire and on the initial height y. Abel's problem is to run the process in reverse: Suppose that we are given a function T. Then find the shape y of a wire that will result in this time-of-descent function T.

What is interesting about this problem, from the point of view of the present section, is that its mathematical formulation leads to an integral equation of the sort that we have just been discussing. And we shall be able to solve it using the Laplace transform.

We begin our analysis with the principle of conservation of energy. Namely,

$$\frac{1}{2}m \left(\frac{ds}{dt} \right)^2 = m \cdot g \cdot (y - v).$$

In this equation, m is the mass of the bead, ds/dt is its velocity (where of course s denotes arc length), and g is the acceleration due to gravity. We assume that $s'(0) = 0$.

We use (u, v) as the coordinates of any intermediate point on the curve. The expression on the left-hand side is the standard one from physics for kinetic energy. And the expression on the right is the potential energy.

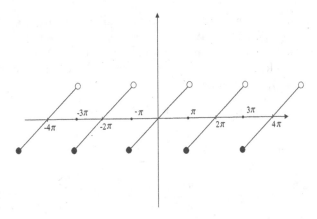

FIGURE 7.1
A smooth curve which a bead will slide down.

We may rewrite the last equation as

$$-\frac{ds}{dt} = \sqrt{2g(y-v)}$$

where the minus sign occurs because the bead is falling. This is equivalent to

$$dt = -\frac{ds}{\sqrt{2g(y-v)}}.$$

Integrating from $v = y$ to $v = 0$ yields

$$T(y) = \int_{v=y}^{v=0} dt = \int_{v=0}^{v=y} \frac{ds}{\sqrt{2g(y-v)}} = \frac{1}{\sqrt{2g}} \int_0^y \frac{s'(v)\,dv}{\sqrt{y-v}}. \qquad (7.4.5)$$

Now we know from calculus how to calculate the length of a curve:

$$s = s(y) = \int_0^y \sqrt{1 + \left(\frac{dx}{dy}\right)^2}\,dy,$$

hence

$$f(y) \equiv s'(y) = \sqrt{1 + \left(\frac{dx}{dy}\right)^2}. \qquad (7.4.6)$$

Substituting this last expression into (7.4.5), we find that

$$T(y) = \frac{1}{\sqrt{2g}} \int_0^y \frac{f(v)\,dv}{\sqrt{y-v}}. \qquad (7.4.7)$$

This formula, in principle, allows us to calculate the total descent time $T(y)$ whenever the curve y is given. From the point of view of Abel's problem, the function $T(y)$ is given, and we wish to find y. We think of $f(y)$ as the unknown. Equation (7.4.7) is called *Abel's integral equation*.

We note that the integral on the right-hand side of Abel's equation is a convolution (of the functions $y^{-1/2}$ and f). Thus, when we apply the Laplace transform to (7.4.7), we obtain

$$L[T(y)] = \frac{1}{\sqrt{2g}} L[y^{-1/2}] \cdot L[f(y)].$$

Now we know from Example 7.2.2 that $L[y^{-1/2}] = \sqrt{\pi/s}$. Hence the last equation may be written as

$$L[f(y)] = \sqrt{2g} \cdot \frac{L[T(y)]}{\sqrt{\pi/s}} = \sqrt{\frac{2g}{\pi}} \cdot s^{1/2} \cdot L[T(y)]. \qquad (7.4.8)$$

When $T(y)$ is given, then the right-hand side of (7.4.8) is completely known, so we can then determine $L[f(y)]$ and hence y (by solving the differential equation (7.4.6)).

EXAMPLE 7.4.9 Analyze the case of Abel's mechanical problem when $T(y) = T_0$, a constant.

Solution: Our hypothesis means that the time of descent is independent of where on the curve we release the bead. A curve with this property (if in fact one exists) is called a *tautochrone*. In this case equation (7.4.8) becomes

$$L[f(y)] = \sqrt{\frac{2g}{\pi}} s^{1/2} L[T_0] = \sqrt{\frac{2g}{\pi}} s^{1/2} \frac{T_0}{s} = b^{1/2} \cdot \sqrt{\frac{\pi}{s}},$$

where we have used the shorthand $b = 2gT_0^2/\pi^2$. Now $L^{-1}[\sqrt{\pi/s}] = y^{-1/2}$, hence we find that

$$f(y) = \sqrt{\frac{b}{y}}. \qquad (7.4.9.1)$$

Now the differential equation (7.4.6) tells us that

$$1 + \left(\frac{dx}{dy}\right)^2 = \frac{b}{y}$$

hence

$$x = \int \sqrt{\frac{b-y}{y}}\, dy.$$

Using the change of variable $y = b\sin^2\phi$, we obtain

$$\begin{aligned}
x &= 2b \int \cos^2\phi\, d\phi \\
&= b \int (1 + \cos 2\phi)\, d\phi \\
&= \frac{b}{2}(2\phi + \sin 2\phi) + C.
\end{aligned}$$

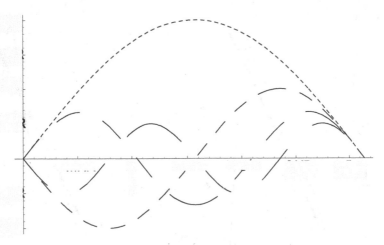

FIGURE 7.2
The cycloid.

In conclusion,

$$x = \frac{b}{2}(2\phi + \sin 2\phi) + C \qquad \text{and} \qquad y = \frac{b}{2}(1 - \cos 2\phi). \qquad (7.4.9.2)$$

The curve must, by the initial mandate, pass through the origin. Hence $C = 0$. If we put $a = b/2$ and $\theta = 2\phi$ then (7.4.9.2) takes the simpler form

$$x = a(\theta + \sin \theta) \qquad \text{and} \qquad y = a(1 - \cos \theta).$$

These are the parametric equations of a cycloid (Figure 7.2). A cycloid is a curve generated by a fixed point on the edge of a disc of radius a rolling along the x-axis. See Figure 7.3. We invite the reader to work from this synthetic definition to the parametric equations that we just enunciated. ∎

Thus the tautochrone turns out to be a cycloid. This problem and its solution is one of the great triumphs of modern mechanics. An additional very interesting property of this curve is that it is the *brachistochrone*. That means that, given two points A and B in space, the curve connecting them down which a bead will slide the *fastest* is the cycloid (Figure 7.4). This last assertion was proved by Isaac Newton, who read the problem as posed in a public challenge by Bernoulli in a periodical. Newton had just come home from a long day at the British Mint (where he worked after he gave up his scientific work). He solved the problem in a few hours, and submitted his solution anonymously. But Bernoulli said he knew it was Newton; he "recognized the lion by his paw."

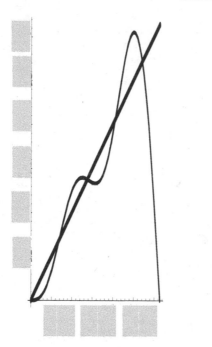
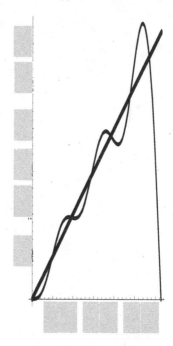

FIGURE 7.3
Generating the cycloid.

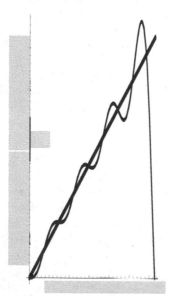
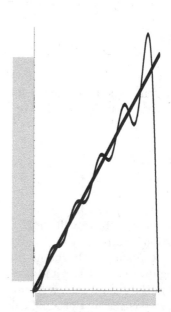

FIGURE 7.4
The brachistrochrone.

Exercises

1. Find $L^{-1}[1/(s^2 + a^2)]$ by using convolution. [**Hint:** Refer to Exercise 1 of the last section.]

2. Solve each of the following integral equations.

   (a) $y(x) = 1 - \int_0^x (x - t)y(t)\, dt$

   (b) $y(x) = e^x \left(1 + \int_0^x e^{-t} y(t)\, dt \right)$

   (c) $e^{-x} = y(x) + 2 \int_0^x \cos(x - t)y(t)\, dt$

   (d) $3 \sin 2x = y(x) + \int_0^x (x - t)y(t)\, dt$

3. Find the equation of the curve of descent if $T(y) = k\sqrt{y}$ for some constant k.

4. Show that the initial value problem

$$y'' + a^2 y = f(x), \qquad y(0) = y'(0) = 0,$$

   has solution

$$y(x) = \frac{1}{a} \int_0^x f(t) \sin a(x - t)\, dt.$$

7.5   The Unit Step and Impulse Functions

In this section our goal is to apply the formula

$$L[f * g] = L[f] \cdot L[g]$$

to study the response of an electrical or mechanical system.

Any physical system that responds to a stimulus can be thought of as a device (or black box) that transforms an *input function* (the stimulus) into an *output function* (the response). If we assume that all initial conditions are zero at the moment $t = 0$ when the input f begins to act, then we may hope to solve the resulting differential equation by application of the Laplace transform.

To be more specific, let us consider solutions of the equation

$$y'' + ay' + by = f$$

satisfying the initial conditions $y(0) = 0$ and $y'(0) = 0$. Notice that, since the equation is nonhomogeneous, these zero initial conditions cannot force the solution to be identically zero. The input f can be thought of as an impressed external force F or electromotive force E that begins to act at time $t = 0$—just as we discussed when we considered forced vibrations.

When the input function happens to be the unit *step function* (or *heaviside function*)

$$u(t) = \begin{cases} 0 & \text{if} \quad t < 0 \\ 1 & \text{if} \quad t \geq 0, \end{cases}$$

then the solution $y(t)$ is denoted by $A(t)$ and is called the *step response* (or indicial response). That is to say,

$$A'' + aA' + bA = u. \tag{7.5.1}$$

Now, applying the Laplace transform to both sides of (7.5.1), and using our standard formulas for the Laplace transforms of derivatives, we find that

$$s^2 L[A] + asL[A] + bL[A] = L[u] = \frac{1}{s}.$$

Here we have calculated that $L[u](s) = 1/s$.

So we may solve for $L[A]$ and obtain that

$$L[A] = \frac{1}{s} \cdot \frac{1}{s^2 + as + b} = \frac{1}{s} \cdot \frac{1}{z(s)}, \tag{7.5.2}$$

where

$$z(s) = s^2 + as + b. \tag{7.5.3}$$

Note that we have just been examining the special case of our differential equation with a step function on the right-hand side. Now let us consider the equation in its general form (with an arbitrary external force function f):

$$y'' + ay' + by = f.$$

Applying the Laplace transform to both sides (and using our zero initial conditions) gives

$$s^2 L[y] + asL[y] + bL[y] = L[f]$$

or

$$L[y] \cdot z(s) = L[f]$$

so

$$L[y] = \frac{L[f]}{z(s)}. \tag{7.5.4}$$

We divide both sides of (7.5.4) by s and use (7.5.2). The result is

$$\frac{1}{s} \cdot L[y] = \frac{1}{sz(s)} \cdot L[f] = L[A] \cdot L[f].$$

This suggests the use of the convolution theorem:

$$\frac{1}{s} \cdot L[y] = L[A * f].$$

As a result,

$$L[y] = s \cdot L\left(\int_0^t A(t-\tau)f(\tau)\,d\tau\right)$$

$$= L\left(\frac{d}{dt}\int_0^t A(t-\tau)f(\tau)\,d\tau\right).$$

Thus we finally obtain that

$$y(t) = \frac{d}{dt}\int_0^t A(t-\tau)f(\tau)\,d\tau. \tag{7.5.5}$$

What we see here is that, once we find the solution A of the differential equation with a step function as an input, then we can obtain the solution for any other input f by convolving A with f and then taking the derivative. With some effort, we can rewrite equation (7.5.5) in an even more appealing way.

In fact we can go ahead and perform the differentiation in (7.5.5) to obtain

$$y(t) = \int_0^t A'(t-\tau)f(\tau)\,d\tau + A(0)f(t).$$

Alternatively, we can use a change of variable to write the convolution as

$$\int_0^t f(t-\sigma)A(\sigma)\,d\sigma.$$

This results in the formula

$$y(t) = \int_0^t f'(t-\sigma)A(\sigma)\,d\sigma + f(0)A(t).$$

Changing variables back again, this gives

$$y(t) = \int_0^t A(t-\tau)f'(\tau)\,d\tau + f(0)A(t). \tag{7.5.6}$$

We notice that the initial conditions force $A(0) = 0$ so our other formula (7.5.6) becomes

$$y(t) = \int_0^t A'(t-\tau)f(\tau)\,d\tau. \tag{7.5.7}$$

Either of (7.5.6) or (7.5.7) is commonly called the *principle of superposition*. They allow us to represent a solution of our differential equation for a general input function in terms of a solution for a step function.

EXAMPLE 7.5.8 Use the principle of superposition to solve the equation

$$y'' + y' - 6y = 2e^{3t}$$

with initial conditions $y(0) = 0$, $y'(0) = 0$.

Solution: We first observe that

$$z(s) = s^2 + s - 6$$

(see the discussion of equation (7.5.3)). Hence

$$L[A] = \frac{1}{s(s^2 + s - 6)}.$$

Now it is a simple matter to apply partial fractions and elementary Laplace transform inversion to obtain

$$A(t) = -\frac{1}{6} + \frac{1}{15}e^{-3t} + \frac{1}{10}e^{2t}.$$

Now $f(t) = 2e^{3t}$, $f'(t) = 6e^{3t}$, and $f(0) = 2$. Thus (7.5.6) gives

$$
\begin{aligned}
y(t) \quad &= \quad \int_0^t \left(-\frac{1}{6} + \frac{1}{15}e^{-3(t-\tau)} + \frac{1}{10}e^{2(t-\tau)} \right) \cdot 6e^{3\tau}\, d\tau \\
&\quad + 2\left(-\frac{1}{6} + \frac{1}{15}e^{-3t} + \frac{1}{10}e^{2t} \right) \\
&= \quad \frac{1}{3}e^{3t} + \frac{1}{15}e^{-3t} - \frac{2}{5}e^{2t}.
\end{aligned}
$$

We invite the reader to confirm that this is indeed a solution to our initial value problem.   ∎

We can use the second principle of superposition, rather than the first, to solve the differential equation. The process is expedited if we first rewrite the equation in terms of an impulse (rather than a step) function.

What is an impulse function? Physicists think of an impulse function as one that takes the value 0 at all points except the origin; at the origin the impulse function takes the value $+\infty$. See Figure 7.5.

In practice, mathematicians think of an impulse function as a limit of functions

$$\varphi_\epsilon(x) = \begin{cases} \dfrac{1}{\epsilon} & \text{if} \quad 0 \le x \le \epsilon \\[2mm] 0 & \text{if} \quad x > \epsilon \end{cases}$$

as $\epsilon \to 0^+$. See Figure 7.6. Observe that, for any $\epsilon > 0$, $\int_0^\infty \varphi_\epsilon(x)\, dx = 1$. It is straightforward to calculate that

$$L[\varphi_\epsilon] = \frac{1 - e^{-s\epsilon}}{s\epsilon}$$

and hence (using l'Hôpital's Rule) that

$$\lim_{\epsilon \to 0} L[\varphi_\epsilon](s) \equiv 1 \qquad \text{for all} \quad s.$$

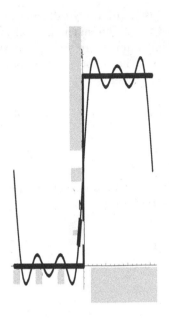

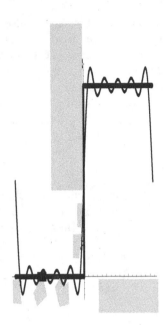

FIGURE 7.5
An impulse function.

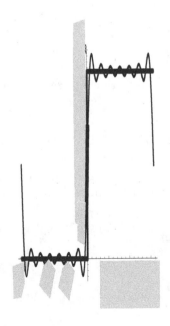

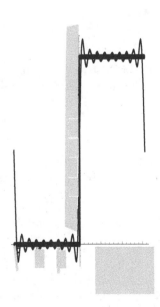

FIGURE 7.6
An impulse function as a limit of tall, thin bumps.

Thus we think of the impulse—intuitively—as an infinitely tall spike at the origin with Laplace transform identically equal to 1. The mathematical justification for the concept of the impulse was outlined in the previous paragraph. A truly rigorous treatment of the impulse requires the theory of distributions (or generalized functions) and we cannot cover it here. We do give a brief introduction to distributions in the next chapter.

It is common to denote the impulse function by $\delta(t)$ (in honor of Paul Dirac, who developed the idea).[3] We have that

$$L[\delta] \equiv 1 .$$

In the special case that the input function for our differential equation is $f(t) = \delta$, then the solution y is called the *impulsive response* and denoted $h(t)$. In this circumstance we have

$$L[h] = \frac{1}{z(s)}$$

hence

$$h(t) = L^{-1}\left(\frac{1}{z(s)}\right) .$$

Now we know that

$$L[A] = \frac{1}{s} \cdot \frac{1}{z(s)} = \frac{L[h]}{s} .$$

As a result,

$$A(t) = \int_0^t h(\tau)\, d\tau .$$

But this last formula shows that $A'(t) = h(t)$, so that our second superposition formula (7.5.7) becomes

$$y(t) = \int_0^t h(t - \tau) f(\tau)\, d\tau . \tag{7.5.8}$$

In summary, the solution of our differential equation with general input function f is given by the convolution of the impulsive response function with f.

EXAMPLE 7.5.9 Solve the differential equation

$$y'' + y' - 6y = 2e^{3t}$$

with initial conditions $y(0) = 0$ and $y'(0) = 0$ using the *second* of our superposition formulas, as rewritten in (7.5.8).

---

[3] It should be noted that, strictly speaking, the Dirac impulse function is not a function. But it is nonetheless useful, from an intuitive point of view, to treat it as a function. Modern mathematical formalism provides rigorous means to handle this object.

Solution: We know that

$$h(t) = L^{-1}\left(\frac{1}{z(s)}\right)$$

$$= L^{-1}\left(\frac{1}{(s+3)(s-2)}\right)$$

$$= L^{-1}\left\{\frac{1}{5}\left(\frac{1}{s-2} - \frac{1}{s+3}\right)\right\}$$

$$= \frac{1}{5}\left(e^{2t} - e^{-3t}\right).$$

As a result,

$$y(t) = \int_0^t \frac{1}{5}\left(e^{2(t-\tau)} - e^{-3(t-\tau)}\right)2e^{3t}\,d\tau$$

$$= \frac{1}{3}e^{3t} + \frac{1}{15}e^{-3t} - \frac{2}{5}e^{2t}.$$

Of course this is the same solution that we obtained in the last example, using the other superposition formula. ∎

To form a more general view of the meaning of convolution, consider a linear physical system in which the effect at the present moment of a small stimulus $g(\tau)\,d\tau$ at any past time τ is proportional to the size of the stimulus. We further assume that the proportionality factor depends only on the elapsed time $t - \tau$, and thus has the form $f(t - \tau)$. The effect at the present time t is therefore

$$f(t - \tau) \cdot g(\tau)\,d\tau.$$

Since the system is linear, the total effect at the present time t due to the stimulus acting throughout the entire past history of the system is obtained by adding these separate effects, and this observation leads to the convolution integral

$$\int_0^t f(t - \tau)g(\tau)\,d\tau.$$

The lower limit is 0 just because we assume that the stimulus started acting at time $t = 0$, i.e., that $g(\tau) = 0$ for all $\tau < 0$. Convolution plays a vital role in the study of wave motion, heat conduction, diffusion, and many other areas of mathematical physics.

Exercises

1. Find the convolution of each of the following pairs of functions:

   (a) 1, $\sin at$

   (b) e^{at}, e^{bt} for $a \neq b$

   (c) t, e^{at}

   (d) $\sin at$, $\sin bt$ for $a \neq b$

2. Verify that the Laplace transform of a convolution is the product of the Laplace transforms for each of the pairs of functions in Exercise 1.

3. Use the methods of Examples 8.4.8, 8.4.9 to solve each of the following differential equations.

   (a) $y'' + 5y' + 6y = 5e^{3t}$, $y(0) = y'(0) = 0$

   (b) $y'' + y' - 6y = t$, $y(0) = y'(0) = 0$

   (c) $y'' - y' = t^2$, $y(0) = y'(0) = 0$

4. When the polynomial $z(s)$ has distinct real roots a and b, so that

$$\frac{1}{z(s)} = \frac{1}{(s-a)(s-b)} = \frac{A}{s-a} + \frac{B}{s-b}$$

   for suitable constants A and B, then

$$h(t) = Ae^{at} + Be^{bt}.$$

   Also equation (7.5.8) takes the form

$$y(t) = \int_0^t f(\tau)[Ae^{a(t-\tau)} + Be^{b(t-\tau)}] \, d\tau.$$

   This formula is sometimes called the *Heaviside expansion theorem*.

   (a) Use this theorem to write the solution of $y'' + 3y' + 2y = f(t)$, $y(0) = y'(0) = 0$.

   (b) Give an explicit evaluation of the solution in (a) for the cases $f(t) = e^{3t}$ and $f(t) = t$.

   (c) Find the solutions in (b) by using the superposition principle.

5. Show that $f * g = g * f$ directly from the definition of convolution, by introducing a new dummy variable $\sigma = t - \tau$. This calculation shows that the operation of convolution is commutative. It is also associative and distributive:

$$f * [g * h] = [f * g] * h$$

   and

$$f * [g + h] = f * g + f * h$$

   and

$$[f + g] * h = f * h + g * h.$$

   Use a calculation to verify each of these last three properties.

6. We know from our earlier studies that the forced vibrations of an undamped spring-mass system are described by the differential equation

$$Mx'' + kx = f(t),$$

where $x(t)$ is the displacement and $f(t)$ is the impressed external force or "forcing function." If $x(0) = x'(0) = 0$, then find the functions A and h and write down the solution $x(t)$ for any $f(t)$.

7. The current I in an electric field with inductance L and resistance R is given (as we saw in Section 1.12) by

$$L\frac{I}{dt} + RI = E.$$

Here E is the impressed electromotive force. If $I(0) = 0$, then use the methods of this section to find I in each of the following cases.

(a) $E(t) = E_0 u(t)$

(b) $E(t) = E_0 \delta(t)$

(c) $E(t) = E_0 \sin \omega t$

Historical Note

Laplace

Pierre Simon de Laplace (1749–1827) was a French mathematician and theoretical astronomer who was so celebrated in his own time that he was sometimes called "the Isaac Newton of France." His main scientific interests were celestial mechanics and the theory of probability.

Laplace's monumental treatise *Mécanique Céleste* (published in five volumes from 1799 to 1825) contained a number of triumphs, including a rigorous proof that our solar system is a stable dynamical system that will not (as Newton feared) degenerate into chaos. Laplace was not always true to standard scholarly dicta; he frequently failed to cite the contributions of his predecessors, leaving the reader to infer that all the ideas were due to Laplace.

Many anecdotes are associated with Laplace's work in these five tomes. One of the most famous concerns an occasion when Napoleon Bonaparte endeavored to get a rise out of Laplace by protesting that he had written a huge book on the system of the world without once making reference to its author (God). Laplace is reputed to have replied, "Sire, I had no need of that hypothesis." Lagrange is reputed to have then said that, "It is a beautiful hypothesis just the same. It explains so many things."

One of the most important features of Laplace's *Mécanique Céleste* is its development of potential theory. Even though he borrowed some of the ideas without attribution from Lagrange, he contributed many of his own. To this

day, the fundamental equation of potential theory is called "Laplace's equation," and the partial differential operator involved is called "the Laplacian."

Laplace's other great treatise was *Théorie Analytique des Probabilités*
(1812). This is a great masterpiece of probability theory, and establishes many
analytic techniques for studying this new subject. It is technically quite sophisticated, and uses such tools as the Laplace transform and generating functions.

Laplace was politically very clever, and always managed to align himself
with the party in power. As a result, he was constantly promoted to ever more
grandiose positions. To balance his other faults, Laplace was quite generous
in supporting and encouraging younger scientists. From time to time he went
to the aid of Gay-Lussac (the chemist), Humboldt (the traveler and natu-
ralist), Poisson (the physicist and mathematician), and Cauchy (the complex
analyst). Laplace's overall impact on modern mathematics has been immense,
and his name occurs frequently in the literature.

7.6   Flow Initiated by an Impulsively Started Flat Plate

Imagine the two-dimensional flow of a semi-infinite extent of viscous fluid,
supported on a flat plate, caused by the motion of the flat plate in its own
plane. Let us use cartesian coordinates with the x-axis lying in the plane of
the plate and the y-axis pointing into the fluid. See Figure 7.7.

Now let $u(x, y, t)$ denote the velocity of the flow *in the x-direction only*.
It can be shown that this physical system is modeled by the boundary value
problem

$$\frac{\partial u}{\partial t} = \nu \frac{\partial^2 u}{\partial y^2}$$

$$u = 0 \quad \text{if} \quad t = 0, y > 0$$

$$u = \mathcal{U} \quad \text{if} \quad t > 0, y = 0$$

$$u \to 0 \quad \text{if} \quad t > 0, y \to \infty.$$

Here ν is a physical constant known as the *kinematic viscosity*. The constant
$\mathcal{U}$ is determined by the initial state of the system. This partial differential
equation is a version of the classical *heat equation*. It is parabolic in form.
It can also be used to model other diffusive systems, such as a semi-infinite
bar of metal, insulated along its sides, suddenly heated up at one end. The
system we are considering is known as *Rayleigh's problem*. This mathematical
model shows that the only process involved in the flow is the diffusion of x-
momentum into the bulk of the fluid (since u represents unidirectional flow in
the x-direction).

In order to study this problem, we shall freeze the y-variable and take
the Laplace transform in the time variable t. We denote this "partial Laplace

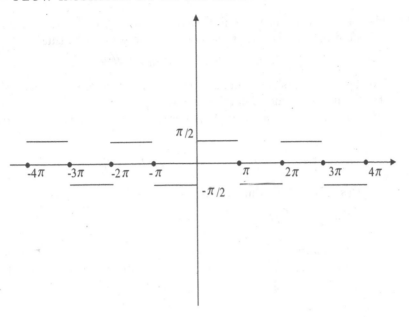

FIGURE 7.7
Flow of a viscous fluid on a flat plate.

transform" by $\tilde{L}$. Thus we write

$$\tilde{L}[u(y,t)] \equiv U(y,s) = \int_0^\infty e^{-st} u(y,t)\, dt\,.$$

We differentiate both sides of this equation twice with respect to y—of course these differentiations commute with the Laplace transform in t. The result is

$$\hat{L}\left(\frac{\partial^2 u}{\partial y^2}\right) = \frac{\partial^2 U}{\partial y^2}\,.$$

But now we use our partial differential equation to rewrite this as

$$\tilde{L}\left(\frac{1}{\nu}\frac{\partial u}{\partial t}\right) = \frac{\partial^2 U}{\partial y^2}\,.$$

But of course we have a formula for the Laplace transform (in the t-variable) of the derivative in t of u. Using the first boundary condition, that formula simplifies to $\tilde{L}[\partial u/\partial t] = sU(y,s)$. Thus the equation becomes

$$\frac{\partial^2 U}{\partial y^2} - \frac{s}{\nu}U = 0\,. \tag{7.6.1}$$

In order to study equation (7.6.1), we think of s as a parameter and of y

as the independent variable. So we now have a familiar second-order ordinary differential equation with constant coefficients. The solution is thus

$$U(y, s) = A(s)e^{\sqrt{s}y/\sqrt{\nu}} + B(s)e^{-\sqrt{s}y/\sqrt{\nu}}.$$

Notice that the "constants" depend on the parameter s. Also $u \to 0$ as $y \to \infty$. Passing the limit under the integral sign, we then see that $U = \tilde{u} \to 0$ as $y \to \infty$. It follows that $A(s) = 0$. We may also use the second boundary condition to write

$$U(0, s) = \int_0^\infty u(0, t)e^{-st}\, dt = \int_0^\infty \mathcal{U}e^{-st}\, dt = \frac{\mathcal{U}}{s}.$$

As a result, $B(s) = \mathcal{U}/s$. We thus know that

$$U(y, s) = \frac{\mathcal{U}}{s} \cdot e^{-\sqrt{s}y/\sqrt{\nu}}.$$

A difficult calculation (see [KBO, pp. 164–167]) now shows that the inverse Laplace transform of U is the important *complementary erf function* erfc. Here we define

$$\mathrm{erfc}(x) = \frac{2}{\sqrt{\pi}} \int_0^x e^{-t^2}\, dt.$$

(This function, as you may know, is modeled on the Gaussian distribution from probability theory.) It can be calculated that

$$u(y, t) = \mathcal{U} \cdot \mathrm{erfc}\left(\frac{y}{2\sqrt{\nu t}}\right).$$

We conclude this discussion by noting that the analysis applies, with some minor changes, to the situation when the velocity of the plate is a function of time. The only change is that the second boundary condition becomes

$$u = \mathcal{U}f(t) \quad \text{if } t > 0, y = 0. \tag{7.6.2}$$

Note the introduction of the function $f(t)$ to represent the dependence on time. The Laplace transform of equation (7.6.2) is $U(0, s) = \mathcal{U}F(s)$, where F is the Laplace transform of f. Now it follows, just as before, that $B(s) = F(s)$. Therefore

$$U(y, s) = \mathcal{U}F(s)e^{-\sqrt{s}y/\sqrt{\nu}} = \mathcal{U}sF(s) \cdot \frac{e^{-\sqrt{s}y/\sqrt{\nu}}}{s}.$$

For simplicity, let us assume that $f(0) = 0$.

Of course $L[f'(t)] = sF(s)$ and so

$$U(y, s) = \mathcal{U} \cdot L[f'(t)] \cdot \tilde{L}\left\{\mathrm{erfc}\left(\frac{y}{2\sqrt{\nu t}}\right)\right\}.$$

Finally, we may use the convolution theorem to invert the Laplace transform and obtain

$$u(y, t) = \mathcal{U}\left\{\int_{\tau=0}^t f'(t - \tau) \cdot \mathrm{erfc}\left(\frac{y}{2\sqrt{\nu \tau}}\right) d\tau\right\}.$$

Problems for Review and Discovery

A. Drill Exercises

1. Calculate the Laplace transforms of each of the following functions.

   (a) $f(t) = 8 + 4e^{3t} - 5\cos 3t$

   (b) $g(t) = \begin{cases} 1 & \text{if} & 0 < t < 4 \\ 0 & \text{if} & 4 \le t \le 8 \\ e^{2t} & \text{if} & 8 < t \end{cases}$

   (c) $h(t) = \begin{cases} 2 - t & \text{if} & 0 < t < 2 \\ 0 & \text{if} & 2 \le t < \infty \end{cases}$

   (d) $f(t) = \begin{cases} 0 & \text{if} & 0 < t \le 4 \\ 3t - 12 & \text{if} & 4 < t < \infty \end{cases}$

   (e) $g(t) = t^5 - t^3 + \cos\sqrt{2}t$

   (f) $h(t) = e^{-2t}\cos 4t + t^2 - e^{-t}$

   (g) $f(t) = e^{-4t}\sin\sqrt{5}t + t^3 e^{-3t}$

   (h) $g(t) = te^t - t^2 e^{-t} + t^3 e^{3t}$

   (i) $h(t) = \sin^2 t$

   (j) $f(t) = \sin 3t \sin 5t$

   (k) $g(t) = (1 - e^{-t})^2$

   (l) $h(t) = \cosh 4t$

   (m) $f(t) = \cos 2t \sin 3t$

2. Find a function f whose Laplace transform is equal to the given expression.

   (a) $\dfrac{4}{s^2 + 16}$

   (b) $\dfrac{s - 2}{s^2 - 3s + 6}$

   (c) $\dfrac{4}{(s + 3)^4}$

   (d) $\dfrac{3s - 2}{s^2 + s + 4}$

   (e) $\dfrac{2s - 5}{(s + 1)(s + 3)(s - 4)}$

   (f) $\dfrac{s^2 + 2s + 2}{(s - 3)^2(s + 1)}$

   (g) $\dfrac{3s^2 + 4s}{(s^2 - s + 2)(s - 1)}$

   (h) $\dfrac{6s^2 - 13s + 2}{s(s - 1)(s - 5)}$

   (i) $\dfrac{s + 1}{s^2 + s + 6}$

   (j) $\dfrac{3}{(s^2 - 4)(s + 2)}$

3. Use the method of Laplace transforms to solve the following initial value problems:

    (a) $y''(t) + 3ty'(t) - 5y(t) = 1,$ $y(0) = y'(0) = 1$

    (b) $y''(t) + 3y'(t) - 2y(t) = -6e^{\pi-t},$ $y(\pi) = 1,\ y'(\pi) = 4$

    (c) $y''(t) + 2y'(t) - y(t) = te^{-t},$ $y(0) = 0,\ y'(0) = 1$

    (d) $y'' - y' + y = 3e^{-t},$ $y(0) = 3, y'(0) = 2$

4. Use the method of Laplace transforms to find the general solution of each of the following differential equations. [**Hint:** Use the boundary conditions $y(0) = A$ and $y'(0) = B$ to introduce the two undetermined constants that you need.]

    (a) $y'' - 5y' + 4y = 0$

    (b) $y'' + 3y' + 3y = 2$

    (c) $y'' + y' + 2y = t$

    (d) $y'' - 7y' + 12y = te^{2t}$

5. Express each of these functions using one or more step functions, and then calculate the Laplace transform.

    (a) $f(t) = \begin{cases} 0 & \text{if } 0 < t < 2 \\ 3 & \text{if } 2 \le t \le 5 \\ 1 & \text{if } 5 < t < 8 \\ -4 & \text{if } 8 \le t < \infty \end{cases}$

    (b) $g(t) = \begin{cases} 0 & \text{if } 0 < t < 3 \\ t-1 & \text{if } 3 \le t < \infty \end{cases}$

    (c) $h(t) = \begin{cases} t & \text{if } 0 < t < 3 \\ 1 & \text{if } 3 \le t \le 6 \\ 1-t & \text{if } 6 < t < \infty \end{cases}$

B. Challenge Problems

1. Solve each equation for $L^{-1}(F)$.

    (a) $s^2 F(s) - 9F(s) = \dfrac{4}{s+1}$

    (b) $pF(s) + 3F(s) = \dfrac{s^2 + 3s + 5}{s^2 - 2s - 2}$

    (c) $pF(s) - F(s) = \dfrac{s+1}{s^2 + 6s + 9}$

    (d) $s^2 F(s) + F(s) = \dfrac{s-1}{s+1}$

2. Use the formula for the Laplace transform of a derivative to calculate the inverse Laplace transforms of these functions.

    (a) $F(s) = \ln\left(\frac{s+3}{s-4}\right)$

    (b) $F(s) = \ln\left(\frac{s-2}{s-1}\right)$

    (c) $F(s) = \ln\left(\frac{s^2+4}{s^2+2}\right)$

3. The current $I(t)$ in a circuit involving resistance, conductance, and capacitance is described by the initial value problem

$$\frac{d^2I}{dt^2} + 2\frac{dI}{dt} + 3I = g(t)$$

$$I(0) = 8 \ , \quad \frac{dI}{dt}(0) = 0 \, ,$$

where

$$g(t) = \begin{cases} 30 & \text{if } 0 < t < 2\pi \\ 0 & \text{if } 2\pi \le t \le 5\pi \\ 10 & \text{if } 5\pi < t < \infty \, . \end{cases}$$

Find the current as a function of time.

In Exercises 4–7, determine the Laplace transform of the function which is described by the given graph.

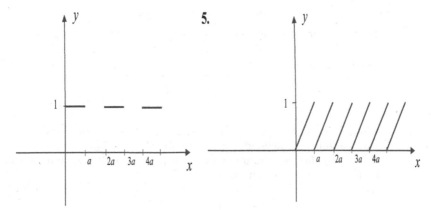

4.

5.

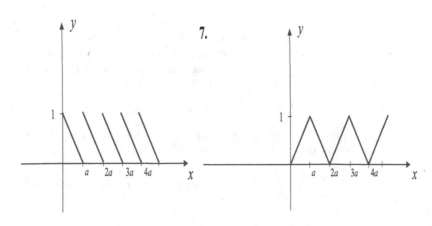

6.

7.

C. Problems for Discussion and Exploration

1. Define, for j a positive integer,

$$\phi_j(x) = \begin{cases} 0 & \text{if } -\infty < x < -\frac{1}{j} \\ 2j & \text{if } -\frac{1}{j} \le x \le \frac{1}{j} \\ 0 & \text{if } \frac{1}{j} < x < \infty. \end{cases}$$

Calculate the Laplace transform of ϕ_j, and verify that it converges to the Laplace transform of a unit impulse function.

2. Derive this formula of Oliver Heaviside. Suppose that P and Q are polynomials with the degree of P less than the degree of q. Assume that $r_1, \ldots, r_n$ are the distinct real roots of Q, and that these are all the roots of Q. Show that

$$L^{-1}\left(\frac{P}{Q}\right)(t) = \sum_{j=1}^{n} \frac{P(r_j)}{Q'(r_j)} e^{r_j t}.$$

3. Let us consider a linear system controlled by the ordinary differential equation

$$ay''(t) + by'(t) + cy(t) = g(t).$$

Here a, b, c are real constants. We call g the *input* function for the system and y the *output* function.

Let $Y = L[y]$ and $G = L[g]$. We set

$$H(s) = \frac{Y(s)}{G(s)}. \qquad (*)$$

Then H is called the *transfer function* for the system. Show that the transfer function depends on the choice of a, b, c but *not* on the input function g. In case the input function g is the unit step function $u(t)$, then equation $(*)$ tells us that

$$L[y](s) = \frac{H(s)}{s}.$$

In these circumstances we call the solution function the *indicial admittance* and denote it by $A(t)$ (instead of the customary $y(t)$).

We can express the general response function $y(t)$ for an arbitrary input $g(t)$ in terms of the special response function $A(t)$ for the step function input $u(t)$. To see this assertion, first show that

$$L[y](s) = sL[A](s)L[g](s).$$

Next apply the fact that the Laplace transform of a convolution is the product of the Laplace transforms to see that

$$y(t) = \frac{d}{dt}\left(\int_0^t A(t-v)g(v)\,dv\right) = \frac{d}{dt}\left(\int_0^t A(v)g(t-v)\,dv\right).$$

Actually carry out these differentiations and make the change of variable $\zeta = t - v$ to obtain *Duhamel's formulas*

$$y(t) \;\; = \;\; \int_0^t A'(\zeta)g(t-\zeta)\,d\zeta\,,$$

$$y(t) \;\; = \;\; \int_0^t A(t-\zeta)g'(\zeta)\,d\zeta + A(t)g(0)\,.$$

8

Distributions

- Schwartz functions
- Schwartz distributions
- Cutoff functions
- Differentiation of distributions
- Fourier transform of distributions
- Other spaces of distributions
- Structure theorem for distributions

8.1 Schwartz Distributions

Thorough treatments of distribution theory may be found in [HOR], [KRA4]. Here we give a quick review.

We define the space of Schwartz functions:

$$\mathcal{S} = \left\{ \phi \in C^\infty(\mathbb{R}^N) : \rho_{\alpha,\beta}(\phi) \equiv \sup_{x \in \mathbb{R}^N} \left| x^\alpha \left(\frac{\partial}{\partial x} \right)^\beta \phi(x) \right| < \infty, \right.$$

$$\left. \alpha = (\alpha_1, \ldots, \alpha_N), \beta = (\beta_1, \ldots, \beta_N) \right\}.$$

Here

$$\left(\frac{\partial}{\partial x} \right)^\beta \equiv \frac{\partial_1^\beta}{\partial x_1^{\beta_1}} \frac{\partial_2^\beta}{\partial x_1^{\beta_2}} \cdots \frac{\partial_N^\beta}{\partial x_1^{\beta_N}}$$

and

$$x^\alpha \equiv x_1^{\alpha_1} \cdot x_2^{\alpha_2} \cdot \cdots \cdot x_N^{\alpha_N}.$$

Observe that $e^{-|x|^2} \in \mathcal{S}$ and $p(x) \cdot e^{-|x|^2} \in \mathcal{S}$ for any polynomial p. Any derivative of a Schwartz function is still a Schwartz function. The Schwartz space is obviously a linear space.

It is worth noting that the space of C^∞ functions with compact support (which we have been denoting by C_c^∞) forms a proper subspace of $\mathcal{S}$. Since as recently as 1930 there was some doubt as to whether C_c^∞ functions are genuine functions, it may be worth seeing how to construct elements of this space.

Let the dimension N equal 1. Define

$$\lambda(x) = \begin{cases} e^{-1/|x|^2} & \text{if } x \geq 0 \\ 0 & \text{if } x < 0 \end{cases}$$

Then one checks, using l'Hôpital's Rule, that $\lambda \in C^\infty(\mathbb{R})$. Set

$$h(x) = \lambda(-x - 1) \cdot \lambda(x + 1) \in C_c^\infty(\mathbb{R}).$$

Moreover, if we define

$$g(x) = \int_{-\infty}^{x} h(t)\, dt$$

then the function

$$f(x) = g(x + 2) \cdot g(-x - 2)$$

lies in C_c^∞ and is identically equal to a constant on $(-1, 1)$. Thus we have constructed a standard "cutoff function" on $\mathbb{R}^1$. On $\mathbb{R}^N$, the function

$$F(x) \equiv f(x_1) \cdots f(x_N)$$

plays a similar role.

Exercise: [**The C^∞ Urysohn lemma**]   Let K and L be disjoint compact sets in $\mathbb{R}^N$. Prove that there is a C^∞ function ϕ on $\mathbb{R}^N$ such that $\phi \equiv 0$ on K and $\phi \equiv 1$ on L. (Details of this sort of construction may be found in [HIR].)

8.1.1   The Topology of the Space $\mathcal{S}$

The functions $\rho_{\alpha,\beta}$ are seminorms on $\mathcal{S}$. A neighborhood basis of 0 for the corresponding topology on $\mathcal{S}$ is given by the sets

$$N_{\epsilon,\ell,m} = \Big\{ \phi : \sum_{\substack{|\alpha| \leq \ell \\ |\beta| \leq m}} \rho_{\alpha,\beta}(\phi) < \epsilon \Big\}.$$

Exercise: The space $\mathcal{S}$ cannot be normed.

Definition 8.1.1 A *Schwartz distribution* α is a continuous linear functional on $\mathcal{S}$. We write $\alpha \in \mathcal{S}'$.

Examples:

1. If $f \in L^1$ then f induces a Schwartz distribution as follows:

$$\mathcal{S} \ni \phi \mapsto \int \phi f \, dx \in \mathbb{C}.$$

We see that this functional is continuous by noticing that

$$\left| \int \phi(x) f(x) \, dx \right| \leq \sup |\phi| \cdot \|f\|_{L^1} = C \cdot \rho_{0,0}(\phi).$$

A similar argument shows that any finite Borel measure induces a distribution.

2. Differentiation is a distribution: On $\mathbb{R}^1$, for example, we have

$$\mathcal{S} \ni \phi \mapsto \phi'(0)$$

satisfies

$$|\phi'(0)| \leq \sup_{x \in \mathbb{R}} |\phi'(x)| = \rho_{0,1}(\phi).$$

3. If $f \in L^p, 1 \leq p \leq \infty$, then f induces a distribution:

$$T_f : \mathcal{S} \ni \phi \mapsto \int \phi f \, dx \in \mathbb{C}.$$

To see that this functional is bounded, we first notice that

$$\left| \int \phi f \right| \leq \|f\|_{L^p} \cdot \|\phi\|_{L^{p'}}, \qquad (8.1.2)$$

where $1/p + 1/p' = 1$. Now notice that

$$(1 + |x|^{N+1})|\phi(x)| \leq C\big(\rho_{0,0}(\phi) + \rho_{N+1,0}(\phi)\big)$$

hence

$$|\phi(x)| \leq \frac{C}{1 + |x|^{N+1}} \big(\rho_{0,0}(\phi) + \rho_{N+1,0}(\phi)\big).$$

Finally,

$$\|\phi\|_{L^{p'}} \leq C \cdot \left[\int \left(\frac{1}{1 + |x|^{N+1}} \right)^{p'} dx \right]^{1/p'} \cdot \big[\rho_{0,0}(\phi) + \rho_{N+1,0}(\phi)\big].$$

As a result, (8.1.2) tells us that

$$T_f(\phi) \leq C\|f\|_{L^p}\big(\rho_{0,0}(\phi) + \rho_{N+1,0}(\phi)\big).$$

8.1.2   Algebraic Properties of Distributions

(i) If $\alpha, \beta \in \mathcal{S}'$ then $\alpha + \beta$ is defined by $(\alpha + \beta)(\phi) = \alpha(\phi) + \beta(\phi)$. Clearly $\alpha + \beta$ so defined is a Schwartz distribution.

(ii) If $\alpha \in \mathcal{S}'$ and $c \in \mathbb{C}$ then $c\alpha$ is defined by $(c\alpha)(\phi) = c[\alpha(\phi)]$. We see that $c\alpha \in \mathcal{S}'$.

(iii) If $\psi \in \mathcal{S}$ and $\alpha \in \mathcal{S}'$ then define $(\psi\alpha)(\phi) = \alpha(\psi\phi)$. It follows that $\psi\alpha$ is a distribution.

(iv) It is a theorem of Laurent Schwartz (see [SCH]) that there is no continuous operation of multiplication on $\mathcal{S}'$. However it is a matter of great interest, especially to mathematical physicists, to have such an operation. Colombeau [CMB] has developed a substitute operation. We shall say no more about it here.

(v) Schwartz distributions may be differentiated as follows: If $\mu \in \mathcal{S}'$ then $(\partial/\partial x)^\beta \mu \in \mathcal{S}'$ is defined, for $\phi \in \mathcal{S}$, by

$$\left[\left(\frac{\partial}{\partial x} \right)^\beta \mu \right] (\phi) = (-1)^{|\beta|} \mu \left(\left(\frac{\partial}{\partial x} \right)^\beta \phi \right).$$

Observe that in case the distribution μ is induced by integration against a C_c^k function f, then the definition is compatible with what integration by parts would yield.

Let us differentiate the distribution induced by integration against the function $f(x) = |x|$ on $\mathbb{R}$. Now, for $\phi \in \mathcal{S}$,

$$
\begin{aligned}
f'(\phi) &\equiv -f(\phi') \\
&= -\int_{-\infty}^{\infty} f\phi' \, dx \\
&= -\int_{0}^{\infty} f(x)\phi'(x) \, dx - \int_{-\infty}^{0} f(x)\phi'(x) \, dx \\
&= -\int_{0}^{\infty} x\phi'(x) \, dx + \int_{-\infty}^{0} x\phi'(x) \, dx \\
&= -\left[x\phi(x) \right]_{0}^{\infty} + \int_{0}^{\infty} \phi(x) \, dx + \left[x\phi(x) \right]_{-\infty}^{0} - \int_{-\infty}^{0} \phi(x) \, dx \\
&= \int_{0}^{\infty} \phi(x) \, dx - \int_{-\infty}^{0} \phi(x) \, dx.
\end{aligned}
$$

Thus f' consists of integration against $b(x) = -\chi_{(-\infty,0]} + \chi_{[0,\infty)}$, where

$$
\chi_S = \begin{cases} 1 & \text{if} \quad x \in S \\ 0 & \text{if} \quad x \notin S \end{cases}
$$

is the characteristic function of the set S. This function is a version of the *heaviside function* that we saw in Section 7.4.

Exercise: Let $\Omega \subseteq \mathbb{R}^N$ be a smoothly bounded domain. Let ν be the unit outward normal vector field to $\partial\Omega$. Prove that $-\nu\chi_\Omega \in \mathcal{S}'$. [**Hint:** Use Green's theorem. It will turn out that $(-\nu\chi_\Omega)(\phi) = \int_{\partial\Omega} \phi \, d\sigma$, where $d\sigma$ is area measure on the boundary.]

8.1.3   The Fourier Transform

The principal importance of the Schwartz distributions as opposed to other distribution theories (more on those below) is that they are well-behaved under the Fourier transform. First we need a lemma:

Lemma 8.1.3 *If $f \in \mathcal{S}$ then $\widehat{f} \in \mathcal{S}$.*

Proof: Recall (see Section 7.6) that the Fourier transform converts multiplication by monomials into differentiation and vice versa. $\square$

Definition 8.1.4 If u is a Schwartz distribution then we define a Schwartz distribution $\widehat{u}$ by
$$\widehat{u}(\phi) = u(\widehat{\phi}).$$

By the lemma, the definition of $\widehat{u}$ makes good sense. Moreover, by 8.1.5 below,
$$|\widehat{u}(\phi)| = |u(\widehat{\phi})| \leq \sum_{|\alpha|+|\beta|\leq M} \rho_{\alpha,\beta}(\widehat{\phi})$$

for some $M > 0$ (by the definition of the topology on $\mathcal{S}$). It is a straightforward exercise with 8.1.3 and 8.1.4 to see that the sum on the right is majorized by the sum
$$C \cdot \sum_{|\alpha|+|\beta|\leq M} \rho_{\alpha,\beta}(\phi).$$

In conclusion, the Fourier transform of a Schwartz distribution is also a Schwartz distribution.

8.1.4   Other Spaces of Distributions

Let $\mathcal{D} = C_c^\infty$ and $\mathcal{E} = C^\infty$. Clearly $\mathcal{D} \subseteq \mathcal{S} \subseteq \mathcal{E}$. On each of the spaces $\mathcal{D}$ and $\mathcal{E}$ we use the semi-norms
$$\eta_{K,\alpha}(\phi) = \sup_K \left| \left(\frac{\partial}{\partial x} \right)^\alpha \phi \right|,$$

where $K \subseteq \mathbb{R}^N$ is a compact set and $\alpha = (\alpha_1, \ldots, \alpha_N)$ is a multi-index. These induce a topology on $\mathcal{D}$ and $\mathcal{E}$ which turn them into topological vector spaces. The spaces $\mathcal{D}'$ and $\mathcal{E}'$ are defined to be the continuous linear functionals on $\mathcal{D}$ and $\mathcal{E}$ respectively. Trivially, $\mathcal{E}' \subseteq \mathcal{S}' \subseteq \mathcal{D}'$. The functional in $\mathbb{R}^1$ given by

$$\mu = \sum_{j=1}^{\infty} 2^j \delta_j,$$

where δ_j is the Dirac mass centered at the integer j, is readily seen to be in $\mathcal{D}'$ but not in $\mathcal{E}'$.

The *support* of a distribution μ is defined to be the complement of the union of all open sets U such that $\mu(\phi) = 0$ for all elements ϕ of C_c^{∞} that are supported in U. As an example, the support of the Dirac mass δ_0 is the origin: when δ_0 is applied to any testing function ϕ with support disjoint from 0 then the result is 0.

Exercise: Let $\mu \in \mathcal{D}'$. Then $\mu \in \mathcal{E}'$ if and only if μ has compact support. The elements of $\mathcal{E}'$ are sometimes referred to as the "compactly supported distributions."

Proposition 8.1.5 *A linear functional L on $\mathcal{S}$ is a Schwartz distribution (tempered distribution) if and only if there is a $C > 0$ and integers m and ℓ such that for all $\phi \in \mathcal{S}$ we have*

$$|L(\phi)| \leq C \cdot \sum_{|\alpha| \leq \ell} \sum_{|\beta| \leq m} \rho_{\alpha,\beta}(\phi). \qquad (8.1.5.1)$$

Sketch of Proof: If an inequality like (8.1.5.1) holds then clearly L is continuous.

For the converse, assume that L is continuous. Recall that a neighborhood basis of 0 in $\mathcal{S}$ is given by sets of the form

$$N_{\epsilon,\ell,m} = \{\phi \in \mathcal{S} : \sum_{\substack{|\alpha| \leq \ell \\ |\beta| \leq m}} \rho_{\alpha,\beta} < \epsilon\}.$$

Since L is continuous, the inverse image of an open set under L is open. Consider

$$L^{-1}\left(\{z \in \mathbb{C} : |z| < 1\}\right).$$

There exist ϵ, ℓ, m such that

$$N_{\epsilon,\ell,m} \subseteq L^{-1}\left(\{z \in \mathbb{C} : |z| < 1\}\right).$$

Thus

$$\sum_{\substack{|\alpha| \leq \ell \\ |\beta| \leq m}} \rho_{\alpha,\beta}(\phi) < \epsilon$$

implies that

$$|L(\phi)| < 1.$$

That is the required result, with $C = 1/\epsilon$. □

Exercise: A similar result holds for $\mathcal{D}'$ and for $\mathcal{E}'$.

Theorem 8.1.6 (Structure Theorem for $\mathcal{D}'$) *If $u \in \mathcal{D}'$ then*

$$u = \sum_{j=1}^{k} D^j \mu_j,$$

where μ_j is a finite Borel measure and each D^j is a differential monomial.

Idea of Proof: For simplicity restrict attention to $\mathbb{R}^1$. We know that the dual of the continuous functions with compact support is the space of finite Borel measures. In a natural fashion, the space of C^1 functions with compact support can be identified with a subspace of the set of ordered pairs of C_c functions: $f \leftrightarrow (f, f')$. Then every functional on C_c^1 extends, by the Hahn-Banach theorem, to a functional on $C_c \times C_c$. But such a functional will be given by a pair of measures. Combining this information with the definition of derivative of a distribution gives that an element of the dual of C_c^1 is of the form $\mu_1 + (\mu_2)'$. In a similar fashion, one can prove that an element of the dual of C_c^k must have the form $\mu_1 + (\mu_2)' + \cdots + (\mu_{k+1})^{(k)}$.

Finally, it is necessary to note that $\mathcal{D}'$ is nothing other than the countable union over k of the dual spaces $(C_c^k)'$. □

The theorem makes explicit the fact that an element of $\mathcal{D}'$ can depend on only finitely many derivatives of the testing function, that is on only finitely many of the $\eta_{K,\alpha}$.

We have already noted that the Schwartz distributions are the most convenient for Fourier transform theory. But the space $\mathcal{D}'$ is often more convenient in the theory of partial differential equations (because of the control on the support of testing functions). It will sometimes be necessary to pass back and forth between the two theories. In any given context, no confusion should result.

Exercise: Use the Paley-Wiener theorem or some other technique to prove that if $\phi \in \mathcal{D}$ then $\hat{\phi} \notin \mathcal{D}$. (This fact is often referred to as the Heisenberg uncertainty principle. In fact it has a number of qualitative and quantitative formulations that are useful in quantum mechanics. See [FEF] for more on these matters.)

8.1.5 More on the Topology of $\mathcal{D}$ and $\mathcal{D}'$

We say that a sequence $\{\phi_j\} \subseteq \mathcal{D}$ converges to $\phi \in \mathcal{D}$ if

1. All the functions ϕ_j have compact support in a single compact set K_0.

2. $\eta_{K,\alpha}(\phi_j - \phi) \to 0$ for each compact set K and for every multi-index α.

The enemy here is the example of the "gliding hump:" On $\mathbb{R}^1$, if ψ is a fixed C^∞ function and $\phi_j(x) \equiv \psi(x - j)$ then we do *not* want to say that the sequence $\{\phi_j\}$ converges to 0.

A functional μ on $\mathcal{D}$ is continuous if $\mu(\phi_j) \to \mu(\phi)$ whenever $\phi_j \to \phi$. This is equivalent to the already noted characterization that there exist a compact K and an $N > 0$ such that

$$|\mu(\phi)| \leq C \sum_{|\alpha| \leq N} \eta_{K,\alpha}(\phi)$$

for every testing function ϕ.

Exercises

1. Show that any derivative of the Dirac mass has support consisting of just the origin.

2. Calculate the Fourier transform of the Dirac mass.

3. Calculate the Fourier transform of the derivative of the Dirac mass.

4. Which distribution has derivative equal to the heaviside function?

5. In $\mathbb{R}^3$, what is the Laplacian (in the sense of distributions) of $|x|^{-1}$?

6. What is the distribution derivative of the characteristic function of the unit interval in $\mathbb{R}$?

7. Let f be an integrable function on $\mathbb{R}$. Why is its derivative equal to a distribution?

8. The function $f(x) = x^2$ is not integrable on $\mathbb{R}$. But it is still a Schwartz distribution. Explain why.

Problems for Review and Discovery

A. Drill Exercises

1. Calculate the distribution derivative of the function $f(x) = x^{-1.3}$.
2. What is the third derivative of the Dirac delta mass?
3. Calculate all second derivatives of the Dirac delta mass in $\mathbb{R}^2$.
4. Calculate the second distribution derivative of $f(x) = |x|$ in $\mathbb{R}^1$.
5. Calculate the distribution derivative of $\log |x|$.
6. Give three distinct examples of distributions that are not functions.
7. Give an example of a function whose distribution derivative is Lebesgue measure dx.
8. Let τ_j be the Dirac mass at the point $1/j \in \mathbb{R}$. Does $\{\tau_j\}$ have a limit in the distribution topology? If so, what is it?

B. Challenge Problems

1. Give an example of an element of $\mathcal{D}'$ that is not an element of $\mathcal{S}'$.
2. Show that if f is continuously differentiable, and is also a distribution, then its calculus derivative and its distribution derivative are equal.
3. Show that if f is a continuous function, and is also a distribution, then its classical Fourier transform and its distribution Fourier transform are equal.
4. In the section of this book on the Fourier transform, we showed how the classical Fourier transform interacts with translations and dilations and rotations. Prove analogous results for the distribution Fourier transform.
5. In the section of this book on the Fourier transform, we showed how the classical Fourier transform interacts with the derivative. Prove analogous results for the distribution Fourier transform.
6. Formulate and prove a version of the fundamental theorem of calculus for distributions.
7. Show that if two distributions on $\mathbb{R}$ have the same derivative then they differ by a constant.
8. Show that the distribution derivative of a measure cannot be a function.

C. Problems for Discussion and Exploration

1. Let $n \geq 3$. Show that the fundamental solution for the Laplacian in $\mathbb{R}^n$ is $\Gamma(x) = c \cdot |x|^{-n+2}$ for a suitable constant c. This means that $\Delta \Gamma(x) = \delta(x)$, where δ is the Dirac function.
2. Refer to Exercise 1. Show that the fundamental solution for the Laplacian in dimension 2 is $c \log |x|$.
3. Inequality (8.1.5.1) suggests a topology on the space of Schwartz functions. This in turn induces a topology on the space of Schwartz distributions. Describe these topologies explicitly.

4. Refer to Exercise 3. Show that the functions

$$\varphi_j(x) = j \cdot \chi_{[0,1/j]}(x)$$

converge to the Dirac delta mass in the given distribution topology.

5. What is the closure of C_c^∞ in the Schwartz space topology?

6. Is the Schwartz space complete in the indicated topology?

9

Wavelets

- Localization in the space and frequency variables

- Building a custom Fourier analysis

- The Haar basis

- A wavelet basis

- The wavelet transform

- Decomposition and reconstruction

- Applications

- Cumulative energy and entropy

9.1   Localization in the Space and Frequency Variables

The premise of the new versions of Fourier analysis that are being developed today is that sines and cosines are not an optimal model for some of the phenomena that we want to study. As an example, suppose that we are developing software to detect certain erratic heartbeats by analysis of an electrocardiogram. (Note that the discussion that we present here is philosophically correct but is over-simplified to facilitate the exposition.) The scheme is to have the software break down the patient's electrocardiogram into component waves. If a wave that is known to be a telltale signal of heart disease is detected, then the software notifies the user.

A good plan, and there is indeed software of this nature in use across America. But let us imagine that a typical electrocardiogram looks like that shown in Figure 9.1. Imagine further that the aberrant heartbeat that we wish to detect is the one in Figure 9.2.

What we want the software to do is to break up the wave in Figure 9.1 into fundamental components, and then to see whether one of those components is the wave in Figure 9.2. Of what utility is Fourier theory in such an analysis? Fourier theory would allow us to break the wave in Figure 9.1 into sines and

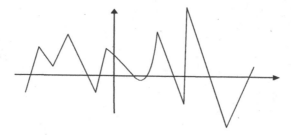

FIGURE 9.1
A heartbeat shown in an electrocardiogram.

FIGURE 9.2
An aberrant heartbeat.

cosines, then break the wave in Figure 9.2 into sines and cosines, and then attempt to match up coefficients. Such a scheme may be dreadfully inefficient, because sines and cosines *have nothing to do* with the waves we are endeavoring to analyze.

The Fourier analysis of sines and cosines arose historically because sines and cosines are eigenfunctions for the wave equation (see Chapter 10). Their place in mathematics became even more firmly secured because they are orthonormal in L^2. They also commute with translations in natural and useful ways. The standard trigonometric relations between the sine and cosine functions give rise to elegant and useful formulas—such as the formulas for the Dirichlet kernel and the Fejér kernel and the Poisson kernel. Sines and cosines have played an inevitable and fundamental historical role in the development of harmonic analysis.

In the same vein, translation-invariant operators have played an important role in our understanding of how to analyze partial differential equations (see [KRA3]), and as a step toward the development of the more natural theory of pseudodifferential operators. Today we find ourselves studying translation *non*-invariant operators—such as those that arise in the analysis on the boundary of a (smoothly bounded) domain in $\mathbb{R}^N$ (see Figure 9.3). The $T(\mathbf{1})$ theorem of David-Journé gives the most natural and comprehensive method

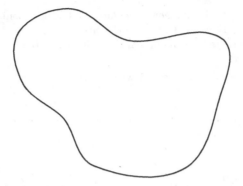

FIGURE 9.3
A smoothly bounded domain.

of analyzing integral operators, and their boundedness on a great variety of spaces.

The next, and current, step in the development of Fourier analysis is to replace the classical sine and cosine building blocks with more flexible units—indeed, with units that can be tailored to the situation at hand. Such units should, ideally, be localizable; in this way they can more readily be tailored to any particular application. This, roughly speaking, is what wavelet theory is all about.

In a book of this nature, we clearly cannot develop the full assemblage of tools that are a part of modern wavelet theory. [See [HERG], [MEY1], [MEY2], [DAU] for more extensive treatments of this beautiful and dynamic subject. The papers [STR] and [WAL] provide nice introductions as well.] What we can do is to give the reader a taste. Specifically, we shall develop a Multi-Resolution Analysis, or MRA; this study will show how Fourier analysis may be carried out with localization in either the space variable or the Fourier transform (frequency) variable. In short, the reader will see how either variable may be localized. Contrast this notion with the classical construction, in which the units are sines and cosines—clearly functions which *do not* have compact support. The exposition here derives from that in [HERG], [STR], and [WAL]. We also thank G. B. Folland and J. Walker for considerable guidance in preparing this chapter.

Exercises

1. Let
$$f(x) = \begin{cases} N & \text{if} & 1 \le x \le 1 + 1/N \\ 0 & \text{if} & x < 1 \text{ or } x > 1 + 1/N \end{cases}$$

for N a large positive integer. Calculate the Fourier coefficients of f using ideas from Section 7.1. Calculate the first five partial sums, and notice that each of these partial sums has a tail that extends across the entire interval $[0, 2\pi]$.

2. Perform the steps of Exercise 1 with

$$f(x) = \begin{cases} 1 & \text{if} & 1 \le x \le 2 \\ 0 & \text{if} & x < 1 \text{ or } x > 2. \end{cases}$$

3. Let $\epsilon > 0$. Perform the steps of Exercise 1 with

$$f(x) = \begin{cases} 1 & \text{if} & \epsilon \le x \le 2\pi - \epsilon \\ 0 & \text{if} & x < \epsilon \text{ or } x > 2\pi - \epsilon. \end{cases}$$

How do the Fourier coefficients behave as $\epsilon \to 0$?

4. Perform the steps of Exercise 1 with

$$f(x) = \begin{cases} \sin x & \text{if} & 1 \le x \le 1 + 1/N \\ 0 & \text{if} & x < 1 \text{ or } x > 1 + 1/N \end{cases}$$

for N a large positive integer. Notice that this Fourier series has a tail on the entire interval $[0, 2\pi]$.

9.2   Building a Custom Fourier Analysis

Typical applications of classical Fourier analysis are

Frequency Modulation: Alternating current, radio transmission

Mathematics: Ordinary and partial differential equations, analysis of linear and nonlinear operators

Medicine: Electrocardiography, magnetic resonance imaging, biological neural systems

Optics and Fiber-Optic Communications: Lens design, crystallography, image processing

Radio, Television, Music Recording: Signal compression, signal reproduction, filtering

Spectral Analysis: Identification of compounds in geology, chemistry, biochemistry, mass spectroscopy

Telecommunications: Transmission and compression of signals, filtering of signals, frequency encoding

In fact, the applications of Fourier analysis are so pervasive that they are part of the very fabric of modern technological life.

The applications that are being developed for wavelet analysis are very similar to those just listed. But the wavelet algorithms give rise to faster and more accurate image compression, faster and more accurate signal compression and analysis, and better denoising techniques that preserve the original signal more completely. The applications in mathematics lead, in many situations, to better and more rapid convergence results.

What is lacking in classical Fourier analysis can be readily seen by examining the Dirac delta mass. Because, if the unit ball of L^1—thought of as a subspace of the dual space of $C(\mathbb{T})$—had any extremal functions (it does not), they would be objects of this sort: the weak-$*$ limit of functions of the form $N^{-1}\chi_{[-1/2N,1/2N]}$ as $N \to +\infty$. That weak-$*$ limit is the Dirac mass. We know the Dirac mass as the functional that assigns to each smooth function with compact support its value at 0:

$$\delta : C_c(\mathbb{R}^N) \ni \phi \longmapsto \phi(0).$$

The point comes through most clearly by way of Fourier series. Consider the Dirac mass δ supported at the origin in the circle group $\mathbb{T}$. Then the Fourier-Stieltjes coefficients of δ are

$$\widehat{\delta}(j) \equiv \frac{1}{2\pi} \int_{-\pi}^{\pi} e^{-ijt}\, d\delta(t) = 1.$$

Thus recovering δ from its Fourier series amounts to finding a way to sum the formal series

$$\sum_{j=-\infty}^{\infty} 1 \cdot e^{ijt}$$

in order to obtain the Dirac mass. Since each exponential is supported on the entire circle group, the imagination is defied to understand how these exponentials could sum to a point mass. (To be fair, the physicists have no trouble seeing this point: at the origin the terms all add up, and away from zero they all cancel out.)

The study of the point mass is not merely an affectation. In a radio signal, noise (in the form of spikes) is frequently a sum of point masses (Figure 9.4). On a phonograph record, the pops and clicks that come from imperfections in the surface of the record exhibit themselves (on an oscilloscope, for instance) as spikes, or point masses.

For the sake of contrast, in the next section we shall generate an *ad hoc* family of wavelet-like basis elements for L^2 and show how these may be used much more efficiently to decompose the Dirac mass into basis elements.

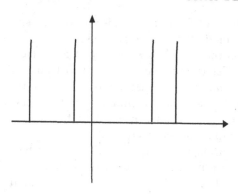

FIGURE 9.4
Noise in a radio signal.

Exercises

1. Define
    $$f_N(x) = \begin{cases} N & \text{if} & 0 \le x \le 1/N \\ 0 & \text{if} & x < 0 \quad \text{or} \quad 1/N < x. \end{cases}$$
    Show that $f_N \to \delta$ (the Dirac mass) in the sense of distributions. This means that, if $\varphi \in C_c^\infty$, then
    $$\int f_N(x)\varphi(x)\,dx \to \varphi(0) = \int \varphi(x)\,d\delta(x).$$

2. Refer to Exercise 1. Let $\varphi \in C_c^\infty$, $\varphi \ge 0$, $\int \varphi\,dx = 1$. Prove that $N\varphi(Nx) \to \delta$ as $N \to +\infty$ in the sense of distributions.

3. Refer to Exercise 2. Calculate the Fourier coefficients of $N\varphi(Nx)$. How do these coefficients behave as $N \to +\infty$?

4. Refer to Exercise 2. Calculate the Fourier coefficients of $\epsilon\varphi(\epsilon x)$ for $\epsilon > 0$ small. How do these coefficients behave as $\epsilon \to 0^+$?

9.3   The Haar Basis

In this section we shall describe the Haar wavelet basis. While the basis elements are not smooth functions (as wavelet basis elements usually are), they will exhibit the other important features of a Multi-Resolution Analysis (MRA). In fact we shall follow the axiomatic treatment as developed by Mallat and exposited in [WAL] in order to isolate the essential properties of an MRA.

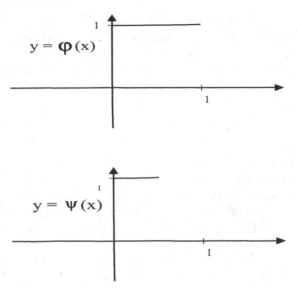

FIGURE 9.5
The functions φ and ψ.

We shall produce a dyadic version of the wavelet theory. Certainly other theories, based on other dilation paradigms, may be produced. But the dyadic theory is the most standard, and quickly gives the flavor of the construction. In this discussion we shall use, as we did in Chapter 6, the notation α_δ to denote the dilate of a function: $\alpha_\delta f(x) \equiv f(\delta x)$.[1] And we shall use the notation τ_a to denote the translate of a function: $\tau_a f(x) \equiv f(x-a)$.

We work on the real line $\mathbb{R}$. Our universe of functions will be $L^2(\mathbb{R})$, the square-integrable functions. Define

$$\varphi(x) = \chi_{[0,1)}(x) = \begin{cases} 1 & \text{if} & x \in [0,1) \\ 0 & \text{if} & x \notin [0,1) \end{cases}$$

and

$$\psi(x) \equiv \varphi(2x) - \varphi(2x-1) = \chi_{[0,1/2)}(x) - \chi_{[1/2,1)}(x).$$

These functions are exhibited in Figure 9.5.

The function φ will be called a *scaling function* and the function ψ will be called the associated *wavelet*. The basic idea is this: translates of φ will generate a space V_0 that can be used to analyze a function f on a large scale— more precisely, on the scale of size 1 (because 1 is the length of the support of φ). But the elements of the space V_0 cannot be used to detect information that is at a scale *smaller* than 1. So we will scale the elements of V_0 down by a factor of 2^j, each $j = 1, 2, \ldots$, to obtain a space that can be used for

---

[1] We use the notation δ in other parts of the book to denote the Dirac delta mass. You should be able to tell from context which meaning of δ is intended.

analysis at the scale 2^{-j} (and we will also scale V_0 up to obtain elements that are useful at an arbitrarily large scale). Let us complete this program now for the specific φ that we have defined above, and then present some axioms that will describe how this process can be performed in a fairly general setting.

Now we use φ to generate a scale of function spaces $\{V_j\}_{j\in\mathbb{Z}}$. We set

$$V_0 = \left\{\sum_{k\in\mathbb{Z}} a_k[\tau_k\varphi] : \sum |a_k|^2 < \infty\right\},$$

for the particular function φ that was specified above. Of course each element of V_0 so specified lies in L^2 (because the functions $\tau_k\varphi$ have disjoint supports). But it would be wrong to think that V_0 is all of L^2, for an element of V_0 is constant on each interval $[k, k+1)$, and has possible jump discontinuities only at the integers. The functions $\{\tau_k\varphi\}_{k\in\mathbb{Z}}$ form an orthonormal basis (with respect to the L^2 inner product) for V_0.

Now let us say that a function g is in V_1 if and only if $\alpha_{1/2}g$ lies in V_0. Thus $g \in V_1$ means that g is constant on the intervals determined by the lattice $(1/2)\mathbb{Z} \equiv \{n/2 : n \in \mathbb{Z}\}$ and has possible jump discontinuities only at the elements of $(1/2)\mathbb{Z}$. It is easy to see that the functions $\{\sqrt{2}\alpha_2\tau_k f : f \in V_0\}$ form an orthonormal basis for V_1.

Observe that $V_0 \subseteq V_1$ since every jump point for elements of V_0 is also a jump point for elements of V_1 (but not conversely). More explicitly, we may write :

$$\tau_k f = \alpha_2\tau_{2k} f + \alpha_2\tau_{2k+1} f,$$

thus expressing an element of V_0 as a linear combination of elements of V_1.

Now that we have the idea down, we may iterate it to define the spaces V_j for any $j \in \mathbb{Z}$. Namely, for $j \in \mathbb{Z}$, V_j will be generated by the functions $\alpha_{2^j}\tau_m\varphi$, all $m \in \mathbb{Z}$. In fact we may see explicitly that an element of V_j will be a function of the form

$$f = \sum_{\ell\in\mathbb{Z}} a_\ell\chi_{(\ell/2^j,[\ell+1]/2^j)}$$

where $\sum |a_\ell|^2 < \infty$. Thus an orthonormal basis for V_j is given by $\{2^{j/2}\alpha_{2^j}\tau_m\varphi\}_{m\in\mathbb{Z}}$.

Now the spaces V_j have no common intersection except the zero function. This is so because, since a function $f \in \cap_{j\in\mathbb{Z}}V_j$ would be constant on arbitrarily large intervals (of length 2^{-j} for j negative), then it can only be in L^2 if it is zero. Also $\cup_{j\in\mathbb{Z}}V_j$ is dense in L^2 because any L^2 function can be approximated by a simple function (i.e., a finite linear combination of characteristic functions), and any characteristic function can be approximated by a sum of characteristic functions of dyadic intervals.

We therefore might suspect that if we combine all the orthonormal bases for all the $V_j, j \in \mathbb{Z}$, then this would give an orthonormal basis for L^2. That supposition is, however, incorrect. For the basis elements $\varphi \in V_0$ and $\alpha_{2^j}\tau_0\varphi \in V_j$ are not orthogonal. This is where the function ψ comes in.

Since $V_0 \subseteq V_1$ we may proceed by trying to complete the orthonormal basis $\{\tau_k \varphi\}$ of V_0 to an orthonormal basis for V_1. Put in other words, we write $V_1 \equiv V_0 \oplus W_0$, and we endeavor to write a basis for W_0. Let $\psi = \alpha_2 \varphi - \alpha_2 \tau_1 \varphi$ be as above, and consider the set of functions $\{\tau_m \psi\}$ for $m \in \mathbb{Z}$. Then this is an orthonormal set. Let us see that it spans W_0.

Let h be an arbitrary element of W_0. So certainly $h \in V_1$. It follows that

$$h = \sum_j b_j \alpha_2 \tau_j \varphi$$

for some constants $\{b_j\}$ that are square-summable. Of course h is constant on the interval $[0, 1/2)$ and also constant on the interval $[1/2, 1)$. We note that

$$\varphi(t) = \frac{1}{2} [\varphi(t) + \psi(t)] \qquad \text{on } [0, 1/2)$$

and

$$\varphi(t) = \frac{1}{2} [\varphi(t) - \psi(t)] \qquad \text{on } [1/2, 1).$$

It follows that

$$h(t) = \left(\frac{b_0 + b_1}{2} \right) \varphi(t) + \left(\frac{b_0 - b_1}{2} \right) \psi(t)$$

on $[0, 1)$. Of course a similar decomposition obtains on every interval $[j, j+1)$. As a result,

$$h = \sum_{j \in \mathbb{Z}} c_j \tau_j \varphi + \sum_{j \in \mathbb{Z}} d_j \tau_j \psi,$$

where

$$c_j = \frac{b_j + b_{j+1}}{2} \quad \text{and} \quad d_j = \frac{b_j - b_{j+1}}{2}.$$

Note that $h \in W_0$ implies that $h \in V_0^\perp$. Also every $\tau_j \varphi$ is orthogonal to every $\tau_k \psi$. Consequently every coefficient $c_j = 0$. Thus we have proved that h is in the closed span of the terms $\tau_j \psi$. In other words, the functions $\{\tau_j \psi\}_{j \in \mathbb{Z}}$ span W_0.

Thus we have $V_1 = V_0 \oplus W_0$, and we have an explicit orthonormal basis for W_0. Of course we may scale this construction up and down to obtain

$$V_{j+1} = V_j \oplus W_j \tag{9.3.1}_j$$

for every j. And we have the explicit orthonormal basis $\{2^{j/2} \alpha_{2^j} \tau_m \psi\}_{m \in \mathbb{Z}}$ for each W_j.

We may iterate the equation $(9.3.1)_j$ to obtain

$$
\begin{aligned}
V_{j+1} &= V_j \oplus W_j = V_{j-1} \oplus W_{j-1} \oplus W_j \\
&= \cdots = V_0 \oplus W_0 \oplus W_1 \oplus \cdots \oplus W_{j-1} \oplus W_j.
\end{aligned}
$$

Letting $j \to +\infty$ yields

$$L^2 = V_0 \oplus \bigoplus_{j=0}^{\infty} W_j. \tag{9.3.2}$$

But a similar decomposition may be performed on V_0, with W_j in descending order:

$$V_0 = V_{-1} \oplus W_{-1} = \cdots = V_{-\ell} \oplus W_{-\ell} \oplus \cdots \oplus W_{-1}.$$

Letting $\ell \to +\infty$, and substituting the result into (9.3.2), now yields that

$$L^2 = \bigoplus_{j \in \mathbb{Z}} W_j.$$

Thus we have decomposed $L^2(\mathbb{R})$ as an orthonormal sum of Haar wavelet subspaces. We formulate one of our main conclusions as a theorem:

Theorem 9.3.1 *The collection*

$$\mathcal{H} \equiv \left\{ 2^{j/2} \alpha_{2^j} \tau_m \psi : m, j \in \mathbb{Z} \right\}$$

is an orthonormal basis for L^2, and will be called a wavelet basis for L^2.

Now it is time to axiomatize the construction that we have just performed in a special instance.

Axioms for a Multi-Resolution Analysis (MRA)

A collection of subspaces $\{V_j\}_{j \in \mathbb{Z}}$ of $L^2(\mathbb{R})$ is called a *Multi-Resolution Analysis* or MRA if

MRA$_1$ (**Scaling**) For each j, the function $f \in V_j$ if and only if $\alpha_2 f \in V_{j+1}$.

MRA$_2$ (**Inclusion**) For each j, $V_j \subseteq V_{j+1}$.

MRA$_3$ (**Density**) The union of the V_j's is dense in L^2:

$$\text{closure} \left\{ \bigcup_{j \in \mathbb{Z}} V_j \right\} = L^2(\mathbb{R}).$$

MRA$_4$ (**Maximality**) The spaces V_j have no non-trivial common intersection:

$$\bigcap_{j \in \mathbb{Z}} V_j = \{0\}.$$

MRA_5 (**Basis**) There is a function φ such that $\{\tau_j\varphi\}_{j\in\mathbb{Z}}$ is an orthonormal basis for V_0.

We invite the reader to review our discussion of $\varphi = \chi_{[0,1)}$ and its dilates and confirm that the spaces V_j that we constructed above do indeed form an MRA. Notice in particular that, once the space V_0 has been defined, then the other V_j are completely and uniquely determined by the MRA axioms.

Exercises

1. What is a typical element of V_2? What is a typical element of V_{-3}?

2. What is a typical element of W_2? What is a typical element of W_{-3}?

3. Give an example of a function f such that $\int f(x)\mu(x)\,dx = 0$ for every $\mu \in V_0$.

4. Give an example of a function g such that $\int g(x)\nu(x)\,dx = 0$ for every $\nu \in W_0$.

5. Verify explicitly that $V_0 \perp W_0$.

6. Verify explicitly that $V_j \perp W_j$ for any index j.

7. If $\int f(x)\mu(x)\,dx = 0$ for every $\mu \in V_j$ for every j, then what can you say about f?

8. If $\int f(x)\nu(x)\,dx = 0$ for every $\nu \in W_j$ for every j, then what can you say about f?

9.4 Some Illustrative Examples

In this section we give two computational examples that provide concrete illustrations of how the Haar wavelet expansion is better behaved—especially with respect to detecting *local* data—than the Fourier series expansion.

EXAMPLE 9.4.1 Our first example is quick and dirty. In particular, we cheat a bit on the topology to make a simple and dramatic point. It is this: If we endeavor to approximate the Dirac delta mass δ with a Fourier series, then the partial sums will always have a *slowly decaying* tail that extends far beyond the highly localized support of δ. By contrast, the partial sums of the Haar series for δ localize rather nicely. We will see that the Haar series has a tail too, but it is small.

Let us first examine the expansion of the Dirac mass in terms of the Haar basis. Properly speaking, what we have just proposed is not feasible because the Dirac mass does not lie in L^2. Instead let us consider, for $N \in \mathbb{N}$, functions

$$f_N = 2^N \chi_{[0,1/2^N)}.$$

The functions f_N each have mass 1, and the sequence $\{f_N \, dx\}$ converges, in the weak-∗ sense of measures (i.e., the weak-∗ topology), to the Dirac mass δ.

First, we invite the reader to calculate the ordinary Fourier series, or Fourier transform, of f_N (see also the calculations at the end of this example). Although (by the Riemann-Lebesgue lemma) the coefficients die out, the fact remains that any finite part of the Fourier transform, or any partial sum of the Fourier series, gives a rather poor approximation to f_N. After all, any partial sum of the Fourier series is a trigonometric polynomial, and any trigonometric polynomial has support on the *entire interval* $[-\pi, \pi)$. In conclusion, whatever the merits of the approximation to f_N by the Fourier series partial sums, they are offset by the unwanted portion of the partial sum that exists *off the support of* f_N. (For instance, if we were endeavoring to construct a filter to remove pops and clicks from a musical recording, then the pop or click (which is mathematically modeled by a Dirac mass) would be replaced by the tail of a trigonometric polynomial—which amounts to undesired low level noise (usually a hiss), as in Figure 9.6 below.)

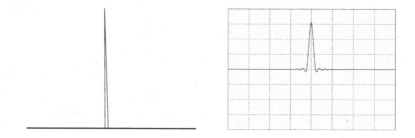

Figure 9.6. Undesired low-level noise.

Now let us do some calculations with the Haar basis. Fix an integer $N > 0$. If $j \geq N$, then any basis element for W_j will integrate to 0 on the support of f_N—just because the basis element will be 1 half the time and -1 half the time on each dyadic interval of length 2^{-j}. If instead $j < N$, then the single basis element μ_j from W_j that has support intersecting the support of f_N is in fact constantly equal to $2^{j/2}$ on the support of f_N. Therefore the coefficient b_j of μ_j in the expansion of f_N is

$$b_j = \int f_N(x)\mu_j(x) \, dx = 2^N \int_0^{2^{-N}} 2^{j/2} \, dx = 2^{j/2}.$$

Thus the expansion for f_N is, for $0 \leq x < 2^{-N}$,

$$
\sum_{j=-\infty}^{N-1} 2^{j/2} \mu_j(x) = \sum_{j=-\infty}^{0} 2^{j/2} \cdot 2^{j/2} + \sum_{j=1}^{N-1} 2^{j/2} \cdot 2^{j/2}
$$
$$
= 2 + (2^N - 2)
$$
$$
= 2^N
$$
$$
= f_N(x).
$$

Notice here that the contribution of terms of negative index in the series—which corresponds to "coarse scale" behavior that is of little interest—is constantly equal to 2 (regardless of the value of N) and is relatively trivial (i.e., small) compared to the interesting part of the series (of size $2^N - 2$) that comes from the terms of positive index.

If instead $2^{-N} \leq x < 2^{-N+1}$, then $\mu_{N-1}(x) = -2^{(N-1)/2}$ and $b_{N-1}\mu_{N-1}(x) = -2^{N-1}$; also

$$
\sum_{j=-\infty}^{N-2} b_j \mu_j(x) = \sum_{j=-\infty}^{N-2} 2^j = 2^{N-1}.
$$

Of course $b_j = 0$ for $j \geq N$. In summary, for such x,

$$
\sum_{j=-\infty}^{\infty} b_j \mu_j(x) = 0 = f_N(x).
$$

A similar argument shows that if $2^{-\ell} \leq x < 2^{-\ell+1}$ for $-\infty < \ell \leq N$, then $\sum b_j \mu_j(x) = 0 = f_N(x)$. And the same result holds if $x < 0$.

Thus we see that the Haar basis expansion for f_N converges pointwise to f_N. More is true: the partial sums of the series give a rather nice approximation to the function f_N. Notice, for instance, that the partial sum $S_{N-1} = \sum_{j=-N+1}^{N-1} b_j \mu_j$ has the following properties:

(9.4.1.1) $S_{N-1}(x) = f_N(x) = 2^{N-1}$ for $0 \leq x < 2^{-N}$.

(9.4.1.2) $S_{N-1}(x) = 0$ for $-2^{-N} \leq x < 0$.

(9.4.1.3) $S_{N-1}(x) = 0$ for $|x| > 2^{-N}$.

It is worth noting that the partial sums of the Haar series for the Dirac mass δ,

$$
H_N(x) = \sum_{|j| \leq N} 2^{j/2} \mu_j(x),
$$

form (almost) a standard family of summability kernels as discussed in [KAT] (the missing feature is that each kernel integrates to 0 rather than 1); but the partial sums of the *Fourier series* for the Dirac mass δ,

$$
D_N(x) = \sum_{|j| \leq N} 1 \cdot e^{ijx},
$$

do *not*. Refer to Figure 9.7, which uses the software FAWAV by J. S. Walker ([WAL]) to illustrate partial sums of both the Fourier series and the Haar series for the Dirac mass.

The perceptive reader will have noticed that the Haar series does not give an entirely satisfactory approximation to our function f_N, just because the partial sums each have mean-value zero (which f_N most certainly does not!). Matters are easily remedied by using the decomposition

$$L^2 = V_0 \oplus \bigoplus_0^\infty W_j \qquad\qquad (9.4.1.5)$$

instead of the decomposition

$$L^2 = \bigoplus_{-\infty}^\infty W_j$$

that we have been using. For, with (9.4.1.5), V_0 takes care of the coarse scale behavior all at once, and also gets the mean-value condition right.

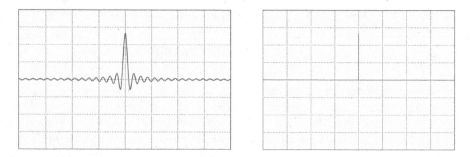

Figure 9.7: Partial sums of the Fourier and Haar series for the Dirac mass.

Thus we see, in the context of a very simple example, that the partial sums of the Haar series for a function that closely approximates the Dirac mass at the origin give a more accurate and satisfying approximation to the function than do the partial sums of the Fourier series. To be sure, the partial sums of the Fourier series of each f_N tend to f_N, but the oscillating error persists no matter how high the degree of the partial sum (in the classical literature this is called *Gibbs's phenomenon*). The situation would be similar if we endeavored to approximate f_N by its Fourier transform.

We close this discussion with some explicit calculations to recap the point that has just been made. It is easy to calculate that the jth Fourier coefficient of the function f_N is

$$\widehat{f_N}(j) = \frac{i2^{N-1}}{j\pi}\left(e^{-ij/2^N} - 1\right).$$

Therefore, with S_M denoting the Mth partial sum of the Fourier series,

$$\|f_N - S_M\|_{L^2}^2 = \sum_{|j|>M} \left(\frac{2^{N-1}}{j\pi}\right)^2 |e^{-ij/2^N} - 1|^2.$$

Imitating the proof of the integral test for convergence of series, it is now straightforward to see that

$$\|f_N - S_M\|_{L^2}^2 \approx \frac{C}{M}.$$

In short, $\|f_N - S_M\|_{L^2} \to 0$, as $M \to \infty$, at a rate comparable to $M^{-1/2}$, and that is quite slow.

By contrast, if we let $H_M \equiv \sum_{|j|\leq M} 2^{j/2}\mu_j$ (where $\mu_j \in W_j$) then, for $M \geq N - 1$, our earlier calculations show that

$$\|f_N - H_M\|_{L^2}^2 = \sum_{j=-\infty}^{-M-1} 2^j = 2^{-M}.$$

Therefore $\|f_N - H_M\|_{L^2} \to 0$, as $M \to \infty$, at a rate comparable to $2^{-M/2}$, or *exponentially fast*. This is a strong improvement over the convergence supplied by classical Fourier analysis. ∎

Our next example shows quite specifically that Haar series can beat Fourier series at their own game. Specifically, we shall approximate the function $g(x) \equiv [\cos \pi x] \cdot \chi_{[0,1]}(x)$ both by Haar series and by using the Fourier transform. The Haar series will win by a considerable margin. [**Note:** A word of explanation is in order here. Instead of the function g, we could consider $h(x) \equiv [\cos \pi x] \cdot \chi_{[0,2]}(x)$. Of course the interval $[0, 2]$ is the natural support for a period of the trigonometric function $\cos \pi x$, and the (suitably scaled) *Fourier series* of this function h is just the single term $\cos \pi x$. In this special circumstance Fourier series is hands down the best method of approximation—just because the support of the function is a good fit to the function. Such a situation is too artificial, and not a good test of the method. A more realistic situation is to chop off the cosine function so that its support does not mesh naturally with the period of cosine. That is what the function g does. We give Fourier every possible chance: by approximating with the Fourier *transform*, we allow all possible frequencies, and let Fourier analysis pick those that will best do the job.]

EXAMPLE 9.4.2 Consider $g(x) = [\cos \pi x] \cdot \chi_{[0,1]}(x)$ as a function on the entire real line. We shall compare and contrast the approximation of g by partial sums using the Haar basis with the approximation of g by "partial sums" of the Fourier transform. Much of what we do here will be traditional hand work; but, at propitious moments, we shall bring the computer to our aid.

Let us begin by looking at the Fourier transform of g. We calculate that

$$\widehat{g}(\xi) = \frac{1}{2}\int_0^1 (e^{i\pi x} + e^{-i\pi x})e^{ix\cdot\xi}\,dx$$

$$= \frac{1}{2}\left[\frac{-e^{i\xi}-1}{i(\xi+\pi)} + \frac{-e^{i\xi}-1}{i(\xi-\pi)}\right]$$

$$= \frac{-e^{i\xi}-1}{i(\xi^2-\pi^2)}\cdot\xi.$$

Observe that the function $\widehat{g}$ is continuous on all of $\mathbb{R}$ and vanishes at ∞. The Fourier inversion formula (last section of Chapter 6) then tells us that g may be recovered from $\widehat{g}$ by the integral

$$\frac{1}{2\pi}\int_{\mathbb{R}}\widehat{g}(\xi)e^{-ix\cdot\xi}\,d\xi.$$

We study this integral by considering the limit of the integrals

$$\eta_N(x) \equiv \frac{1}{2\pi}\int_{-N}^{N}\widehat{g}(\xi)e^{-ix\cdot\xi}\,d\xi \qquad (9.4.2.1)$$

as $N \to +\infty$. Elementary calculations show that (9.4.2.1) equals

$$\eta_N(x) = \frac{1}{2\pi}\int_{-N}^{N}\int_{-\infty}^{\infty}g(t)e^{i\xi t}\,dt\,e^{-ix\xi}\,d\xi$$

$$= \frac{1}{2\pi}\int_0^1 g(t)\int_{-N}^{N}e^{i(t-x)\xi}\,d\xi\,dt$$

$$= \frac{1}{2\pi i}\int_0^1 g(t)\frac{1}{t-x}e^{i\xi(t-x)}\Big]_{\xi=-N}^{\xi=N}\,dt$$

$$= \frac{1}{2\pi i}\int_0^1 g(t)\frac{1}{t-x}\left[e^{iN(t-x)} - e^{i(-N)(t-x)}\right]\,dt$$

$$= \frac{1}{2\pi i}\int_0^1 g(t)\frac{1}{t-x}2i\sin N(t-x)\,dt$$

$$= \frac{1}{\pi}\int_0^1 g(t)\frac{\sin N(x-t)}{x-t}\,dt$$

$$= \frac{1}{\pi}\int_0^1 \cos\pi t\frac{\sin N(x-t)}{x-t}\,dt. \qquad (9.4.2.2)$$

We see, by inspection of (9.4.2.2), that η_N is a continuous, indeed an analytic function. Thus it is supported on the entire real line (not on any compact set). Notice further that it could not be the case that $\eta_N = \mathcal{O}(|x|^{-r})$ for some $r > 1$; if it were, then η_N would be in $L^1(\mathbb{R})$ and then $\widehat{\eta_N}$ would be continuous (which it is certainly not). It turns out (we omit the details) that in fact $\eta_N = \mathcal{O}(|x|^{-1})$. This statement says, in a quantitative way, that η_N has a tail.

We can rewrite the far right expression in formula (9.4.2.2) (the last item in our long calculation) in the form

$$\eta_N(x) = \frac{1}{\pi} \int_{\mathbb{R}} g(t) \widetilde{D}_N(x - t) \, dt,$$

where

$$\widetilde{D}_N(t) = \frac{\sin Nt}{\pi t}.$$

The astute reader will realize that the kernel $\widetilde{D}_N$ is quite similar to the Dirichlet kernel that arises in connection with Fourier series. A proof analogous to ones we considered in Chapter 6 will show that $\eta_N(x) \to g(x)$ pointwise as $N \to \infty$.

Our calculations confirm that the Fourier transform of g can be "Fourier-inverted" (in the L^2 sense) back to g. But they also show that, for any particular $N > 0$ large, the expression

$$\eta_N(x) \equiv \frac{1}{2\pi} \int_{-N}^{N} \widehat{g}(\xi) e^{-ix \cdot \xi} \, d\xi \qquad (9.4.2.3)$$

is supported *on the entire real line*. (Of course this must be so since, if we replace the real variable x with a complex variable z, then (9.4.2.3) defines an entire function.) Thus, for practical applications, the convergence of η_N to g on the support $[0, 1]$ of g is seriously offset by the fact that η_N has a "tail" that persists no matter how large N. And the key fact is that the tail is *not small*. This feature is built in just because the function we are expanding has discontinuities.

We now contrast the preceding calculation of the Fourier transform of the function $g(x) = [\cos \pi x] \cdot \chi_{[0,1]}(x)$ with the analogous calculation using the Haar basis (but we shall perform these new calculations with the aid of a computer). The first thing that we will notice is that the only Haar basis elements that end up being used in the expansion of g are *those basis elements that are supported in the interval* $[0, 1]$. For the purposes of signal processing, this is already a dramatic improvement.

Figure 9.8 shows the Fourier series approximation to the function g. Specifically, this is a graph of the sum of 64 terms of the Fourier series created with Walker's software FAWAV. Figure 9.9 shows the improved approximation attained by η_{128}. Figures 9.10 and 9.11, respectively, superimpose the approximations η_{64} and η_{128} against the graph of g. Notice that, while the approximations are reasonable *inside*—and away from the endpoints of—the unit interval, the "inverse" of the Fourier transform goes out of control as x moves left across 0 or as x moves right across 1. By contrast, the Haar series for g is quite tame and gives a good approximation. Figure 9.12 shows the 256-term Haar series approximation—an even more dramatic improvement.

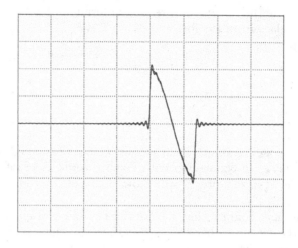

FIGURE 9.8
Fourier series approximation with 64 terms to the function g.

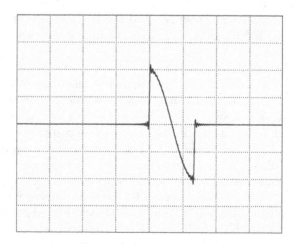

FIGURE 9.9
Improved approximation with 128 terms to the function g.

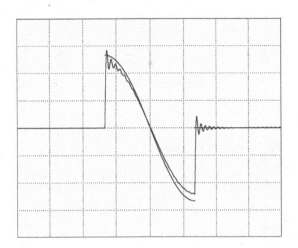

FIGURE 9.10
The graph of η_{128} superimposed on the graph of g.

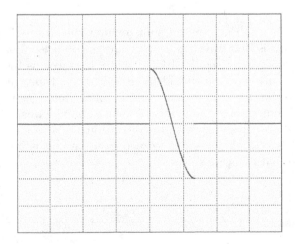

FIGURE 9.11
The approximation by η_{128}.

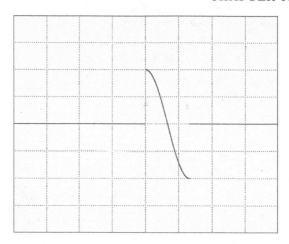

FIGURE 9.12
The 256-term Haar series partial sum.

More precisely, the Haar series partial sums are supported on $[0,1]$ (just like the function g) and they converge uniformly on $[0,1)$ to g (exercise). Of course the Haar series is not the final solution either. It has good quantitative behavior, but its qualitative behavior is poor because the partial sums are piecewise constant (i.e., *jagged*) functions. We thus begin to see the desirability of smooth wavelets. ∎

Part of the reason that wavelet sums exhibit this dramatic improvement over Fourier sums is that wavelets provide an "unconditional basis" for many standard function spaces (see [HERG, p. 233 ff.], as well as the discussion in the next section, for more on this idea). Briefly, the advantage that wavelets offer is that we can select only those wavelet basis functions whose supports overlap with the support of the function being approximated. This procedure corresponds, roughly speaking, with the operation of rearranging a series; such rearrangement is possible for series formed from an unconditional basis, but not (in general) with Fourier series.

Exercises

1. Calculate the Fourier series expansion of $f(x) = \chi_{[0,1]}(x) \cdot e^x$. Also calculate the Haar basis expansion of f. Which is a more accurate approximation? Why?

2. Refer to Exercise 1. Calculate the Fourier *transform* of f. Use Fourier

inversion to recover f from $\hat{f}$. How good an approximation to you get? How does this compare with the results from Exercise 1?

3. Calculate the Fourier series expansion of $f(x) = \chi_{[0,1]}(x)$. Also calculate the Haar basis expansion of f. Which is a more accurate approximation? Why?

4. If f is continuously differentiable, then the Fourier series of f converges uniformly to f. Can you explain why—at least intuitively? Can you say something similar about the Haar basis expansion?

5. Can you write a formula for the Nth partial sum of the Haar basis expansion?

6. What is the orthogonal complement of W_0 in L^2?

7. What is the orthogonal complement of V_0 in L^2?

8. Let λ be a linear functional on V_0. Can it be represented by an integral formula?

9. Let λ be a linear functional on W_0. Can it be represented by an integral formula?

9.5 Construction of a Wavelet Basis

There exist examples of an MRA for which the scaling function φ is smooth and compactly supported. It is known—for reasons connected with the uncertainty principle (Section 9.1)—that there do not exist C^∞ (infinitely differentiable) scaling functions which are compactly supported, or which satisfy the weaker condition that they decay exponentially at infinity—see [HERG, p. 197] for a proof. But there *do exist* compactly supported C^k (k times continuously differentiable) scaling functions for each k. In this section we will give an indication of I. Daubechies's construction of such scaling functions. We begin, however, by first describing the properties of such a scaling function, and how the function might be utilized.

So suppose that φ is a scaling function that is compactly supported and is C^k. By the axioms of an MRA, the functions $\{\tau_k\varphi\}_{k\in\mathbb{Z}}$ form a basis for V_0. It follows then that the functions $\{\sqrt{2}\alpha_2\tau_k\varphi\}_{k\in\mathbb{Z}}$ form an orthonormal basis for V_1. Written more explicitly, these functions have the form $\sqrt{2}\varphi(2x - k)$, and they span V_1. Since $\varphi \in V_0 \subseteq V_1$, we may expand φ itself in terms of the functions $\sqrt{2}\varphi(2x - k)$. Thus

$$\varphi(x) = \sum_k c_k \sqrt{2}\varphi(2x - k), \qquad (9.5.1)$$

where

$$c_k = \int \varphi(x)\sqrt{2}\,\varphi(2x - k)\,dx.$$

If we set

$$\psi(x) = \sum_k (-1)^k c_{1-k} \sqrt{2} \varphi(2x - k), \tag{9.5.2}$$

then the functions $\psi(x - \ell)$, $\ell \in \mathbb{Z}$, will be orthogonal and will span W_0. To see the first assertion, we calculate the integral

$$\int \psi(x - k)\psi(x - \ell)\,dx$$

$$= \int \left[\sum_k (-1)^k c_{1-k} \sqrt{2}\varphi(2x - k) \right] \times \left[\sum_\ell (-1)^\ell c_{1-\ell}\sqrt{2}\varphi(2x-\ell) \right] dx$$

$$= \sum_{k,\ell} 2 c_{1-k} c_{1-\ell} (-1)^{k+\ell} \int \varphi(2x - k) \cdot \varphi(2x - \ell)\,dx.$$

Of course the kth integral in this last sum will be zero if $k \neq \ell$ because of Axiom 5 of an MRA. If instead $k = \ell$, then the integral evaluates to $1/2$ by a simple change of variable. If we mandate in advance that $\int |\varphi|^2 = 1$, then $\sum_k |c_k|^2 = 1$ and the result follows.

As for the functions $\psi(x - \ell)$ spanning W_0, it is slightly more convenient to verify that $\{\varphi(x - m)\}_{m\in\mathbb{Z}} \cup \{\psi(x - \ell)\}_{\ell\in\mathbb{Z}}$ spans $V_1 = V_0 \oplus W_0$. Since the functions $\varphi(2x - n)$ already span V_1, it is enough to express each of them as a linear combination of functions $\varphi(x - m)$ and $\psi(x - \ell)$. If this is to be so, then the coefficient $a_n(m)$ of $\varphi(2x - m)$ will have to be

$$a_n(m) = \int \varphi(2x - n)\varphi(x - m)\,dx$$

$$= \int \varphi(2x - n) \sum_k c_k \sqrt{2}\,\varphi(2x - 2m - k)\,dx$$

$$= \sum_k \sqrt{2}\,c_k \int \varphi(2x - n)\varphi(2x - 2m - k)\,dx.$$

The summand can be non-zero only when $n = 2m + k$, that is, when $k = n - 2m$. Hence

$$a_n(m) = \frac{1}{\sqrt{2}} c_{n-2m}.$$

Likewise, the coefficient $b_n(m)$ of $\psi(2x - \ell)$ will have to be

$$b_n(m) = \int \varphi(2x - n)\psi(x - \ell)\,dx$$

$$= \int \varphi(2x - n) \sum_k (-1)^k c_{1-k}\sqrt{2}\varphi(2x - 2\ell - k)\,dx$$

$$= \sum_k \sqrt{2}\,c_{1-k} \int \varphi(2x - n)\varphi(2x - 2\ell - k)\,dx.$$

Of course this integral can be non-zero only when $2\ell + k = n$, that is, when $k = n - 2\ell$. Thus the ℓth coefficient is

$$b_n(m) = \frac{(-1)^n}{\sqrt{2}} c_{1-n+2\ell}.$$

Thus our task reduces to showing that

$$\varphi(2x - n) = \frac{1}{\sqrt{2}} \left[\sum_{m \in \mathbb{Z}} c_{n-2m}\varphi(x - m) + (-1)^n \sum_{\ell \in \mathbb{Z}} c_{1-n+2\ell}\psi(x - \ell) \right].$$

If we plug (9.5.1) and (9.5.2) into this last equation, we end up with an identity in the functions $\varphi(2x - p)$, which in turn reduces to an algebraic identity on the coefficients. This algebraic lemma is proved in [STR, p. 546]; it is similar in spirit to the calculations that precede Theorem 9.3.1. We shall not provide the details here.

The Haar basis, while elementary and convenient, has several shortcomings. Chief among these is the fact that each basis element is discontinuous. One consequence is that the Haar basis does a poor job of approximating continuous functions. A more profound corollary of the discontinuity is that the Fourier transform of a Haar wavelet dies like $1/x$ at infinity, hence is not integrable. It is desirable to have smooth wavelets, for as we know (Chapter 6), the Fourier transform of a smooth function dies rapidly at infinity. The Daubechies wavelets are important partly because they are as smooth as we wish; for a thorough discussion of these see [WAL] or [HERG].

Our discussion of the Haar wavelets (or MRA) already captures the spirit of wavelet analysis. In particular, it generates a complete orthonormal basis for L^2 with the property that finite sums of the basis elements give a good approximation (better than partial sums of Fourier series exponentials) to the Dirac mass δ. Since any L^2 function f can be written as $f = f * \delta$, it follows (subject to checking that Haar wavelets interact nicely with convolution) that any L^2 function with suitable properties will have a good approximation by wavelet partial sums.

The Haar wavelets are particularly effective at encoding information coming from a function that is constant on large intervals. The reason is that the function ψ integrates to zero—we say that it has "mean value zero." Thus integration against ψ annihilates constants. If we want a wavelet that compresses more general classes of functions, then it is natural to mandate that the wavelet annihilate first linear functions, then quadratic functions, and so forth. In other words, we typically demand that our wavelet satisfy

$$\int \psi(x)x^j \, dx = 0 \,, \qquad j = 0, 1, 2, \ldots, L - 1 \tag{9.5.3}$$

for some pre-specified positive integer L. In this circumstance we say that ψ has "L vanishing moments". Of course it would be helpful, although it is not necessary, in achieving these vanishing moment conditions to have a wavelet ψ that is smooth.

It is a basic fact that smooth wavelets must have vanishing moments. More precisely, if ψ is a C^k wavelet such that

$$|\psi(x)| \leq C(1+|x|)^{-k-2},$$

then it must be that

$$\int x^j \psi(x)\, dx = 0, \qquad 0 \leq j \leq k.$$

Here is a sketch of the reason:

Let $\{\psi_k^j\}_{k \in \mathbb{Z}}$ be the basis generated by ψ for the space W_j. Let $j \gg j'$. Then ψ_k^j lives on a *much smaller scale* than does $\psi_{k'}^{j'}$. Therefore $\psi_{k'}^{j'}$ is (essentially) a Taylor polynomial on the interval where ψ_k^j lives. Hence the orthogonality

$$\int \psi_k^j \overline{\psi_{k'}^{j'}} = 0$$

is essentially equivalent to

$$\int \psi_k^j(x)\overline{x^m}\, dx = 0$$

for appropriate m. This is the vanishing-moment condition.

It is a basic fact about calculus that a function with many vanishing moments must oscillate a great deal. For instance, if a function f is to integrate to 0 against both 1 and x, then f integrates to zero against all linear functions. So f itself cannot be linear; it must be at least quadratic. That gives one oscillation. Likewise, if f is to integrate to 0 against $1, x, x^2$, then f must be at least cubic. That gives two oscillations. And so forth.

Remark: It is appropriate at this point to offer an aside about why there cannot exist a C^∞ wavelet with compact support. First, a C^∞ wavelet ψ must have vanishing moments of all orders. Passing to the Fourier transform, we see therefore that $\widehat{\psi}$ vanishes to infinite order at the origin (i.e., $\widehat{\psi}$ and all its derivatives vanish at 0). If ψ were compactly supported, then $\widehat{\psi}$ would be analytic (see [KRA3, §2.4]); the infinite-order vanishing then forces $\widehat{\psi}$, and hence ψ, to be identically zero.

9.5.1 A Combinatorial Construction of the Daubechies Wavelets

With these thoughts in mind, let us give the steps that explain how to use Daubechies's construction to create a *continuous* wavelet. (Constructing a C^1

or smoother wavelet follows the same lines, but is much more complicated.) We begin by nearly repeating the calculations at the beginning of this section, but then we add a twist.

Imagine functions φ and ψ, both continuous, and satisfying

$$\int_{-\infty}^{+\infty} \varphi(x)\,dx = 1\;, \quad \int_{-\infty}^{+\infty} |\varphi(x)|^2\,dx = 1\;, \quad \int_{-\infty}^{+\infty} |\psi(x)|^2\,dx = 1. \qquad (9.5.4)$$

We know from the MRA axioms that the function φ must generate the basic space V_0. Moreover, we require that $V_0 \subseteq V_1$. It follows that

$$\varphi(x) = \sum_{j\in\mathbb{Z}} c_j \sqrt{2}\varphi(2x - j) \qquad (9.5.5)$$

for some constants c_j. The equation

$$\psi(x) = \sum_{j\in\mathbb{Z}} (-1)^j c_{1-j} \sqrt{2}\varphi(2x - j) \qquad (9.5.6)$$

defines a wavelet ψ such that $\{\tau_k\psi\}$ spans the subspace W_0. Notice that equations (9.5.5) and (9.5.6) generalize the relations that we had between φ and ψ for the Haar basis.

Equation (9.5.5), together with the first two integrals in (9.5.4), shows that

$$\sum_{j\in\mathbb{Z}} c_j = \sqrt{2} \quad \text{and} \quad \sum_{j\in\mathbb{Z}} |c_j|^2 = 1. \qquad (9.5.7)$$

If, for specificity, we take $L = 2$, then equation (9.5.3) combined with (9.5.6) implies that

$$\sum_{j\in\mathbb{Z}} (-1)^j c_j = 0 \quad \text{and} \quad \sum_{j\in\mathbb{Z}} j(-1)^j c_j = 0. \qquad (9.5.8)$$

In fact one can solve the equations in (9.5.7) and (9.5.8); one standard solution is

$$c_0 = \frac{1 + \sqrt{3}}{4\sqrt{2}}\;, \quad c_1 = \frac{3 + \sqrt{3}}{4\sqrt{2}}\;, \quad c_2 = \frac{3 - \sqrt{3}}{4\sqrt{2}}\;, \quad c_3 = \frac{1 - \sqrt{3}}{4\sqrt{2}} \qquad (9.5.9)$$

and all other $c_j = 0$.

Now here comes the payoff. Using these values of c_j, we may define

$$\begin{aligned}
\varphi_0(x) &= \chi_{[0,1)}(x) \\
\varphi_j(x) &= \sum_{\ell\in\mathbb{Z}} c_\ell \sqrt{2}\varphi_{j-1}(2x - \ell) \quad \text{when } j \geq 1.
\end{aligned}$$

The functions φ_j, iteratively defined as above, converge to a continuous function φ that is supported in $[0, 3]$. It can then be seen from (9.5.6) that the corresponding function ψ is continuous and supported in $[-1, 2]$.

Now that suitable φ and ψ have been found, we may proceed step by step as we did with the construction of the Haar wavelet basis. We shall not provide the details, but instead refer the reader to [WAL], from which this particular presentation of wavelet ideas derives.

9.5.2 The Daubechies Wavelets from the Point of View of Fourier Analysis

We now give a last loving look at the Daubechies wavelet construction—this time from the point of view of Fourier analysis. Observe that if φ is smooth and of compact support, then the sum in (9.5.5) must be finite. Using ideas from Chapter 6, we may calculate that

$$\left(\sqrt{2}\varphi(2\cdot(x-j))\right)\widehat{}(\xi) = \left(\sqrt{2}\alpha_2\tau_j\varphi\right)\widehat{}(\xi) = \frac{\sqrt{2}}{2}\left(\tau_j\varphi\right)\widehat{}(\xi/2) = \frac{1}{\sqrt{2}}e^{ij\xi/2}\widehat{\varphi}(\xi/2).$$

If we set

$$m(\xi) = \frac{1}{\sqrt{2}}\sum_j c_j e^{ij\xi},$$

where the c_j are as in (9.5.5), then we may write (applying the Fourier transform to the sum in (9.5.5))

$$\widehat{\varphi}(\xi) = m(\xi/2)\widehat{\varphi}(\xi/2). \tag{9.5.10}$$

We call the function m a *low-pass filter*.

Iterating this last identity yields

$$\widehat{\varphi}(\xi) = m(\xi/2)m(\xi/4)\widehat{\varphi}(\xi/4)$$

$$\cdots$$

$$\widehat{\varphi}(\xi) = m(\xi/2)m(\xi/4)\cdots m(\xi/2^p)\widehat{\varphi}(\xi/2^p).$$

Since $\widehat{\varphi}(0) = \int \varphi(x)\,dx = 1$, we find in the limit that

$$\widehat{\varphi}(\xi) = \prod_{p=1}^{\infty} m(\xi/2^p). \tag{9.5.11}$$

Now the orthonormality of the $\{\varphi(x-j)\}$ implies the identity

$$|m(\xi)|^2 + |m(\xi+\pi)|^2 \equiv 1. \tag{9.5.12}$$

To wit, we calculate using Plancherel's theorem (and with $\delta_{j,k}$ denoting the

Kronecker delta) that

$$
\begin{aligned}
\delta_{j,0} &= \int_{\mathbb{R}} \varphi(x)\overline{\varphi(x-j)}\,dx \quad . \\
&= \frac{1}{2\pi}\int_{\mathbb{R}} |\widehat{\varphi}(\xi)|^2 e^{-ij\xi}\,d\xi \\
&= \frac{1}{2\pi}\sum_{\ell=-\infty}^{\infty}\int_{2\ell\pi}^{2(\ell+1)\pi} |\widehat{\varphi}(\xi)|^2 e^{-ij\xi}\,d\xi \\
&= \frac{1}{2\pi}\sum_{\ell=-\infty}^{\infty}\int_{0}^{2\pi} |\widehat{\varphi}(\lambda+2\ell\pi)|^2 e^{-ij\lambda}\,d\lambda \\
&= \frac{1}{2\pi}\int_{0}^{2\pi}\left(\sum_{\ell\in\mathbb{Z}} |\widehat{\varphi}(\lambda+2\ell\pi)|^2\right) e^{-ij\lambda}\,d\lambda.
\end{aligned}
$$

This calculation tells us that the 2π-periodic function $\sum_\ell |\widehat{\varphi}(\mu + 2\ell\pi)|^2$ has Fourier coefficient 1 at the frequency 0 and all other Fourier coefficients 0. In other words,

$$\sum_\ell |\widehat{\varphi}(\mu + 2\ell\pi)|^2 = 1\,.$$

Using (9.5.10), we find that

$$\sum_\ell |\widehat{\varphi}(\xi + \ell\pi)|^2 |m(\xi + \ell\pi)|^2 = 1\,.$$

Now separating into sums over even and odd indices ℓ, and using the 2π-periodicity of m, yields (9.5.12).

Running our last arguments backwards, it can be shown that if m is a trigonometric polynomial satisfying (9.5.12) and such that $m(0) = 1$, then the product in (9.5.11) converges uniformly on compact sets to a function $\widehat{\varphi} \in L^2$. Also, if $\widehat{\varphi}$ decays sufficiently rapidly ($|\widehat{\varphi}| = \mathcal{O}(1 + |\xi|)^{-1-\epsilon}$ will do), then its inverse Fourier transform φ is the scaling function of an MRA.

In summary, if we can find a trigonometric polynomial m satisfying (9.5.12) with $m(0) = 1$ and so that the resulting $\widehat{\varphi}$ satisfies $|\widehat{\varphi}(\xi)| = \mathcal{O}(1+|\xi|)^{-k-1-\epsilon}$ for some integer $k \geq 0$, then φ will be a compactly supported wavelet of class C^k. It should be noted that *finding such a trigonometric polynomial m is hard work.*

As a final note, if φ_0 is a "nice" function with $\widehat{\varphi_0}(0) = 1$ and if we define φ_K inductively by

$$\widehat{\varphi_K}(\xi) = m(\xi/2)\widehat{\varphi_{K-1}}(\xi/2) \tag{9.5.13}$$

then, by (9.5.11),

$$\lim_{K\to\infty}\widehat{\varphi_K}(\xi) = \lim_{K\to\infty}\widehat{\varphi_0}(\xi/2^K)\prod_{k=1}^{K} m(\xi/2^k) \equiv \widehat{\varphi}(\xi).$$

This last equation *defines* $\widehat{\varphi}$, and therefore φ itself. Unraveling the Fourier transform in (9.5.13), we conclude that $\varphi = \lim \varphi_K$, where φ_K is defined inductively by

$$\varphi_K(x) = \sum_j c_j \sqrt{2} \varphi_{K-1}(2x - j).$$

The analysis that we have just given shows that wavelet theory is firmly founded on invariance properties of the Fourier transform. In other words, wavelet theory does not displace the classical Fourier theory; rather, it builds on those venerable ideas.

Reflective Remarks

The iterative procedure that we have used to construct the scaling function φ has some interesting side effects. One is that φ has certain self-similarity properties that are reminiscent of fractals.

We summarize the very sketchy presentation of the present chapter by pointing out that an MRA (and its generalizations to wavelet packets and to the local cosine bases of Coifman and Meyer [HERG]) gives a "designer" version of Fourier analysis that retains many of the favorable features of classical Fourier analysis, but also allows the user to adapt the system to problems at hand. We have given a construction that is particularly well adapted to detecting spikes in a sound wave, and therefore is useful for denoising. Other wavelet constructions have proved useful in signal compression, image compression, and other engineering applications.

In what follows are two noteworthy mathematical applications of wavelet theory. They have independent interest, but are also closely connected to each other (by way of wavelet theory) and to ideas in the rest of the book. We shall indicate some of these connections. One of these applications is to see that wavelets give a natural unconditional basis for many of the classical Banach spaces of analysis. The other is to see that a Calderón–Zygmund operator is essentially diagonal when expressed as a bi-infinite matrix with respect to a wavelet basis.

We sketch some of the ideas adherent to the previous paragraph, and refer the reader to [DAU] for the details.

9.5.3   Wavelets as an Unconditional Basis

Recall that a set of vectors $\{e_0, e_1, \dots\}$ in a Banach space (a complete, normed, linear space) X is called an *unconditional basis* if it has the following properties:

(9.5.14) For each $x \in X$ there is a unique sequence of scalars $\alpha_0, \alpha_1, \dots$ such that

$$x = \sum_{j=0}^{\infty} \alpha_j e_j,$$

in the sense that the partial sums $S_N \equiv \sum_{j=0}^{N} \alpha_j e_j$ converge to x in the topology of the Banach space.

(9.5.15) There exists a constant C such that, for each integer m, for each sequence $\alpha_0, \alpha_1, \ldots$ of coefficients as in **(9.5.14)**, and for any sequence $\beta_0, \beta_1, \ldots$ satisfying $|\beta_k| \leq |\alpha_k|$ for all $0 \leq k \leq m$, we have

$$\left\| \sum_{k=0}^{m} \beta_k e_k \right\| \leq C \cdot \left\| \sum_{k=0}^{m} \alpha_k e_k \right\|. \tag{9.5.15.1}$$

We commonly describe property **(9.5.14)** with the phrase "$\{e_j\}$ is a *Schauder basis* for X." The practical significance of property **(9.5.15)** is that we can decide whether a given formal series $\beta_0 e_0 + \beta_1 e_1 + \cdots$ converges to an element $y \in X$ simply by checking the sizes of the coefficients.

Let us consider the classical L^p spaces on the real line. We know by construction that the wavelets form an orthonormal basis for $L^2(\mathbb{R})$. In particular, the partial sums are dense in L^2. So they are also dense in $L^2 \cap L^p$, $2 < p < \infty$, in the L^2 topology. It follows that they are dense in L^p in the L^p topology for this range of p, since the L^p norm then dominates the L^2 norm (the argument for $p < 2$ involves some extra tricks which we omit). Modulo some technical details, this says in effect that the wavelets form a Schauder basis for L^p, $1 < p < \infty$. Now let us address the "unconditional" aspect.

It can be shown that **(9.5.15)** holds for all sequences $\{\beta_j\}$ if and only if it holds in all the special cases $\beta_j = \pm \alpha_j$. Suppose that

$$L^p \ni f = \sum_{j,k} \alpha_{j,k} \psi_k^j;$$

then of course it must be that $\alpha_{j,k} = \int f(x) \psi_k^j(x) \, dx = \langle f, \psi_k^j \rangle$ (by the orthonormality of the wavelets). So we need to show that, for any choice of $\mathbf{w} \equiv \{w_{j,k}\} = \{\pm 1\}$, the operator $T_{\mathbf{w}}$ defined by

$$T_{\mathbf{w}} f = \sum_{j,k} w_{j,k} \langle f, \psi_k^j \rangle \psi_k^j$$

is a bounded operator on L^p. We certainly know that $T_{\mathbf{w}}$ is bounded on L^2, for

$$\|T_{\mathbf{w}} f\|_{L^2}^2 = \sum_{j,k} |w_{j,k} \langle f, \psi_k^j \rangle|^2 = \sum_{j,k} |\langle f, \psi_k^j \rangle|^2 = \|f\|_{L^2}^2.$$

The L^p boundedness will then follow from the Calderón–Zygmund theorem provided that we can prove suitable estimates for the integral kernel of $T_{\mathbf{w}}$.

The necessary estimates are these (which should look familiar):

$$|k(x,y)| \leq \frac{C}{|x-y|}$$

and

$$\left|\frac{\partial}{\partial x}k(x,y)\right| + \left|\frac{\partial}{\partial y}k(x,y)\right| \leq \frac{C}{|x-y|^2}.$$

These are proved in Lemma 8.1.5 on p. 296 of [DAU]. We shall not provide the details here.

9.5.4   Wavelets and Almost Diagonalizability

Now let us say something about the "almost diagonalizability" of Calderón–Zygmund operators with respect to a wavelet basis. In fact we have already seen an instance of this phenomenon: the kernel K for the operator $T_{\mathbf{w}}$ that we just considered must have the form

$$K(x,y) = \sum_{j,k} w_{j,k}\psi_k^j(x)\overline{\psi_k^j(y)}. \tag{9.5.16}$$

Do not be confused by the double indexing! If we replace the double index (j,k) by the single index ℓ, then the kernel becomes

$$K(x,y) = \sum_{\ell} w_\ell \psi_\ell(x)\overline{\psi_\ell(y)}.$$

We see that the kernel, an instance of a singular integral kernel, is plainly diagonal.

In fact it is easy to see that an operator given by a kernel that is diagonal with respect to a wavelet basis induces an operator that is bounded on L^p. For let

$$K(x,y) = \sum_{j,k} \alpha_{j,k}\psi_k^j(x)\overline{\psi_k^j(y)}.$$

Then the operator

$$T_K : f \mapsto \int K(x,y)f(y)\,dy$$

satisfies (at least at a computational level)

$$T_K f(x) = \sum_{j,k} \alpha_{j,k}\langle f, \psi_k^j\rangle \psi_k^j(x).$$

Since $\{\psi_k^j\}$ forms an orthonormal basis for L^2, we see that if each $\alpha_{j,k} = 1$, then the last line is precisely f. If instead the $\alpha_{j,k}$ form a bounded sequence, then the last displayed line represents a bounded operator on L^2. In fact it turns out that T_K must be (essentially) the sort of operator that is being described in the $T(1)$ theorem. A few details follow:

A translation-invariant operator T, with kernel k, is a Calderón–Zygmund operator if and only if it is "essentially diagonal" with respect to a wavelet basis, in the sense that the matrix entries die off rapidly away from the diagonal of the matrix. To see the "only if" part of this assertion, one calculates

$$\langle T\psi_k^j, \psi_{k'}^{j'}\rangle = \int_{I'}\int_I k(x-y)\psi_k^j(x)\psi_{k'}^{j'}(y)\,dxdy,$$

where I is the interval that is the support of ψ and I' is the interval that is the support of ψ'. One then exploits the mean-value-zero properties of ψ_k^j and $\psi_{k'}^{j'}$, together with the estimates

$$|k(x,y)| \le \frac{C}{|x-y|}$$

and

$$\left|\frac{\partial}{\partial x}k(x,y)\right| + \left|\frac{\partial}{\partial y}k(x,y)\right| \le \frac{C}{|x-y|^2}.$$

After some calculation, the result is that

$$|\langle T\psi_k^j, \psi_{k'}^{j'}\rangle| \le C \cdot e^{-c\cdot\rho(\zeta,\zeta')},$$

where ρ is the hyperbolic (Poincaré) distance between the points ζ, ζ' that are associated with I, I'. See Figure 9.13. We shall not provide the proofs of these assertions, but instead refer the reader to [MEY1], [DAU].

Exercise: Calculate the matrix of the Hilbert transform

$$f \longmapsto \text{P.V.} \int \frac{f(t)}{x-t}\,dt$$

with respect to the Haar basis. Conclude that the Hilbert transform is bounded on $L^2(\mathbb{R})$.

Exercises

1. Provide the details of the derivation of the formula

$$\varphi(2x-n) = \frac{1}{\sqrt{2}}\left[\sum_{m\in\mathbb{Z}}\overline{c_{n-2m}}\varphi(x-m) + (-1)^n\sum_{\ell\in\mathbb{Z}}\overline{c_{1-n+2\ell}}\psi(x-\ell)\right]$$

as given in the text.

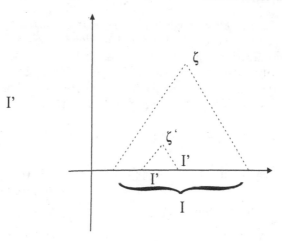

FIGURE 9.13
The geometry of the hyperbolic plane.

2. Explain why a function that integrates to 0 against $1, x, x^2, \ldots, x^k$ must have a graph that oscillates at least k times. [**Hint:** Use induction on k.]

3. Give an explicit example of a nonzero function that, on the interval $[0, 1]$, integrates to 0 against $1, x, x^2$. [**Hint:** Think of a polynomial of degree at least 3.]

4. Prove that if f is a continuous function such that

$$\int_0^1 f(x)x^j \, dx = 0 \quad \text{for all} \quad j = 0, 1, 2, \ldots$$

    then $f \equiv 0$.

5. Show that the functions

$$\varphi_j(x) = \sum_{\ell \in \mathbb{Z}} c_\ell \sqrt{2}\varphi_{j-1}(2x - \ell),$$

    as discussed in the text, converge as $j \to +\infty$ to a continuous function φ that is supported in $[0, 3]$.

6. Show that if m is a trigonometric polynomial satisfying (9.5.12) and such that $m(0) = 1$, then the product in (9.5.11) converges uniformly on compact sets to a function $\widehat{\varphi} \in L^2$.

7. Refer to Exercise 6. Show that if $\widehat{\varphi}$ decays sufficiently rapidly, then its inverse Fourier transform φ is the scaling function of an MRA.

8. Show that the Hilbert transform is bounded on L^2 by calculating the Fourier transform of the kernel $1/x$.

9. Refer to the calculations at the end of the section. Show that

$$|\langle T\psi_k^j, \psi_{k'}^{j'} \rangle| \leq C \cdot e^{-c\rho(\zeta,\zeta')},$$

    where ρ is the hyperbolic metric.

10. Which nonnegative integers k have the property that

$$\int_{-\pi}^{\pi} (\sin x) x^k \, dx = 0 \, ?$$

9.6 The Wavelet Transform

The material presented up until now in this chapter should be considered to be an informal introduction to what wavelet theory is all about. Now we shall engage in some detailed calculations in order to turn wavelets into a useful tool. We shall conclude with some concrete applications.

We now develop a sequence of results that lead up to a powerful tool called the wavelet transform. Afterward we present some compelling applications. Recall the basic function φ that we have treated in detail above.

Proposition 9.6.1 *Define*

$$\varphi_{j,k}(t) = 2^{j/2} \cdot \varphi(2^j t - k) \qquad \text{for } j, k \in \mathbb{Z}.$$

Then the $\{\varphi_{j,k}\}$ form an orthonormal basis for V_j.

Proof: The assertion is obvious from the discussion in Section 9.3. $\qquad\square$

Definition 9.6.2 Let $j, k \in \mathbb{Z}$. Define the interval

$$I_{j,k} = \left[\frac{k}{2^j}, \frac{k+1}{2^j} \right).$$

We refer to the first index j as the *level* of the interval. The level specifies the size of the interval.

Proposition 9.6.3 *Let $j \in \mathbb{Z}$. Then*

$$\mathbb{R} = \cdots \cup I_{j,-2} \cup I_{j,-1} \cup I_{j,0} \cup I_{j,1} \cup I_{j,2} \cup \cdots.$$

Proof: Obvious. $\qquad\square$

Proposition 9.6.4 Let $j, k, m \in \mathbb{Z}$ and assume that $k \neq m$. Then
$$I_{j,k} \cap I_{j,m} = \emptyset.$$

Proof: Obvious.   □

Proposition 9.6.5 Let $j, k, \ell, m \in \mathbb{Z}$. Assume that $\ell > j$. Then either $I_{j,k} \cap I_{\ell,m} = \emptyset$ or $I_{\ell,m} \subseteq I_{j,k}$. In the latter case, $I_{\ell,m}$ is contained in either the left half of $I_{j,k}$ or the right half of $I_{j,k}$.

Proof: Obvious.   □

Now we want to consider the projection of a function $g \in L^2(\mathbb{R})$ into V_j. Denote this projection by $P_j[g](t)$. Just as in finite-dimensional linear algebra, we have that
$$P_j[g](t) = \sum_{k \in \mathbb{Z}} \langle \varphi_{j,k}(t), g(t) \rangle \cdot \varphi_{j,k}(t).$$
We can use Proposition 9.6.1 above to rewrite this last formula as
$$
\begin{aligned}
P_j[g](t) &= \sum_{k \in \mathbb{Z}} \langle 2^{j/2} \varphi(2^j t - k), g(t) \rangle \cdot 2^{j/2} \varphi(2^j t - k) \\
&= 2^j \cdot \sum_{k \in \mathbb{Z}} \langle \varphi(2^j t - k), g(t) \rangle \cdot \varphi(2^j t - k).
\end{aligned}
$$

EXAMPLE 9.6.6 Let $g(t) = e^{-|t|}$. Let us calculate $P_2[g](t)$ and $P_{-1}[g](t)$.
    Now
$$P_2[g](t) = 4 \sum_{k \in \mathbb{Z}} \langle \varphi(4t - k), g(t) \rangle \cdot \varphi(4t - k).$$
We see that
$$\text{supp}\,(\varphi_{2,k}) = \left[\frac{k}{4}, \frac{k+1}{4} \right].$$
Hence the inner produce can be written as
$$
\langle \varphi(4t - k), g(t) \rangle =
\begin{cases}
\int_{k/4}^{(k+1)/4} e^{-t}\, dt & \text{if} \quad k \geq 0 \\[2mm]
\int_{k/4}^{(k+1)/4} e^{t}\, dt & \text{if} \quad k < 0
\end{cases}
$$
$$
=
\begin{cases}
e^{-k/4}(1 - e^{-1/4}) & \text{if} \quad k \geq 0 \\[2mm]
e^{k/4}(e^{1/4} - 1) & \text{if} \quad k < 0.
\end{cases}
$$

As a result,

$$P_2[g](t) = 4(1 - e^{-1/4}) \sum_{k=0}^{\infty} e^{-k/4} \varphi(4t - k) + 4(e^{1/4} - 1) \sum_{k=-\infty}^{-1} e^{k/4} \varphi(4t - k).$$

For the second projection, we have

$$P_{-1}[g](t) = \frac{1}{2} \sum_{k \in \mathbb{Z}} \left\langle \varphi\left(\frac{t}{2} - k\right), g(t) \right\rangle \cdot \varphi\left(\frac{t}{2} - k\right).$$

We know that

$$\operatorname{supp}(\varphi_{-1,k}) = [2k, 2(k+1)]$$

and hence $\varphi_{-1,k}$ equals 0 outside that interval. The inner product is then

$$\langle \varphi(t/2 - k), g(t) \rangle = \begin{cases} \int_{2k}^{2k+2} e^{-t}\, dt & \text{if} \quad k \geq 0 \\ \int_{2k}^{2k+2} e^{t}\, dt & \text{if} \quad k < 0 \end{cases}$$

$$= \begin{cases} e^{-2k}(1 - e^{-2}) & \text{if} \quad k \geq 0 \\ e^{2k}(e^2 - 1) & \text{if} \quad k < 0. \end{cases}$$

Thus we can write the projection as

$$P_{-1}[g](t) = \frac{1}{2}(1 - e^{-2}) \sum_{k=0}^{\infty} e^{-2k} \varphi\left(\frac{t}{2} - k\right) + \frac{1}{2}(e^2 - 1) \sum_{k=-\infty}^{-1} e^{2k} \varphi\left(\frac{t}{2} - k\right).$$

■

Proposition 9.6.7 *The V_j spaces are nested as follows:*

$$\cdots \subseteq V_{-2} \subseteq V_{-1} \subseteq V_0 \subseteq V_1 \subseteq V_2 \subseteq \cdots.$$

Proof: This is immediate from the definition of V_j. □

Proposition 9.6.8 *A function $f \in V_j$ if and only if $f(2t)$ is an element of V_{j+1}.*

Proof: This is immediate from the definitions. □

Proposition 9.6.9 *The spaces V_j satisfy these properties:*

$$\bigcap_{j\in\mathbb{Z}} V_j = \{0\}$$

and

$$\overline{\bigcup_{j\in\mathbb{Z}} V_j} = L^2(\mathbb{R}).$$

Proof: Already discussed in Section 9.3. $\square$

Proposition 9.6.10 *Let $f_1 \in V_1$ be given by*

$$f_1(t) = \sum_{k\in\mathbb{Z}} a_k \varphi_{1,k}(t).$$

Then the projection $P_0[f_1](t)$ of f_1 into V_0 is given by

$$f_0(t) = \sum_{k\in\mathbb{Z}} b_k \varphi(t-k),$$

where

$$b_k = \frac{\sqrt{2}}{2} \cdot (a_{2k} + a_{2k+1}). \qquad (9.6.10.1)$$

Proof: Write

$$f_0(t) = P_0[f_1](t) = \sum_{k\in\mathbb{Z}} b_k \varphi(t-k).$$

By the definition of the projection we have

$$b_k = \langle \varphi(t-k), f_1(t) \rangle = \int_{\mathbb{R}} \varphi(t-k) f_1(t)\, dt.$$

We know that $\operatorname{supp} \varphi(t-k) = [k, k+1]$ and, on this interval, the value of the function is 1. Hence the last integral reduces to

$$b_k = \int_k^{k+1} f_1(t)\, dt.$$

The function f_1 is piecewise constant with possible discontinuities at points of $(1/2)\mathbb{Z}$. Notice that there are two relevant basis functions:

$$\varphi_{1,m}(t) = \sqrt{2}\varphi(2t-m) \qquad \text{with} \quad \operatorname{supp} \varphi_{1,m} = [k, k+1/2]$$

and

$$\varphi_{1,m+1}(t) = \sqrt{2}\varphi(2t - (m+1)) \qquad \text{with} \ \ \operatorname{supp} \varphi_{1,m+1} = [k + 1/2, k + 1].$$

We need to determine the correct choice for m.

Observe that the compact support of $\varphi(2t - m)$ is $[m/2, (m+1)/2]$. We need the left endpoint of this interval to equal the left endpoint of the support of $\varphi(t - k)$. Thus $m/2 = k$ or $m = 2k$. As a result, $\varphi_{1,2k}(t) = \varphi(2t - 2k)$ is the leftmost function whose support overlaps $[k, k+1]$. We find the other function by translating this last function to the right by $1/2$ unit. This is

$$\varphi(2(t - 1/2) - 2k) = \varphi(2t - (2k + 1)) = \varphi_{1,2k+1}.$$

We conclude that, on the interval $[k, k + 1]$,

$$f_1(t) = a_{2k}\varphi(2t - 2k) + a_{2k+1}\varphi(2t - (2k + 1)).$$

Here $\operatorname{supp} \varphi(2t - 2k) = [k, k + 1/2]$ and $\operatorname{supp} \varphi(2t - (2k+1)) = [k + 1/2, k + 1]$. The inner product thus becomes

$$
\begin{aligned}
b_k &= \int_k^{k+1} f_1(t)\, dt \\
&= \int_k^{k+1} \left(a_{2k}\sqrt{2}\varphi(2t - 2k) + a_{2k+1}\sqrt{2}\varphi(2t - (2k + 1)) \right) dt \\
&= \sqrt{2} \int_k^{k+1} a_{2k}\varphi(2t - 2k)\, dt + \sqrt{2} \int_k^{k+1} a_{2k+1}\varphi(2t - (2k + 1))\, dt.
\end{aligned}
$$

Of course $\varphi(2t - 2k) = 1$ on the interval $[k, k+1/2]$ and $\varphi(2t - (2k+1)) = 1$ on the interval $[k + 1/2, k + 1]$. The inner product then becomes

$$
\begin{aligned}
b_k &= \sqrt{2} \int_0^{k+1/2} a_{2k}\varphi(2t - 2k)\, dt + \sqrt{2} \int_{k+1/2}^{k+1} a_{2k+1}\varphi(2t - (2k + 1))\, dt \\
&= \sqrt{2}a_{2k} \int_0^{k+1/2} 1\, dt + \sqrt{2}a_{2k+1} \int_{k+1/2}^{k+1} 1\, dt \\
&= \frac{\sqrt{2}}{2}(a_{2k} + a_{2k+1}).
\end{aligned}
$$

That completes the proof. $\qquad\qquad\qquad\qquad\qquad\qquad\qquad\qquad\qquad\square$

The last proposition is an important calculation that will shape what we do below.

Exercise for the Reader: Let

$$f_1(t) = 5\varphi_{1,0}(t) + 3\varphi_{1,1}(t) + 5\varphi_{1,2}(t) - \varphi_{1,3}(t) + 5\varphi_{1,4}(t) + 7\varphi_{1,5}(t). \qquad (9.6.11)$$

Calculate that the projection f_0 of f_1 into V_0 is

$$f_0 = 4\sqrt{2}\varphi_{0,0}(t) + 2\sqrt{2}\varphi_{0,1}(t) + 6\sqrt{2}\varphi_{0,2}(t).$$

We have constructed f_0 as an approximation in V_0 of $f_1 \in V_1$. Let us now change our point of view and assume that we are given $f_0 \in V_0$ and $f_1 \in V_1$ and we want to produce a residual function g_0 so that $f_0(t) + g_0(t) = f_1(t)$.

Let us find a way to describe g_0 explicitly.

Of course $g_0 = f_1 - f_0$. Since we know exactly what the functions on the right look like, we see that g_0 is a piecewise constant function with discontinuities at $1/2, 1, 3/2, 2, 5/2$. We conclude that $g_0 \in V_1$ so g_0 can be written as a linear combination of the $\varphi_{1,k}$.

Refer to the **Exercise for the Reader** above. Let us focus on those particular f_0, f_1. Since $\operatorname{supp} f_0 = \operatorname{supp} f_1 = [0,3]$, we need only consider $\varphi_{1,k}$ for $k = 0,1,2,\ldots,5$.

Consider now f_0 and f_1 on the interval $[0,1)$. On this interval f_0 equals the average of the two constant values of f_1. Hence g_0 equals the directed distance from f_0 to f_1 on this interval. That directed distance is $1/2$ on $[0,1/2)$ and it is $-\sqrt{2}$ on $[1/2,1)$.

We can repeat this last analysis on the intervals $[1,2)$ and $[2,3)$. We obtain

$$\begin{aligned}
g_0(t) &= f_1(t) - f_0(t) \\
&= \sqrt{2}\varphi(2t) - \sqrt{2}\varphi(2t-1) + 3\sqrt{2}\varphi(2t-2) - 3\sqrt{2}\varphi(2t-3) \\
&\quad - \sqrt{2}\varphi(2t-4) + \sqrt{2}\varphi(2t-5) \\
&= \sqrt{2}(\varphi(2t) - \varphi(2t-1)) + 3\sqrt{2}(\varphi(2t-2) - \varphi(2t-3)) \\
&\quad - \sqrt{2}(\varphi(2t-4) - \varphi(2t-5)).
\end{aligned} \qquad (9.6.12)$$

Now we want to take a closer look at this function g_0. Let us restrict attention to g_0 on the interval $[0,1]$. On that interval, g_0 is $\sqrt{2}$ times the function ψ, where

$$\psi(t) = \varphi(2t) - \varphi(2t-1) = \begin{cases} 1 & \text{if} & 0 \le t < \frac{1}{2} \\ -1 & \text{if} & \frac{1}{2} \le t < 1 \\ 0 & \text{if} & t \le 0 \text{ or } 1 \le t < \infty. \end{cases} \qquad (9.6.13)$$

Next let us look at g_0 on the interval $[1,2)$. We can view this function as the one-unit right-translate of the ψ discussed in the last paragraph multiplied by $3\sqrt{2}$. As a result, on $[1,2)$, $g_0(t) = 3\sqrt{2}\psi(t-1)$. Similar reasoning shows that, on the interval $[2,3)$, $g_0(t) = -\psi(t-2)$. We conclude that

$$g_0(t) = \sqrt{2}\psi(t) + 3\sqrt{2}\psi(t-1) - \sqrt{2}\psi(t-2).$$

We see now that there is a relationship between the coefficients of $\psi(t-k)$,

$k = 0, 1, 2, \ldots$ in the discussion following (9.6.13) and the coefficients of f_1 in (9.6.11). If we write

$$g_0(t) = \sum_{k=0}^{2} c_k \psi(t - k)$$

where $c_0 = \sqrt{2}$, $c_1 = 3\sqrt{2}$, and $c_2 = -\sqrt{2}$, then the c_k obey the relation

$$c_k = \frac{\sqrt{2}}{2} \cdot (a_{2k} - a_{2k+1}).$$

Based on this experimental evidence, we now have the following more general result.

Proposition 9.6.14 *Suppose that the function f_1, an element of V_1, is given by*

$$f_1(t) = \sum_{k \in \mathbb{Z}} a_k \varphi_{1,k}(t).$$

Further assume that f_0 is the projection of $P_0[f_1](t)$ of f_1 into V_0. If $g_0(t) = f_1(t) - f_0(t)$ is the residual function in V_1, then g_0 is given by

$$g_0(t) = \sum_{k \in \mathbb{Z}} c_k \psi(t - k)$$

where ψ is given by (9.6.13) and

$$c_k = \frac{\sqrt{2}}{2}(a_{2k} - a_{2k+1})$$

for $k \in \mathbb{Z}$.

Proof: Exercise. □

In the past we have been interested in the linear span of φ and its integer translates. That is the space V_0. In the last proposition we saw that there is interest in the linear span of ψ defined in (9.6.13) and its integer translates. This gives rise to the space W_0 (which we have already had a glimpse of in Section 9.3). Thus we have

Definition 9.6.15 Let ψ be as above. We define

$$W_0 = \text{span} \{\ldots, \psi(t+1), \psi(t), \psi(t-1), \ldots\} \cap L^2(\mathbb{R}) = \text{span} \{\psi(t-k)\}_{k \in \mathbb{Z}} \cap L^2(\mathbb{R}).$$

We call W_0 the *Haar wavelet space* generated by the Haar wavelet function ψ.

The etymology of the word "wavelet" is now evident. Because of the properties of the approximation f_0 to f_1 (averaging consecutive values a_{2k} and a_{2k+1}), it makes sense to model the error using small waves such as ψ and its integer translates.

Proposition 9.6.16 *The set $\{\psi(t-k)\}_{k \in \mathbb{Z}}$ forms an orthonormal basis for W_0.*

Equation (9.6.10.1) tells us that the vector

$$\mathbf{h} = {}^t[h_0, h_1] = {}^t\left[\frac{\sqrt{2}}{2}, \frac{\sqrt{2}}{2}\right] \tag{9.6.17}$$

has an important role in projecting a function $f_1 \in V_1$ into V_0. In fact we can rewrite the projection coefficients in (9.6.10.1) using formula (9.6.17). For $k \in \mathbb{Z}$ we have

$$b_k = \mathbf{h} \cdot \mathbf{a}^k,$$

where

$$\mathbf{a}^k = {}^t[a_{2k}, a_{2k+1}].$$

We can use $\mathbf{h}$ to rewrite φ in terms of basis functions of V_1. To wit,

$$\begin{aligned}
\varphi(t) &= \varphi(2t) + \varphi(2t-1) \\
&= \left(\frac{\sqrt{2}}{2}\right)\sqrt{2}\varphi(2t) + \left(\frac{\sqrt{2}}{2}\right)\sqrt{2}\varphi(2t-1) \\
&= h_0\varphi_{1,0}(t) + h_1\varphi_{1,1}(t).
\end{aligned} \tag{9.6.18}$$

In a similar manner, we can use the vector

$$\mathbf{g} = {}^t[g_0, g_1] = {}^t\left[\frac{\sqrt{2}}{2}, -\frac{\sqrt{2}}{2}\right] \tag{9.6.19}$$

to project a function $f_1 \in V_1$ into W_0. As we see from (9.6.13),

$$\begin{aligned}
\psi(t) &= \varphi(2t) - \varphi(2t-1) \\
&= \left(\frac{\sqrt{2}}{2}\right)\sqrt{2}\varphi(2t) + \left(-\frac{\sqrt{2}}{2}\right)\sqrt{2}\varphi(2t-1) \\
&= g_0\varphi_{1,0}(t) + g_1\varphi_{1,1}(t).
\end{aligned} \tag{9.6.20}$$

Equations (9.6.18) and (9.6.20) will play a major role in our development of wavelet theory.

Definition 9.6.21 We shall call the equations

$$\varphi(t) = \frac{\sqrt{2}}{2}\varphi_{1,0}(t) + \frac{\sqrt{2}}{2}\varphi_{1,1}(t) = \varphi(2t) + \varphi(2t-1)$$

and

$$\psi(t) = \frac{\sqrt{2}}{2}\varphi_{1,0}(t) - \frac{\sqrt{2}}{2}\varphi_{1,1}(t) = \varphi(2t) - \varphi(2t-1)$$

dilation equations. The Haar function φ which is used to generate these dilation equations is usually called a *scaling function*.

The vectors **h** and **g** in equations (9.6.17) and (9.6.19) play a significant role in applications involving discrete data. In fact we shall see that **h** and **g** can be used to form a matrix that allows us to decompose a signal (or a digital image) into an approximation of the original signal together with the details needed to recover the original signal from that approximation.

It is straightforward for you to check that $\langle \varphi(t-k), \psi(t-m)\rangle = 0$ for any integers k and m. The next proposition then follows immediately.

Proposition 9.6.22 *Suppose that $f \in V_0$ and $g \in W_0$. Then $\langle f, g\rangle = 0$.*

Definition 9.6.23 Let V and W be linear subspaces of $L^2(\mathbb{R})$. We say that V and W are *orthogonal* or *mutually perpendicular* if and only if $\langle f, g\rangle = 0$ for all $f \in V$ and $g \in W$. We write $V \perp W$.

Definition 9.6.24 Let V and W be linear subspaces of $L^2(\mathbb{R})$. We define the *direct sum* of V and W to be the space

$$V \oplus W \equiv \{f(t) + g(t) : f \in V, g \in W\}.$$

If it happens that V is orthogonal to W, then we call this the *direct orthogonal sum*.

We can think of V as a subspace of $V \oplus W$ via the mapping $v \mapsto v + 0$ and similarly for W.

Since $V_0 \perp W_0$, we can consider the direct orthogonal sum $V_0 \oplus W_0$. We have the following.

Proposition 9.6.25 *Suppose that V_0, V_1, and W_0 are defined as usual. Then*

$$V_1 = V_0 \oplus W_0.$$

Proof: We know that a function $f \in V_1$ can be written as the sum of an

$f_0 \in V_0$ and a $g_0 \in W_0$. Since f_0 and g_0 are both elements of V_1, we thus see that $V_1 = V_0 \oplus W_0$. To finish the proof we need to show that if $f_1 \in V_1$ and if f_1 is orthogonal to all the elements of V_0, then $f_1 \in W_0$.

Write $f_1 = f_0 + g_0$, with $f_0 \in V_0$ and $g_0 \in W_0$. Let $h \in V_0$ be an arbitrary element. Then

$$
\begin{aligned}
0 &= \langle f_1, h \rangle \\
  &= \langle f_0, h \rangle + \langle g_0, h \rangle \\
  &= \langle f_0, h \rangle + 0 \\
  &= \langle f_0, h \rangle .
\end{aligned}
$$

We conclude that $f_0 = 0$. Therefore $f_1 = g_0 \in W_0$. That proves the result. $\sqcup$

We now have a well-established connection among V_0, V_1, and W_0. We would like to have a similar result at the jth level. So, given a function $f_j \in V_j$, we wish to approximate it with a function f_{j-1} from V_{j-1}. We could certainly do so by projection, but we need a way to measure the difference. Since V_j is constructed from functions $\varphi(2^j t - k)$, it makes sense that the difference function will be constructed from translates and dilates of the function ψ. Thus we have the following definition.

Definition 9.6.26 Let ψ be the Haar wavelet function. For $j, k \in \mathbb{Z}$ we define

$$\psi_{j,k}(t) = 2^{j/2} \cdot \psi(2^j t - k) .$$

Proposition 9.6.27 *We have that*

$$\|\psi_{j,k}\| = 1 .$$

Proof: This is a straightforward calculation. $\square$

It is also easy to see that $\int \psi_{j,k} \, dt = 0$.

EXAMPLE 9.6.28 Let us describe the support of each of these functions:

    (a) $\psi_{1,-2}$,

    (b) $\psi_{-2,4}$,

    (c) $\psi_{5,-5}$.

For part **(a)** we see that

$$\psi_{1,-2} = 2^{1/2} \psi(2t + 2) = \sqrt{2} \cdot \psi(2(t + 1)) .$$

We conclude then that we are contracting the function ψ by a factor of 2 and

then shifting one unit to the left. Now supp $\psi = [0,1]$. Thus we have that
supp $\psi(2(t+1)) = [-1,-1/2]$.

A similar calculation shows that supp $\psi_{-2,4} = [16,20]$.

Finally, we can calculate that supp $\psi_{4,-5} = [-5/16,-1/4]$.   ∎

Recall that

$$\psi_{j,k}(t) = \begin{cases} 2^{j/2} & \text{if} & 0 \le 2^j t - k < \frac{1}{2} \\ -2^{j/2} & \text{if} & \frac{1}{2} \le 2^j t - k < 1 \\ 0 & \text{if} & t < 0 \text{ or } t \ge . \end{cases}$$

And this equals

$$\begin{cases} 2^{j/2} & \text{if} & 2^{-j}k \le t < 2^{-j}k + 2^{-(j+1)} \\ -2^{j/2} & \text{if} & 2^{-j}k + 2^{-(j+1)} \le t < 2^{-j}k + 2^{-k} \\ 0 & \text{if} & t < 2^{-j}k \text{ or } t \ge 2^{-j}k + 2^{-j}. \end{cases}$$

As a result we have the following.

Proposition 9.6.29 *The compact support for the function*
$\psi_{j,k}$ *is the interval* $[2^{-j}k, 2^{-j}(k+1)]$.

Proof: From the discussion preceding the statement of the proposition we
see that $\psi_{j,k}$ takes the value $2^{j/2}$ on the interval $[2^{-j}k, 2^{-j}k + 2^{-(j+1)}]$ and
takes the value $2^{-j/2}$ on the interval $[2^{-j}k + 2^{-(j+1)}, 2^{-j}k + 2^{-j}]$. It is zero
elsewhere.   □

Definition 9.6.30 Let ψ be the usual Haar function. We define the vector
space

$$W_j = \text{span}\left\{\ldots, \psi(2^j t + 1), \psi(2^j t), \psi(2^j t - 1), \ldots\right\} \cap L^2(\mathbb{R})$$

$$= \text{span}\left\{\psi(2^j t - k)\right\}_{k \in \mathbb{Z}} \cap L^2((R)).$$

We term this space the *Haar wavelet space* W_j.

Proposition 9.6.31 *Let* W_j *be as in the preceding defi-
nition. Then the set* $\{\psi_{j,k}\}_{k \in \mathbb{Z}}$ *is an orthonormal basis for*
W_j.

Proof: The proof is similar to that for Proposition 9.6.1.   □

So we certainly know that $\psi_{j,k}$ is orthogonal to $\psi_{j,m}$ for $k \neq m$. The next result gives even more orthogonality properties.

Proposition 9.6.32 *Let j, k, ℓ, m be integers with $\ell \geq j$. Then $\langle \varphi_{j,k}, \psi_{\ell,m} \rangle = 0$.*

Proof: First we treat the case $\ell = j$. We see that $\varphi_{j,k}(t) = 2^{j/2}$ on $I_{j,k}$ and $\psi_{j,m}(t) = \pm 2^{j/2}$ on $I_{j,m}$. If $k \neq m$, then we know that $I_{j,k} \cap I_{j,m} = \emptyset$. So the supports of the two functions are disjoint hence the inner product is 0. If instead $k = m$, then

$$
\begin{aligned}
\langle \varphi_{j,k}(t), \psi_{j,k}(t) \rangle &= \int_{I_{j,k}} \varphi_{j,k}(t)\psi_{j,k}(t)\, dt \\
&= 2^{j/2} \int_{I_{j,k}} \psi_{j,k}\, dt \\
&= 0.
\end{aligned}
$$

Now consider the case that $\ell > j$. First note that $\varphi_{j,k}$ and $\psi_{\ell,m}$ are nonzero on $I_{j,k}$ and $I_{\ell,m}$ respectively. There are two cases:

(a) If $I_{j,k} \cap I_{\ell,m} = \emptyset$, then the inner product is automatically 0.

(b) If instead $I_{j,k} \cap I_{\ell,m} \neq \emptyset$, then $I_{\ell,m}$ is either entirely contained in the left half of $I_{j,k}$ or entirely contained in the right half of $I_{j,k}$. Since $\varphi_{j,k})(t) = 2^{j/2}$ on $I_{j,k}$, it therefore equals $2^{j/2}$ on $I_{\ell,m}$. Hence

$$
\langle \varphi_{j,k}(t), \psi_{\ell,m} \rangle = 2^{j/2} \int_{I_{\ell,m}} \psi_{\ell,m}(t)\, dt = 0.
$$

That completes the proof.   □

A related idea is the following.

Proposition 9.6.33 *Let j, ℓ be integers with $\ell \geq j$. Then $V_j \perp W_\ell$.*

Proof: Let $f \in V_j$ and $g \in W_\ell$. Since $\{\varphi_{j,k}\}$ is a basis for V_j and $\{\psi_{\ell,m}\}$ is a basis for W_ℓ, we can write

$$
f(t) = \sum_{k \in \mathbb{Z}} c_k \varphi_{j,k}(t) \qquad \text{and} \qquad g(t) = \sum_{m \in \mathbb{Z}} d_m \psi_{\ell,m}(t).
$$

Now we calculate

$$\langle f(t), g(t) \rangle = \int_{\mathbb{R}} f(t) g(t) \, dt$$

$$= \int_{\mathbb{R}} \sum_{k \in \mathbb{Z}} c_k \varphi_{j,k}(t) \cdot \sum_{m \in \mathbb{Z}} d_m \psi_{\ell,m}(t) \, dt$$

$$= \sum_{k \in \mathbb{Z}} c_k \sum_{m \in \mathbb{Z}} d_m \int_{\mathbb{R}} \varphi_{j,k}(t) \psi_{\ell,m}(t) \, dt \, .$$

But the preceding proposition tells us that each of these last integrals is 0. So f and g are orthogonal, as was to be proved. □

Next we have a strengthening of Proposition 9.6.27.

Proposition 9.6.34 *Let* j, k, ℓ, m *be integers. Then*

$$\langle \psi_{j,k}(t), \psi_{\ell,m} \rangle = \begin{cases} 1 & \text{if} \quad j = \ell \text{ and } k = m \,, \\ 0 & \text{if} \quad j \neq \ell \text{ or } k \neq m \,. \end{cases}$$

Proof: We first treat the case $j = \ell$ and $k = m$. Then we calculate that

$$\langle \psi_{j,k}(t), \psi_{\ell,m}(t) \rangle = \int_{\mathbb{R}} \psi_{j,k}^2(t) \, dt = \| 2^{j/2} \psi(2^j t - k) \|^2 = 2^j \| \psi(2^j t - k) \|^2 \, .$$

But of course $\| \psi(2^j t - k) \|^2 = 2^{-j}$. So this establishes the first half of our assertion.

Now if $k \neq m$, then there are two cases.

(a) If $j = \ell$, then the inner product is zero because the functions $\{\psi_{j,k}\}$ form an orthonormal basis of W_j.

(b) Say that $j \neq \ell$. We may as well suppose that $\ell > j$. Then $I_{\ell,m}$ is either entirely contained in the left half of $I_{j,k}$ or else it is entirely contained in the right half of $I_{j,k}$. If it is in the left half, then $\psi_{j,k}(t) = \pm 2^{j/2}$. But then

$$\langle \psi_{j,k}(t), \psi_{\ell,m}(t) \rangle = \pm 2^{j/2} \int_{\mathbb{R}} \psi_{\ell,m} \, dt \, .$$

We know that this last integral equals 0.

The case $k = m$ and $j \neq \ell$ is handled similarly. □

> **Proposition 9.6.35** *Let* $j, \ell \in \mathbb{Z}$ *with* $j \neq \ell$. *Then* $W_j \perp W_\ell$.

Proof: Exercise.   □

> **Proposition 9.6.36** *The function* $\varphi_{j,0}$ *satisfies the dilation equation*
>
> $$\varphi_{j,0}(t) = \frac{\sqrt{2}}{2} \cdot \varphi_{j+1,0}(t) + \frac{\sqrt{2}}{2} \cdot \varphi_{j+1,1}(t). \qquad (9.6.36.1)$$

Proof: We know that

$$\varphi(t) = \varphi(2t) + \varphi(2t - 1).$$

We replace t by $2^j t$ in this equation to arrive at

$$\varphi(2^j t) = \varphi(2^{j+1} t) + \varphi(2^{j+1} t - 1).$$

Let us multiply this last equation by $2^{j/2}$. We obtain

$$2^{j/2} \varphi(2^j t) = \varphi_{j,0}(t) = 2^{j/2} \varphi(2^{j+1} t) + 2^{j/2} \varphi(2^{j+1} t - 1).$$

Now we rewrite the very last equality as

$$\begin{aligned}
\varphi_{j,0}(t) &= 2^{-1/2} \cdot 2^{1/2} \cdot 2^{j/2} \varphi(2^{j+1} t) + 2^{-1/2} \cdot 2^{1/2} \cdot 2^{j/2} \varphi(2^{j+1} t - 1) \\
&= 2^{-1/2} \cdot 2^{(j+1)/2} \varphi(2^{j+1} t) + 2^{-1/2} \cdot 2^{(j+1)/2} \varphi(2^{j+1} t - 1) \\
&= \frac{\sqrt{2}}{2} \varphi_{j+1,0}(t) + \frac{\sqrt{2}}{2} \varphi_{j+1,1}(t).
\end{aligned}$$

                                                                                        □

Exercises

1. Let $g(t) = t^2$. Calculate $P_2[g](t)$ and $P_{-1}[g](t)$.

2. Let

$$f_1(t) = 3\varphi_{1,0}(t) - \varphi_{1,1}(t) + 2\varphi_{1,2}(t) - 4\varphi_{1,3}(t) + 6\varphi_{1,4}(t) + 2\varphi_{1,5}(t) \,.$$

Calculate that the projection f_0 of f_1 into V_0

3. Describe the support of each of these functions:

(a) $\psi_{-2,3}$

(b) $\psi_{4,-1}$

(c) $\psi_{3,-3}$

4. Describe the Haar wavelet space W_j in words.

5. What does the formula

$$\varphi_{j,0}(t) = 2^{-1/2} \cdot 2^{1/2} \cdot 2^{j/2} \varphi(2^{j+1}t) + 2^{-1/2} \cdot 2^{1/2} \cdot 2^{j/2} \varphi(2^{j+1}t - 1)$$
$$= 2^{-1/2} \cdot 2^{(j+1)/2} \varphi(2^{j+1}t) + 2^{-1/2} \cdot 2^{(j+1)/2} \varphi(2^{j+1}t - 1)$$
$$= \frac{\sqrt{2}}{2} \varphi_{j+1,0}(t) + \frac{\sqrt{2}}{2} \varphi_{j+1,1}(t)$$

have to do with the vector **h**?

6. The spaces W_j are mutually orthogonal but the spaces V_j are not. Explain the significance of this fact.

7. Explain why the moment conditions $\int \psi(x)x^j \, dx = 0$ arise and why they are significant.

8. What particular properties set the Daubechies wavelet apart from the Haar wavelet?

9. Is it important that there is no C^∞ Daubechies wavelet? Can you use physical reasoning?

10. Explain why the space W_j is needed to complete the space V_j to V_{j+1}. What is V_j lacking?

9.7 More on the Wavelet Transform

The next result generalizes the last proposition in the last section.

Proposition 9.7.1 *For $j, k \in \mathbb{Z}$ we have*

$$\varphi_{j,k}(t) = \frac{\sqrt{2}}{2} \cdot \varphi_{j+1,2k}(t) + \frac{\sqrt{2}}{2} \cdot \varphi_{j+1,2k+1}(t).$$

Proof: We know that

$$
\begin{aligned}
\varphi_{j,k}(t) &= 2^{j/2}\varphi(2^j t - k) \\
&= 2^{j/2}\varphi\left(2^j \left(t - \frac{k}{2^j}\right)\right) \\
&= \varphi_{j,0}\left(t - \frac{k}{2^j}\right).
\end{aligned}
$$

Now we shall replace t by $t - k/2^j$ in equation (9.6.36.1) to obtain

$$\varphi_{j,k}(t) = \varphi_{j,0}\left(t - \frac{k}{2^j}\right) = \frac{\sqrt{2}}{2}\varphi_{j+1,0}\left(t - \frac{k}{2^j}\right) + \frac{\sqrt{2}}{2}\varphi_{j+1,1}\left(t - \frac{k}{2^j}\right).$$

$$(9.7.1.1)$$

The next step is to expand the term $\varphi_{j+1,0}(t - k/2^j)$:

$$
\begin{aligned}
\varphi_{j+1,0}(t - k/2^j) &= 2^{(j+1)/2}\varphi\left(2^{j+1}\left(t - \frac{k}{2^j}\right)\right) \\
&= 2^{(j+1)/2}\varphi\left(2^{j+1}t - \frac{2^{j+1}k}{2^j}\right) \\
&= 2^{(j+1)/2}\varphi\left(2^{j+1}t - 2k\right) \\
&= \varphi_{j+1,2k}(t).
\end{aligned}
$$

Expanding the term $\varphi_{j+1,1}(t - k/2^j)$ in a similar fashion yields

$$
\begin{aligned}
\varphi_{j+1,1}(t - k/2^j) &= 2^{(j+1)/2}\varphi\left(2^{j+1}\left(t - \frac{k}{2^j}\right) - 1\right) \\
&= 2^{(j+1)/2}\varphi\left(2^{j+1}t - \frac{2^{j+1}k}{2^j} - 1\right) \\
&= 2^{(j+1)/2}\varphi\left(2^{j+1}t - 2k - 1\right) \\
&= 2^{(j+1)/2}\varphi\left(2^{j+1}t - (2k + 1)\right) \\
&= \varphi_{j+1,2k+1}(t).
\end{aligned}
$$

Now substituting these last two calculations into (9.7.1.1) gives the desired result. $\square$

Next we have some analogous results for ψ. The proofs of these results are left as an exercise.

Proposition 9.7.2 *Let* $j, k \in \mathbb{Z}$. *Then*

$$\psi_{j,k}(t) = \frac{\sqrt{2}}{2} \cdot \varphi_{j+1,2k}(t) - \frac{\sqrt{2}}{2} \cdot \varphi_{j+1,2k+1}(t).$$

In particular, or $k = 0$, *we have*

$$\psi_{j,0}(t) = \frac{\sqrt{2}}{2} \cdot \varphi_{j+1,0}(t) - \frac{\sqrt{2}}{2} \cdot \varphi_{j+1,1}(t).$$

We have formerly considered projecting functions from V_1 into V_0. Now we generalize these ideas to projecting functions from V_{j+1} into V_j.

Proposition 9.7.3 *Let* $f_{j+1} \in V_{j+1}$ *be defined by*

$$f_{j+1}(t) = \sum_{m \in \mathbb{Z}} a_m \varphi_{j,m}(t).$$

Further assume that **h** *is the vector defined in (9.6.17). Then the projection vector* $f_j = P_j[f_{j+1}](t)$ *of* f_{j+1} *into* V_j *is given by*

$$f_j(t) = \sum_{k \in \mathbb{Z}} b_k \varphi_{j,k}(t) = \sum_{k \in \mathbb{Z}} \langle f_{j+1}(t), \varphi_{j,k}(t) \rangle \varphi_{j,k}(t),$$

where b_k *is given by*

$$b_k = \frac{\sqrt{2}}{2} \cdot (a_{2k} + a_{2k+1}) = \mathbf{h} \cdot \mathbf{a}^k. \qquad (9.7.3.1)$$

Here we have $\mathbf{a}^k = {}^t[a_{2k}, a_{2k+1}].$

Proof: We write

$$
\begin{aligned}
b_k &= \langle f_{j+1}(t), \varphi_{j,k}(t) \rangle \\
&= \left\langle \sum_{m \in \mathbb{Z}} a_m \varphi_{j+1,m}(t), \varphi_{j,k}(t) \right\rangle \\
&= \sum_{m \in \mathbb{Z}} a_m \langle \varphi_{j+1,m}(t), \varphi_{j,k}(t) \rangle.
\end{aligned}
$$

We now use the dilation equation for $\varphi_{j,k}$ to write

$$b_k = \sum_{m \in \mathbb{Z}} a_m \langle \varphi_{j+1,m}(t), \varphi_{j,k}(t) \rangle$$

$$= \sum_{m \in \mathbb{Z}} a_m \left\langle \varphi_{j+1,m}(t), \frac{\sqrt{2}}{2} \varphi_{j+1,2k}(t) + \frac{\sqrt{2}}{2} \varphi_{j+1,2k+1}(t) \right\rangle$$

$$= \frac{\sqrt{2}}{2} \sum_{m \in \mathbb{Z}} a_m \langle \varphi_{j+1,m}(t), \varphi_{j+1,2k}(t) \rangle + \frac{\sqrt{2}}{2} \sum_{m \in \mathbb{Z}} a_m \langle \varphi_{j+1,m}(t), \varphi_{j+1,2k+1}(t) \rangle.$$

The functions $\varphi_{j+1,m}(t)$ and $\varphi_{j+1,2k}(t)$ in the inner product in the third line above are elements of an orthonormal basis for V_{j+1}. Thus the only nonzero inner product occurs when $m = 2k$. Thus the first term in that line reduces to $(\sqrt{2}/2) \cdot a_{2k}$. Similar reasoning shows that the second term in that line reduces to $(\sqrt{2}/2) \cdot a_{2k+1}$. That ends the proof. □

Next let us think about projecting an element of V_{j+1} into W_j.

Proposition 9.7.4 *Assume that $f_{j+1} \in V_{j+1}$ is given by*

$$f_{j+1}(t) = \sum_{m \in \mathbb{Z}} a_m \varphi_{j+1,m}(t).$$

Further suppose that $f_j = P_j[f_{j+1}](t)$ is the projection of f_{j+1} into V_j. Let $\mathbf{g}$ be given by (9.6.19). If $g_j(t) = f_{j+1}(t) - f_j(t)$ is the error term in V_{j+1}, then $g_j \in W_j$ and is given by

$$g_j(t) = \sum_{k \in \mathbb{Z}} c_k \psi_{j,k}(t).$$

Furthermore,

$$c_k = \frac{\sqrt{2}}{2}(a_{2k} - a_{2k+1}) = \mathbf{g} \cdot \mathbf{a}^k. \tag{9.7.4.1}$$

Here $\mathbf{a}^k = {}^t[a_{2k}, a_{2k+1}]$.

Proof: This proof hinges on the compact support properties of $\varphi_{j,k}$ and $\varphi_{j+1,k}$. We shall analyze $f_{j+1} - f_j$ on an interval-by-interval basis.

The basis elements $\varphi_{j,k}$ that are used to build f_j are nonzero on the intervals $I_{j,k} = [k/2^j, (k+1)/2^j)$. Such an interval has length 2^{-j}. The basis elements $\varphi_{j+1,k}$ that are used to build f_{j+1} are nonzero on intervals that are half as long. If we are going to analyze $f_{j+1} - f_j$, then we must consider the larger intervals $[k/2^j, (k+1)/2^j)$ and we also must consider the two subintervals $[k/2^j, (k+1/2)/2^j)$ and $[(k+1/2)/2^j, (k+1)/2^j)$.

We next must identify the corresponding basis functions. The basis functions for the interval $[k/2^j, (k+1/2)/2^j)$ will assume the value $2^{(j+1)/2}$ for t satisfying

$$\frac{k}{2^j} \le t < \frac{k+1/2}{2^j}.$$

Let us multiply these inequalities through by 2^j to obtain

$$k \le 2^j t < k + \frac{1}{2}.$$

Now multiply this new string of inequalities through by 2 and then subtrace $2k$ from each term. The result is

$$0 \le 2^{j+1} t - 2k < 1.$$

It is now apparent that the basis function we seek is $\varphi_{j+1,2k}$. In a similar manner we may see that the basis function that equals $2^{(j+1)/2}$ on the interval $[(k+1/2)/2, (k+1)/2)$ is $\varphi_{j+1,2k+1}$. We also know that $\varphi_{j,k}$ takes the value

$$\frac{2^{j/2} \cdot \sqrt{2}}{2} \cdot (a_{2k} + a_{2k+1}) = 2^{(j-1)/2}(a_{2k} + a_{2k+1}) \qquad (9.7.4.2)$$

on the interval $[k/2^j, (k+1)/2^j)$. In conclusion, on the left half-interval $[k/2^j, (k+1/2)/2^j)$, the function $f_{j+1} - f_j$ has the value

$$2^{(j+1)/2} a_{2k} - 2^{(j-1)/2}(a_{2k} + a_{2k+1}) = 2^{j/2} a_{2k}(2^{1/2} - 2^{-1/2}) - 2^{(j-1)/2} a_{2k+1}$$
$$= 2^{(j-1)/2}(a_{2k} - a_{2k+1}).$$

By contrast, on the right half-interval $[(k+1/2)/2^j, (k+1)/2^j)$, the function $f_{j+1} - f_j$ assumes the value

$$2^{(j+1)/2} a_{2k+1} - 2^{(j-1)/2}(a_{2k} + a_{2k+1}) = 2^{j/2} a_{2k+1}(2^{1/2} - 2^{-1/2}) - 2^{(j-1)/2} a_{2k}$$
$$= -2^{(j-1)/2}(a_{2k} - a_{2k+1}). \qquad (9.7.4.3)$$

Recall that $\operatorname{supp} \psi_{j,k} = [k/2^j, (k+1)/2^j]$. Write

$$\psi_{j,k}(t) = 2^{j/2} \psi(2^j t - k) = \begin{cases} 2^{j/2} & \text{if} & k/2^j \le t < (k+1/2)/2^j \\ -2^{j/2} & \text{if} & (k+1/2)/2^j \le t < (k+1)/2^j. \end{cases}$$

Finally, we can use (9.7.4.2) and (9.7.4.3) to summarize our findings on $[k/2, (k+1)/2)$. We have

$$f_{j+1}(t) - f_j(t)$$
$$= \begin{cases} 2^{(j-1)/2}(a_{2k} - a_{2k+1} & \text{if} & k/2^j \le t < (k+1/2)/2^j \\ -2^{(j-1)/2}(a_{2k} - a_{2k+1} & \text{if} & (k+1/2)/2^j \le t < (k+1)/2^j \end{cases}$$
$$= \frac{\sqrt{2}}{2}(a_{2k} - a_{2k+1}) \begin{cases} 2^{j/2} & \text{if} & k/2^j \le t < (k+1/2)/2^j \\ -2^{j/2} & \text{if} & (k+1/2)/2^j \le t < (k+1)/2^j \end{cases}$$
$$= \frac{\sqrt{2}}{2} \cdot (a_{2k} - a_{2k+1})\psi_{j,k}(t).$$

That completes the proof.   □

It is an easily verified fact that g_j, as defined in this last proposition, is the projection of f_{j+1} into W_j. It follows that any function $f_{j+1} \in V_{j+1}$ can be written as the sum of an element $f_j \in V_j$ and an element $g_j \in W_j$. Furthermore, f_j is the projection of f_{j+1} into V_j and g_j is the projection of f_{j+1} into W_j. As the next proposition shows, V_{j+1} can be constructed from V_j and W_j.

Proposition 9.7.5 *Suppose that V_j, W_j, and V_{j+1} are as usual. Then*

$$V_{j+1} = V_j \oplus W_j.$$

Proof: Similar to the proof of Proposition 9.6.25.   □

9.7.1   Summary

We now know that every $f_{j+1} \in V_{j+1}$ can be written as the sum of an *approximation function* $f_j \in V_j$ and a *residual function* $g_j \in W_j$. The functions f_j and g_j are orthogonal to each other. Also, if

$$f_{j+1}(t) = \sum_{k \in \mathbb{Z}} b_k \varphi_{j+1,k}(t),$$

$$f_j(t) = \sum_{k \in \mathbb{Z}} a_k \varphi_{j+1,k}(t),$$

and

$$g_j(t) = \sum_{k \in \mathbb{Z}} c_k \psi_{j+1,k}(t).$$

Also, for $k \in \mathbb{Z}$,

$$\mathbf{h} = {}^t\left[\frac{\sqrt{2}}{2}, \frac{\sqrt{2}}{2}\right],$$

$$\mathbf{g} = {}^t\left[\frac{\sqrt{2}}{2}, -\frac{\sqrt{2}}{2}\right],$$

and

$$\mathbf{a}^k = {}^t[a_{2k}, a_{2k+1}],$$

we have

$$b_k = \mathbf{h} \cdot \mathbf{a}^k \qquad \text{and} \qquad c_k = \mathbf{g} \cdot \mathbf{a}^k.$$

The very last two formulas will be the basis for creating a *discrete Haar wavelet transformation* that can be used to process digital signals and images.

Exercises

1. Show that $\int_{\mathbb{R}} \psi_{\ell,m}(t)\,dt = \int_{I_{\ell,m}} \psi_{\ell,m}(t)\,dt = 0$.

2. Plot each of these functions. Give a verbal description of the compact support of each function.

   (a) $\psi_{2,-4}(t)$

   (b) $\psi_{-3,5}(t)$

   (c) $\psi_{4,8}(t)$

3. Plot each of these functions.

   (a) $f(t) = 2\psi(t/8 - 2) - 4\psi(t/8 + 1) + 3\psi(t/8 + 1)$

   (b) $g(t) = \sum_{j=1}^{4} j \cdot \psi(16t + j)$

   (c) $h(t) = \sum_{j=-3}^{3} \sin(\pi j)\psi(t/2 + j)$

   (d) $k(t) = \sum_{j=0}^{\infty} \langle e^{-t}, \psi(2t + j)\rangle \cdot \psi(2t + j)$

4. We know that $V_j \subseteq V_{j+1}$ for each j. Is something similar true for the W_j?

5. Let j be an integer and let W_j be as usual. Prove that the set $\{\psi(2^j t + k)\}_{k \in \mathbb{Z}}$ is a linearly independent set in W_j.

6. Use the u-substitution from calculus to establish that

   (a) $\langle \psi(2^j t + j), \psi(2^j t + k)\rangle = 0$ for $j \neq k$

   (b) $\|\psi(2^j t + k)\| = 2^{-j/2}$

7. Show that the error term in Proposition 9.7.4 is simply the projection of f_{j+1} into W_j.

8. Prove Proposition 9.7.2.

9. Prove Proposition 9.7.5.

9.8   Decomposition and Its Obverse

In this section we learn how to iterate the decomposition process described in the last section. The key is repeated use of the last proposition.

As an illustration, consider the space V_5. We can write

$$V_5 = V_4 \oplus W_4 .$$

But $V_4 = V_3 \oplus W_3$ hence

$$V_5 = V_3 \oplus W_3 \oplus W_4 .$$

Continuing this reasoning, we can write

$$
\begin{aligned}
V_5 &= V_4 \oplus W_4 \\
&= V_3 \oplus W_3 \oplus W_4 \\
&= V_2 \oplus W_2 \oplus W_3 \oplus W_4 \\
&= V_1 \oplus W_1 \oplus W_2 \oplus W_3 \oplus W_4 \\
&= V_0 \oplus W_0 \oplus W_1 \oplus W_2 \oplus W_3 \oplus W_4 \,.
\end{aligned}
$$

EXAMPLE 9.8.1 Suppose that $f_3 \in V_3$ is given by

$$
\begin{aligned}
f_3(t) &= \sum_{k=0}^{7} a_k \varphi_{3,k}(t) \\
&= 3\varphi_{3,0}(t) + \varphi_{3,1}(t) - 2\varphi_{3,2}(t) + 4\varphi_{3,3}(t) + 5\varphi_{3,4}(t) + \varphi_{3,5}(t) \\
&\quad - 2\varphi_{3,6}(t) - 4\varphi_{3,7}(t) \,.
\end{aligned}
$$

Recall that $\varphi_{3,k}(t) = 2^{3/2}\varphi(8t - k)$. Hence the coefficients are each multiplied by $2^{3/2}$. Our goal here is to decompose f_3 into an approximation function $f_0 \in V_0$ and details functions $g_0 \in W_0$, $g_1 \in W_1$, $g_2 \in W_2$.

We know that $V_3 = V_2 \oplus W_2$. Hence $f_3 = f_2 + g_2$ with $f_2 \in V_2$ and $g_2 \in W_2$. Also f_2 is the projection of f_3 into V_2 and g_2 is the projection of f_3 into W_2. Thus we use Proposition 9.7.3 to obtain f_2 and we use Proposition 9.7.4 to obtain g_2. Since only $a_0, a_1, \ldots, a_7$ are nonzero in our expression for f_3, formula (9.7.3.1) tells us that the only nonzero coefficients in

$$
f_2(t) = \sum_{k \in \mathbb{Z}} b_k \varphi_{2,k}(t)
$$

are

$$
\begin{aligned}
b_0 &= \frac{\sqrt{2}}{2}(a_0 + a_1) = \frac{\sqrt{2}}{2}(3 + 1) = 2\sqrt{2} \,, \\
b_1 &= \frac{\sqrt{2}}{2}(a_2 + a_3) = \frac{\sqrt{2}}{2}(-2 + 4) = \sqrt{2} \,, \\
b_2 &= \frac{\sqrt{2}}{2}(a_4 + a_5) = \frac{\sqrt{2}}{2}(5 + 1) = 3\sqrt{2} \,, \\
b_3 &= \frac{\sqrt{2}}{2}(a_6 + a_7) = \frac{\sqrt{2}}{2}(-2 - 4) = -3\sqrt{2} \,,
\end{aligned}
$$

In a similar fashion, we can use (9.7.4.1) to calculate the nonzero coefficients c_k for

$$
g_2(t) = \sum_{k \in \mathbb{Z}} c_k \psi_{2,k}(t) \,.
$$

We obtain

$$c_0 = \frac{\sqrt{2}}{2}(a_0 - a_1) = \frac{\sqrt{2}}{2}(3 - 1) = \sqrt{2},$$

$$c_1 = \frac{\sqrt{2}}{2}(a_2 - a_3) = \frac{\sqrt{2}}{2}(-2 - 4) = -3\sqrt{2},$$

$$c_2 = \frac{\sqrt{2}}{2}(a_4 - a_5) = \frac{\sqrt{2}}{2}(5 - 1) = 2\sqrt{2},$$

$$c_3 = \frac{\sqrt{2}}{2}(a_6 - a_7) = \frac{\sqrt{2}}{2}(-2 + 4) = \sqrt{2},$$

Next we use Propositions 9.7.4 and 9.7.5 to create $f_1 \in V_1$ and $g_1 \in W_1$ from $f_2 \in V_2$. The nonzero coefficients for f_1 are

$$b_0 = \frac{\sqrt{2}}{2}(2\sqrt{2} + \sqrt{2}) = 3 \qquad \text{and} \qquad b_1 = \frac{\sqrt{2}}{2}(3\sqrt{2} - 3\sqrt{2}) = 0.$$

The nonzero coefficients for g_1 are

$$c_0 = \frac{\sqrt{2}}{2}(2\sqrt{2} - \sqrt{2}) = 1 \qquad \text{and} \qquad c_1 = \frac{\sqrt{2}}{2}(3\sqrt{2} - (-3\sqrt{2})) = 6.$$

At last we project $f_1 \in V_1$ into V_0 and W_0. We have

$$b_0 = \frac{\sqrt{2}}{2}(3 + 0) = \frac{3}{2}\sqrt{2} \qquad \text{and} \qquad c_0 = \frac{\sqrt{2}}{2}(3 - 0) = \frac{3}{2}\sqrt{2}.$$

Hence

$$f_0(t) = \frac{3\sqrt{2}}{2}\varphi(t) \qquad \text{and} \qquad g_0(t) = \frac{3\sqrt{2}}{2}\psi(t).$$

We have decomposed f_3 as follows:

$$f_3(t) = f_0(t) + g_0(t) + g_1(t) + g_2(t). \qquad \blacksquare$$

In applications it is often useful to begin with a function $f_{j+L} \in V_{j+L}$ and then perform L decompositions so that we can write

$$f_{j+L} = f_j + g_j + g_{j+1} + \cdots + g_{j+L-1}.$$

In other circumstances we might be given the decomposition (on the right) and be asked to recover f_{j+L}. We now develop some formulas for performing these operations.

Proposition 9.8.2 *Let*

$$f_{j+1}(t) = \sum_{k \in \mathbb{Z}} a_k \varphi_{j+1,k}(t) .$$

Assume that

$$f_j(t) = \sum_{k \in \mathbb{Z}} b_k \varphi_{j,k}(t) \qquad and \qquad g_j(t) = \sum_{k \in \mathbb{Z}} c_k \psi_{j,k}(t)$$

are the projections of f_{j+1} into V_j and W_j respectively. Then

$$a_{2k} = \frac{\sqrt{2}}{2}(b_k + c_k) = \mathbf{h} \cdot \mathbf{d}^k \qquad and \qquad a_{2k+1} = \frac{\sqrt{2}}{2}(b_k - c_k) = \mathbf{g} \cdot \mathbf{d}^k .$$

Here $\mathbf{h}$, $\mathbf{g}$ are as usual and $\mathbf{d}^k = {}^t[b_k, c_k]$.

Proof: Suppose that $f_j \in V_j$ and $g_j \in W_j$ are obtained by projecting $f_{j+1} \in V_{j+1}$ into V_j and W_j respectively. Propositions 9.7.3 and 9.7.4 tell us that we can write $f_{j+1} = \sum_k a_k \varphi_{j+1,k}$ as

$$f_{j+1}(t) = \sum_{k \in \mathbb{Z}} b_k \varphi_{j,k}(t) + c_k \psi_{j,k}(t) . \tag{9.8.2.1}$$

In order to reconstruct f_{j+1} from f_j and g_j, we need to be able to write the a_k in terms of the b_k and the c_k.

The two indicated propositions tell us that

$$b_k = \frac{\sqrt{2}}{2}(a_{2k} + a_{2k+1}) \qquad and \qquad c_k = \frac{\sqrt{2}}{2}(a_{2k} - a_{2k+1}) .$$

Let us look at a single term from (9.8.2.1). Both $\varphi_{j,k}$ and $\psi_{j,k}$ are nonzero on $[k/2^j, (k+1)/2^j)$. Indeed, on the half-interval $[k/2^j, (k+1/2)/2^j)$, both $\varphi_{j,k}$ and $\psi_{j,k}$ assume the value $2^{j/2}$. So simplification of these two functions on the half-interval gives

$$b_k \varphi_{j,k}(t) + c_k \psi_{j,k}(t) = b_k 2^{j/2} + c_k 2^{j/2}$$
$$= 2^{j/2}(b_k + c_k) . \tag{9.8.2.2}$$

We have already seen that the only function at level $j + 1$ with support on the interval $[k/2^j, (k + 1/2)/2^j)$ is $\varphi_{j+1,2k}$. On this interval, this function takes the value $2^{(j+1)/2}$. Hence

$$f_{j+1}(t) = 2^{(j+1)/2} a_{2k} \quad \text{for} \quad t \in \left[\frac{k}{2^j}, \frac{k + 1/2}{2^j} \right) . \tag{9.8.2.3}$$

Comparing (9.8.2.2) and (9.8.2.3), we find that

$$2^{j/2}(b_k + c_k) = 2^{(j+1)/2}a_{2k}.$$

As a result,

$$a_{2k} = \frac{\sqrt{2}}{2}(b_k + c_k).$$

A similar analysis, which we leave to you, shows that, on the right-half interval $[(k+1/2)/2^j, (k+1)/2^j)$, we have the identity

$$a_{2k+1} = \frac{\sqrt{2}}{2}(b_k - c_k).$$

That completes the argument. $\square$

We close this discussion by doing a reconstruction.

EXAMPLE 9.8.3 In the last example we decomposed a function $f_3 \in V_3$ into components $f_0 \in V_0$, $g_0 \in W_0$, $g_1 \in W_1$, and $g_2 \in W_2$. We found that

$$f_0(t) = \frac{3\sqrt{2}}{2}\varphi(t) \quad \text{and} \quad g_0(t) = \frac{3\sqrt{2}}{2}\psi(t).$$

If now we add and subtract $b_0 = c_0 = 3\sqrt{2}/2$ and scale the result by $\sqrt{2}/2$, then we obtain 3 and 0. These are exactly the coefficients of f_1.

If instead we take $b_0 = 3$ and $b_1 = 0$ and combine these with the coefficients $c_0 = 1$ and $c_1 = 6$ of g_1, then we get

$$\frac{\sqrt{2}}{2}(b_0 + c_0) = \frac{\sqrt{2}}{2}(3 + 1) = 2\sqrt{2}.$$

$$\frac{\sqrt{2}}{2}(b_0 - c_0) = \frac{\sqrt{2}}{2}(3 - 1) = \sqrt{2}.$$

$$\frac{\sqrt{2}}{2}(b_1 + c_1) = \frac{\sqrt{2}}{2}(0 + 6) = 3\sqrt{2}.$$

$$\frac{\sqrt{2}}{2}(b_1 - c_1) = \frac{\sqrt{2}}{2}(0 - 6) = -3\sqrt{2}.$$

These numbers are the coefficients of $f_2 \in V_2$. If we combine these with the coefficients of $g_2 \in W_2$, then we have

$$
\frac{\sqrt{2}}{2}(b_0 + c_0) = \frac{\sqrt{2}}{2}(2\sqrt{2} + \sqrt{2}) = 3.
$$

$$
\frac{\sqrt{2}}{2}(b_0 - c_0) = \frac{\sqrt{2}}{2}(2\sqrt{2} - \sqrt{2}) = 1.
$$

$$
\frac{\sqrt{2}}{2}(b_1 + c_1) = \frac{\sqrt{2}}{2}(\sqrt{2} + (-3\sqrt{2})) = -2.
$$

$$
\frac{\sqrt{2}}{2}(b_1 - c_1) = \frac{\sqrt{2}}{2}(\sqrt{2} - (-3\sqrt{2})) = 4.
$$

$$
\frac{\sqrt{2}}{2}(b_2 + c_2) = \frac{\sqrt{2}}{2}(3\sqrt{2} + 2\sqrt{2}) = 5.
$$

$$
\frac{\sqrt{2}}{2}(b_2 - c_2) = \frac{\sqrt{2}}{2}(3\sqrt{2} - 2\sqrt{2}) = 1.
$$

$$
\frac{\sqrt{2}}{2}(b_3 + c_3) = \frac{\sqrt{2}}{2}(-3\sqrt{2} + \sqrt{2}) = -2.
$$

$$
\frac{\sqrt{2}}{2}(b_3 - c_3) = \frac{\sqrt{2}}{2}(-3\sqrt{2} - \sqrt{2}) = -4.
$$

These last, of course, are the coefficients of our original $f_3 \in V_3$. ∎

Exercises

1. Calculate the sum $f_0(t) + g_0(t) + g_1(t) + g_2(t)$ in Example 9.8.1 and verify that it equals f_3. [**Hint:** Consider this sum on each of the subintervals $[j, j + 1/8)$ for $j = 0, 1, 2, \ldots, 7$. On each of these intervals, compute the sum and check that it equals a_j.]

2. Refer again to Example 9.7.1. Decompose each of the following functions in V_2, into elements f_0, g_0, g_1 from the spaces V_0, W_0, W_1 respectively.

    (a) $f = 2\varphi_{3,0} - 5\varphi_{2,1} + 2\varphi_{2,5}$

    (b) $g = \sum_{j=0}^{7} \varphi_{2,j}$

    (c) $h = \sum_{j=0}^{7} (j + 1)\varphi_{2,j}$

3. Complete the proof of Proposition 9.8.2 by verifying that

$$
a_{2k+1} = \frac{\sqrt{2}}{2}(b_k - c_k).
$$

    [**Hint:** Equate the terms given by $b_k\varphi_{j,k} + c_k\psi_{j,k}$ and f_{j+1}.]

4. For each of the functions in Exercise 2, use Proposition 9.8.2 and the ideas in Example 9.8.3 to reproduce f_2 from the functions f_0, g_0, and g_1.

   5.   Find f_0 and g_0 for each of the following functions from V_1:

   (a) $f_1 = 5\varphi_{1,0} - 2\varphi_{2,2} + 5\varphi_{2,4}$

   (b) $f_1 = 10\varphi_{1,0} - 10\varphi_{1,1} - 5\varphi_{1,3} + 4\varphi_{1,5}$

   (c) $f_1 = 4\varphi_{1,0} + 3\varphi_{1,1} + 8\varphi_{1,2} - 2\varphi_{1,5}$

   (d) $f_1 = \sum_{j=1}^{\infty} \frac{1}{j}\varphi_{1,j}$

   6.   Prove that the set $\{\psi(t - j)\}_{j \in \mathbb{Z}}$ is a linearly independent set in W_0.

   7.   Show that

   (a) $\int_{\mathbb{R}} \psi(t)\, dt = 0$

   (b)

   $$\langle \psi(t - k), \psi(t - j) \rangle = \begin{cases} 0 & \text{if} \quad k \neq j \\ 1 & \text{if} \quad k = j \end{cases}$$

   (c) $\langle \varphi(t - k), \psi(t - j) \rangle = \int_{\mathbb{R}} \varphi(t - k)\psi(t - j)\, dt = 0$

   8.   Show that the function g_j in Proposition 9.7.4 is actually the projection of f_{j+1} into W_j.

9.9   Some Applications

We have developed considerable wavelet machinery in the preceding pages, and now it is time to see how these tools can be used.

   Recall that

$$\mathbf{h} = {}^t[h_0, h_1] = {}^t\!\left[\frac{\sqrt{2}}{2}, \frac{\sqrt{2}}{2}\right]$$

and

$$\mathbf{g} = {}^t[g_0, g_1] = {}^t\!\left[\frac{\sqrt{2}}{2}, -\frac{\sqrt{2}}{2}\right].$$

EXAMPLE 9.9.1 Consider the function f_3 from Example 9.8.1. We saw that $a_k = 0$ for $k < 0$ or $k \geq 8$. Furthermore,

$$a_0 = 3\,, \qquad a_1 = 1\,, \qquad a_2 = -2\,, \qquad a_3 = 4$$
$$a_4 = 5\,, \qquad a_5 = 1\,, \qquad a_6 = 4\,, \qquad a_7 = -4\,.$$

To project f_3 into V_2, we use Proposition 9.7.3 to write

$$b_0 = \mathbf{h} \cdot \mathbf{a}^0 = [h_0, h_1] \cdot \begin{bmatrix} a_0 \\ a_1 \end{bmatrix} = \left[\frac{\sqrt{2}}{2}, \frac{\sqrt{2}}{2}\right] \cdot \begin{bmatrix} 3 \\ 1 \end{bmatrix} = 2\sqrt{2}\,,$$

$$b_1 = \mathbf{h} \cdot \mathbf{a}^1 = [h_0, h_1] \cdot \begin{bmatrix} a_2 \\ a_3 \end{bmatrix} = \left[\frac{\sqrt{2}}{2}, \frac{\sqrt{2}}{2}\right] \cdot \begin{bmatrix} -2 \\ 4 \end{bmatrix} = \sqrt{2}\,,$$

$$b_2 = \mathbf{h} \cdot \mathbf{a}^2 = [h_0, h_1] \cdot \begin{bmatrix} a_4 \\ a_5 \end{bmatrix} = \left[\frac{\sqrt{2}}{2}, \frac{\sqrt{2}}{2} \right] \cdot \begin{bmatrix} 5 \\ 1 \end{bmatrix} = 3\sqrt{2},$$

$$b_3 = \mathbf{h} \cdot \mathbf{a}^3 = [h_0, h_1] \cdot \begin{bmatrix} a_6 \\ a_7 \end{bmatrix} = \left[\frac{\sqrt{2}}{2}, \frac{\sqrt{2}}{2} \right] \cdot \begin{bmatrix} 4 \\ -4 \end{bmatrix} = 0.$$

We can naturally formulate these computations as the matrix product $H_4 \mathbf{a} = \mathbf{b}$ or

$$\begin{bmatrix} h_0 & h_1 & 0 & 0 & 0 & 0 & 0 & 0 \\ 0 & 0 & h_0 & h_1 & 0 & 0 & 0 & 0 \\ 0 & 0 & 0 & 0 & h_0 & h_1 & 0 & 0 \\ 0 & 0 & 0 & 0 & 0 & 0 & h_0 & h_1 \end{bmatrix} \cdot \begin{bmatrix} a_0 \\ a_1 \\ a_2 \\ a_3 \\ a_4 \\ a_5 \\ a_6 \\ a_7 \end{bmatrix} = \begin{bmatrix} b_0 \\ b_1 \\ b_2 \\ b_3 \end{bmatrix}$$

or

$$\begin{bmatrix} \frac{\sqrt{2}}{2} & \frac{\sqrt{2}}{2} & 0 & 0 & 0 & 0 & 0 & 0 \\ 0 & 0 & \frac{\sqrt{2}}{2} & \frac{\sqrt{2}}{2} & 0 & 0 & 0 & 0 \\ 0 & 0 & 0 & 0 & \frac{\sqrt{2}}{2} & \frac{\sqrt{2}}{2} & 0 & 0 \\ 0 & 0 & 0 & 0 & 0 & 0 & \frac{\sqrt{2}}{2} & \frac{\sqrt{2}}{2} \end{bmatrix} \cdot \begin{bmatrix} 3 \\ 1 \\ -2 \\ 4 \\ 5 \\ 1 \\ 4 \\ -4 \end{bmatrix} = \begin{bmatrix} 2\sqrt{2} \\ \sqrt{2} \\ 3\sqrt{2} \\ 0 \end{bmatrix}.$$

$$(9.9.1.1)$$

In the same fashion we can compute the detail coefficients (formerly known as residual or error terms) using the matrix equation $G_4 \mathbf{a} = \mathbf{c}$ or

$$\begin{bmatrix} g_0 & g_1 & 0 & 0 & 0 & 0 & 0 & 0 \\ 0 & 0 & g_0 & g_1 & 0 & 0 & 0 & 0 \\ 0 & 0 & 0 & 0 & g_0 & g_1 & 0 & 0 \\ 0 & 0 & 0 & 0 & 0 & 0 & g_0 & g_1 \end{bmatrix} \cdot \begin{bmatrix} a_0 \\ a_1 \\ a_2 \\ a_3 \\ a_4 \\ a_5 \\ a_6 \\ a_7 \end{bmatrix} = \begin{bmatrix} c_0 \\ c_1 \\ c_2 \\ c_3 \end{bmatrix}$$

or

$$
\begin{bmatrix}
\frac{\sqrt{2}}{2} & -\frac{\sqrt{2}}{2} & 0 & 0 & 0 & 0 & 0 & 0 \\
0 & 0 & \frac{\sqrt{2}}{2} & -\frac{\sqrt{2}}{2} & 0 & 0 & 0 & 0 \\
0 & 0 & 0 & 0 & \frac{\sqrt{2}}{2} & -\frac{\sqrt{2}}{2} & 0 & 0 \\
0 & 0 & 0 & 0 & 0 & 0 & \frac{\sqrt{2}}{2} & -\frac{\sqrt{2}}{2}
\end{bmatrix}
\cdot
\begin{bmatrix}
3 \\ 1 \\ -2 \\ 4 \\ 5 \\ 1 \\ 4 \\ -4
\end{bmatrix}
=
\begin{bmatrix}
\sqrt{2} \\ -3\sqrt{2} \\ 2\sqrt{2} \\ 4\sqrt{2}
\end{bmatrix}.
$$

(9.9.1.2)

We can concatenate the matrix equations (9.9.1.1) and (9.9.1.2) to write

$$
Q_8 \mathbf{a} = \begin{bmatrix} b \\ \hline c \end{bmatrix}.
$$

Here

$$
Q_8 = \begin{bmatrix} H_4 \\ \hline G_4 \end{bmatrix}
$$

$$
= \begin{bmatrix}
h_0 & h_1 & 0 & 0 & 0 & 0 & 0 & 0 \\
0 & 0 & h_0 & h_1 & 0 & 0 & 0 & 0 \\
0 & 0 & 0 & 0 & h_0 & h_1 & 0 & 0 \\
0 & 0 & 0 & 0 & 0 & 0 & h_0 & h_1 \\
\hline
g_0 & g_1 & 0 & 0 & 0 & 0 & 0 & 0 \\
0 & 0 & g_0 & g_1 & 0 & 0 & 0 & 0 \\
0 & 0 & 0 & 0 & g_0 & g_1 & 0 & 0 \\
0 & 0 & 0 & 0 & 0 & 0 & g_0 & g_1
\end{bmatrix}
$$

$$
= \begin{bmatrix}
\frac{\sqrt{2}}{2} & \frac{\sqrt{2}}{2} & 0 & 0 & 0 & 0 & 0 & 0 \\
0 & 0 & \frac{\sqrt{2}}{2} & \frac{\sqrt{2}}{2} & 0 & 0 & 0 & 0 \\
0 & 0 & 0 & 0 & \frac{\sqrt{2}}{2} & \frac{\sqrt{2}}{2} & 0 & 0 \\
0 & 0 & 0 & 0 & 0 & 0 & \frac{\sqrt{2}}{2} & \frac{\sqrt{2}}{2} \\
\hline
\frac{\sqrt{2}}{2} & -\frac{\sqrt{2}}{2} & 0 & 0 & 0 & 0 & 0 & 0 \\
0 & 0 & \frac{\sqrt{2}}{2} & -\frac{\sqrt{2}}{2} & 0 & 0 & 0 & 0 \\
0 & 0 & 0 & 0 & \frac{\sqrt{2}}{2} & -\frac{\sqrt{2}}{2} & 0 & 0 \\
0 & 0 & 0 & 0 & 0 & 0 & \frac{\sqrt{2}}{2} & -\frac{\sqrt{2}}{2}
\end{bmatrix}.
$$

■

Now let us take a closer look at the matrix Q_8. We see that the inner product of the first row with itself is 1. In fact the inner product of *any* row with itself is 1. And the rows are orthogonal to each other. We conclude that Q_8 is an orthogonal matrix, so that $Q_8^{-1} = {}^t Q_8$.

If we form the vector

$$\mathbf{y} = \left[\frac{\mathbf{b}}{\mathbf{c}} \right],$$

where $\mathbf{b}$ and $\mathbf{c}$ are defined as in the last example, then we see that

$${}^t Q_8 \mathbf{y} \; = \; {}^t Q_8 \left[\frac{\mathbf{b}}{\mathbf{c}} \right]$$

$$= \left[\begin{array}{cccc|cccc} \frac{\sqrt{2}}{2} & 0 & 0 & 0 & \frac{\sqrt{2}}{2} & 0 & 0 & 0 \\ \frac{\sqrt{2}}{2} & 0 & 0 & 0 & -\frac{\sqrt{2}}{2} & 0 & 0 & 0 \\ 0 & \frac{\sqrt{2}}{2} & 0 & 0 & 0 & \frac{\sqrt{2}}{2} & 0 & 0 \\ 0 & \frac{\sqrt{2}}{2} & 0 & 0 & 0 & -\frac{\sqrt{2}}{2} & 0 & 0 \\ 0 & 0 & \frac{\sqrt{2}}{2} & 0 & 0 & 0 & \frac{\sqrt{2}}{2} & 0 \\ 0 & 0 & \frac{\sqrt{2}}{2} & 0 & 0 & 0 & -\frac{\sqrt{2}}{2} & 0 \\ 0 & 0 & 0 & \frac{\sqrt{2}}{2} & 0 & 0 & 0 & \frac{\sqrt{2}}{2} \\ 0 & 0 & 0 & \frac{\sqrt{2}}{2} & 0 & 0 & 0 & -\frac{\sqrt{2}}{2} \end{array} \right].$$

$$\left[\begin{array}{c} b_0 \\ b_1 \\ b_2 \\ b_3 \\ \hline c_0 \\ c_1 \\ c_2 \\ c_3 \end{array} \right]$$

$$= \left[\begin{array}{c} \frac{\sqrt{2}}{2} b_0 + \frac{\sqrt{2}}{2} c_0 \\ \frac{\sqrt{2}}{2} b_0 - \frac{\sqrt{2}}{2} c_0 \\ \frac{\sqrt{2}}{2} b_1 + \frac{\sqrt{2}}{2} c_1 \\ \frac{\sqrt{2}}{2} b_1 - \frac{\sqrt{2}}{2} c_1 \\ \hline \frac{\sqrt{2}}{2} b_2 + \frac{\sqrt{2}}{2} c_2 \\ \frac{\sqrt{2}}{2} b_2 - \frac{\sqrt{2}}{2} c_2 \\ \frac{\sqrt{2}}{2} b_3 + \frac{\sqrt{2}}{2} c_3 \\ \frac{\sqrt{2}}{2} b_3 - \frac{\sqrt{2}}{2} c_3 \end{array} \right]$$

$$
= \begin{bmatrix}
\frac{\sqrt{2}}{2}(2\sqrt{2}) + \frac{\sqrt{2}}{2}(\sqrt{2}) \\
\frac{\sqrt{2}}{2}(2\sqrt{2}) - \frac{\sqrt{2}}{2}(\sqrt{2}) \\
\frac{\sqrt{2}}{2}(\sqrt{2}) + \frac{\sqrt{2}}{2}(-3\sqrt{2}) \\
\frac{\sqrt{2}}{2}(\sqrt{2}) - \frac{\sqrt{2}}{2}(-3\sqrt{2}) \\
\hline
\frac{\sqrt{2}}{2}(3\sqrt{2}) + \frac{\sqrt{2}}{2}(2\sqrt{2}) \\
\frac{\sqrt{2}}{2}(3\sqrt{2}) - \frac{\sqrt{2}}{2}(2\sqrt{2}) \\
\frac{\sqrt{2}}{2}(0) + \frac{\sqrt{2}}{2}(4\sqrt{2}) \\
\frac{\sqrt{2}}{2}(0) - \frac{\sqrt{2}}{2}(4\sqrt{2})
\end{bmatrix}
$$

$$
= \begin{bmatrix}
3 \\
1 \\
-2 \\
4 \\
\hline
5 \\
1 \\
4 \\
-4
\end{bmatrix}
$$

$$
= \begin{bmatrix}
a_0 \\
a_1 \\
a_2 \\
a_3 \\
\hline
a_4 \\
a_5 \\
a_6 \\
a_7
\end{bmatrix}
$$

We see that the matrix Q_8 gives us a way to take a vector of finite length and decompose it. Since Q_8 is orthogonal, we can use ${}^tQ_8$ to recover the original vector. Now we can define the discrete Haar wavelet transformation matrix.

Definition 9.9.2 Let N be an even positive integer. We define the *discrete*

Haar wavelet transformation to be

$$
Q_N = \left[\frac{H_{N/2}}{G_{N/2}}\right] = \left[\begin{array}{cccccccc}
\frac{\sqrt{2}}{2} & \frac{\sqrt{2}}{2} & 0 & 0 & & & 0 & 0 \\
0 & 0 & \frac{\sqrt{2}}{2} & \frac{\sqrt{2}}{2} & & & 0 & 0 \\
\vdots & & & & \ddots & & & \vdots \\
0 & 0 & 0 & 0 & \cdots & & \frac{\sqrt{2}}{2} & \frac{\sqrt{2}}{2} \\
\hline
\frac{\sqrt{2}}{2} & -\frac{\sqrt{2}}{2} & 0 & 0 & & & 0 & 0 \\
0 & 0 & \frac{\sqrt{2}}{2} & -\frac{\sqrt{2}}{2} & & & 0 & 0 \\
\vdots & & & & \ddots & & & \vdots \\
0 & 0 & 0 & 0 & \cdots & & \frac{\sqrt{2}}{2} & -\frac{\sqrt{2}}{2}
\end{array}\right] \qquad (9.9.2.1)
$$

The $N/2 \times N$ block, which we have called $H_{N/2}$, is called the *averages block* and the $N/2 \times N$ block, which we have called $G_{N/2}$, is called the *details block*.

We can apply the matrix $H_{N/2}$ to a vector $\mathbf{a}$ to see why it is called the averages block. We calculate

$$
\begin{aligned}
H_{N/2}\mathbf{a} &= \left[\begin{array}{cccccccc}
\frac{\sqrt{2}}{2} & \frac{\sqrt{2}}{2} & 0 & 0 & & & 0 & 0 \\
0 & 0 & \frac{\sqrt{2}}{2} & \frac{\sqrt{2}}{2} & & & 0 & 0 \\
\vdots & & & & \ddots & & & \vdots \\
0 & 0 & 0 & 0 & \cdots & & \frac{\sqrt{2}}{2} & \frac{\sqrt{2}}{2}
\end{array}\right] \cdot \left[\begin{array}{c}
a_0 \\ a_1 \\ \vdots \\ a_{N-2} \\ a_{N-1}
\end{array}\right] \\[2ex]
&= \sqrt{2} \cdot \left[\begin{array}{cccccccc}
\frac{1}{2} & \frac{1}{2} & 0 & 0 & & & 0 & 0 \\
0 & 0 & \frac{1}{2} & \frac{1}{2} & & & 0 & 0 \\
\vdots & & & & \ddots & & & \vdots \\
0 & 0 & 0 & 0 & \cdots & & \frac{1}{2} & \frac{1}{2}
\end{array}\right] \cdot \left[\begin{array}{c}
a_0 \\ a_1 \\ \vdots \\ a_{N-2} \\ a_{N-1}
\end{array}\right] \\[2ex]
&= \sqrt{2} \cdot \left[\begin{array}{c}
\frac{a_0 + a_1}{2} \\[1ex] \frac{a_2 + a_3}{2} \\[1ex] \vdots \\[1ex] \frac{a_{N-2} + a_{N-1}}{2}
\end{array}\right].
\end{aligned}
$$

We see then that $H_{N/2}\mathbf{a}$ calculates pairwise averages of consecutive values of $\mathbf{a}$ and then weights the result by $\sqrt{2}$.

Now, at last, we shall study the application of the Haar wavelet transform to a problem of noise-level estimation.

EXAMPLE 9.9.3 Let

$$\mathbf{y} = \mathbf{v} + \mathbf{e}.$$

Here $\mathbf{v}$ is the true signal and $\mathbf{e}$ is a noise vector. In practice we do not know

v or **e**; instead we are given **y** and we want to estimate **v**. A key step in this process is estimate the noise level σ that occurs from **e**. In this example, **e** is composed of independent samples from a normal distribution[2] with mean 0 and variance σ^2. Such noise is called *Gaussian white noise*.

Now assume that $\mathbf{y} \in \mathbb{R}^N$. We apply the discrete Haar wavelet transform to obtain

$$\left[\begin{array}{c} \mathbf{s} \\ \mathbf{d} \end{array} \right] = Q_N \mathbf{y} = Q_N(\mathbf{v} + \mathbf{e}) = Q_N \mathbf{v} + Q_N \mathbf{e}.$$

Here $\mathbf{s} = H_{N/2}\mathbf{y}$ is the approximation part of the transform and $\mathbf{d} = G_{N/2}\mathbf{y}$ is the detail part of the transform. Because Q_N is an orthogonal matrix, it can be seen that the elements of the vector $Q_N\mathbf{e}$ are normally distributed with mean 0 and variance σ^2. As Donoho and Johnstone point out in [DOJ], the main portion of the transformed noise $Q_N\mathbf{e}$ ends up in the detail portion **d**. So the vector **d** is an excellent candidate for estimating σ.

Hampel [HAM] showed that the median absolute deviation (MAD) of a sample can be used to estimate σ. Let a_{med} denote the median of samples $\mathbf{a} = {}^t[a_1, a_2, \ldots, a_N]$. Then we define the MAD to be

$$\mathrm{MAD}(\mathbf{a}) = \mathrm{median}(|a_1 - a_{\mathrm{med}}|, |a_2 - a_{\mathrm{med}}|, \ldots, |a_N - a_{\mathrm{med}}|). \quad (9.9.3.1)$$

Hampel showed that

$$\mathrm{MAD}(\mathbf{a}) \to 0.6745 \cdot \sigma$$

as the sample size tends to $+\infty$. Combining this result with the fact that most of the noise in $Q_N\mathbf{y}$ resides in **d** gives us the following estimate $\widetilde{\sigma}$ fo σ:

$$\widetilde{\sigma} = \frac{\mathrm{MAD}(\mathbf{d})}{0.6745}.$$

We now illustrate this wavelet-based method for estimating the noise level of a signal by considering the signal formed by evaluating the heaviside function and adding some white noise to it.[3] We define

$$h(t) = 4\sin(4\pi t) - \mathrm{sgn}(t - 0.3) - \mathrm{sgn}(0.72 - t) \quad \text{for } t \in [0, 1].$$

Here we recall that

$$\mathrm{sgn}(t) = \left\{ \begin{array}{rcl} 1 & \text{if} & t > 0 \\ 0 & \text{if} & t = 0 \\ -1 & \text{if} & t < 0 \end{array} \right.$$

We form the vector $\mathbf{v} \in \mathbb{R}^{2048}$ using the formula

$$v_k = h\left(\frac{k}{2048} \right) \quad \text{for } k = 0, 1, 2, \ldots, 2047.$$

---

[2] If this terminology from statistics is unfamiliar to you then you should consult a basic statistics textbook like [DES].

[3] Recall that the heaviside function is that function which is equal to 0 on the interval $(-\infty, 0]$ and equal to 1 on the interval $(0, +\infty)$.

Next we create a noise vector $\mathbf{e}$ whose components e_k, for $k = 0, 1, 2, \ldots, 2047$, are independent samples from a normal distribution with mean 0 and variance σ^2. Just to be specific, we shall take $\sigma = 0.5$.

The first thing we do is to calculate the discrete Haar wavelet transform of $\mathbf{y}$. We use (9.9.2.1) for this purpose. The 1024-vector $\mathbf{s} = H_{1024}\mathbf{y}$ (that is, the averages) and the detail vector $\mathbf{d} = G_{1024}\mathbf{y}$ result.

Using (9.9.3.1) we calculate the median absolute deviation of $\mathbf{d}$ and find that $\mathrm{MAD}(\mathbf{d}) = 0.356043$. By Hampel's result, cited above, the noise level σ can be estimated by dividing this number by 0.6745. Thus

$$\sigma \approx \widetilde{\sigma} = \frac{0.356043}{0.6745} = 0.537862 \, .$$

The absolute error in this approximation is $|\sigma - \widetilde{\sigma}| = 0.027862$ and the percentage error is about 05.57%.   ∎

Exercises

1. Compute the discrete Haar wavelet transformations for each of these vectors:

   (a) $\mathbf{u} = {}^t\langle 1, 1, 1, 1, 1, 1, 1, 1 \rangle$
   (b) $\mathbf{v} = {}^t\langle 1, 2, 3, 4, 5, 6, 7, 8 \rangle$
   (c) $\mathbf{w} = {}^t\langle 1, 4, 9, 16, 25, 36, 49, 64 \rangle$

2. Compute three iterated discrete Haar wavelet transforms for each of the vectors in Exercise 1.

3. Let $\mathbf{v} \in \mathbb{R}^N$, where N is an even, positive integer. Show that

$$\mathbf{y} = Q_N \mathbf{v} = \begin{bmatrix} X\mathbf{h} \\ X\mathbf{g} \end{bmatrix},$$

   where $\mathbf{h}$, $\mathbf{g}$ are as usual and

$$X = \begin{bmatrix} v_1 & v_2 \\ v_3 & v_4 \\ & \vdots \\ v_{N-1} & v_N \end{bmatrix}.$$

   These ideas lead to an efficient algorithm for calculating $Q_N\mathbf{v}$.

4. Suppose that $\mathbf{y} = Q_N\mathbf{v}$, where N is a positive, even integer. Then it is possible to recover $\mathbf{v}$ using the inverse transform $\mathbf{v} = {}^tQ_N\mathbf{y}$. Prove this assertion.

Can you reformulate this matrix product in terms of the vectors **h**, **g** and

$$
\mathbf{Y} = \begin{bmatrix} y_1 & y_{N/2+1} \\ y_2 & y_{N/2+2} \\ & \vdots \\ y_{N/2} & y_N \end{bmatrix} ?
$$

5. Define

$$
Q_{N,j} = \mathrm{diag}[Q_{N/2^j}, I_{N/2^j}, I_{N/2^{j-1}}, \ldots, I_{N/4}, I_{N/2}].
$$

Show that the jth iteration of the discrete Haar wavelet transform is given by

$$
\mathbf{y}^j = Q_{N,j-2}Q_{N,j-2} \cdots\cdots Q_{N,1}Q_{N,0}\mathbf{v}.
$$

6. Refer to Exercises 5 for notation. We define

$$
\mathcal{Q}_N = Q_{N,2}Q_{N,1}Q_{N,0},
$$

where $N = 2^p$, $p \geq 3$ is a positive integer. Let $\mathbf{v} \in \mathbb{R}^8$. We wish to apply three iterations of the discrete Haar wavelet transform to $\mathbf{v}$. We may write

$$
\mathbf{y}^3 = \mathcal{Q}_8\mathbf{v} = Q_{8,2}Q_{8,1}Q_{8,0}\mathbf{v}.
$$

(a) Show that

$$
\mathbf{y}^3 = \mathcal{Q}_8 \cdot \mathbf{v} = \begin{bmatrix}
\frac{\sqrt{2}}{4} & \frac{\sqrt{2}}{4} & \frac{\sqrt{2}}{4} & \frac{\sqrt{2}}{4} & \frac{\sqrt{2}}{4} & \frac{\sqrt{2}}{4} & \frac{\sqrt{2}}{4} & \frac{\sqrt{2}}{4} \\
-\frac{\sqrt{2}}{4} & -\frac{\sqrt{2}}{4} & -\frac{\sqrt{2}}{4} & -\frac{\sqrt{2}}{4} & \frac{\sqrt{2}}{4} & \frac{\sqrt{2}}{4} & \frac{\sqrt{2}}{4} & \frac{\sqrt{2}}{4} \\
-\frac{1}{2} & -\frac{1}{2} & \frac{1}{2} & \frac{1}{2} & 0 & 0 & 0 & 0 \\
0 & 0 & 0 & 0 & -\frac{1}{2} & -\frac{1}{2} & \frac{1}{2} & \frac{1}{2} \\
-\frac{\sqrt{2}}{2} & \frac{\sqrt{2}}{2} & 0 & 0 & 0 & 0 & 0 & 0 \\
0 & 0 & -\frac{\sqrt{2}}{2} & \frac{\sqrt{2}}{2} & 0 & 0 & 0 & 0 \\
0 & 0 & 0 & 0 & -\frac{\sqrt{2}}{2} & \frac{\sqrt{2}}{2} & 0 & 0 \\
0 & 0 & 0 & 0 & 0 & 0 & -\frac{\sqrt{2}}{2} & \frac{\sqrt{2}}{2}
\end{bmatrix} \cdot \mathbf{v}
$$

(b) Redo part (a) where now we take $\mathbf{v} \in \mathbb{R}^{16}$ and we calculate four iterations of the discrete Haar wavelet transform. What is $\mathcal{Q}_{16}$ in this case?

(c) Describe the general form of $\mathcal{Q}_N$, where $N = 2^p$.

7. For the vectors **g** and **h** defined as usual, show the following.

(a) The Fourier series constructed from **h** is given by

$$
\mathcal{H}(\omega) = \frac{\sqrt{2}}{2} + \frac{\sqrt{2}}{2}e^{i\omega} = \sqrt{2}e^{i\omega}\cos(\omega/2).
$$

(b) Using part (a), show that

$$|\mathcal{H}(\omega)| = \sqrt{2}\cos(\omega/2) \quad \text{for} \quad -\pi \leq \omega \leq \pi.$$

(c) The Fourier series constructed from **g** is given by

$$\mathcal{G}(\omega) = \frac{\sqrt{2}}{2} - \frac{\sqrt{2}}{2}e^{i\omega} = -\sqrt{2}ie^{i\omega}\sin(\omega/2).$$

(d) We have that

$$|\mathcal{G}(\omega)| = \sqrt{2}|\sin(\omega/2)| \quad \text{for} \quad -\pi \leq \omega \leq \pi.$$

(e) We may use parts (b) and (d) to show that

$$|\mathcal{H}(\omega)|^2 + |\mathcal{H}(\omega + \pi)|^2 = |\mathcal{G}(\omega)|^2 + |\mathcal{G}(\omega + \pi)|^2 = 2$$

and

$$\mathcal{H}(\omega) \cdot \overline{\mathcal{G}(\omega)} + \mathcal{H}(\omega + \pi) \cdot \overline{\mathcal{G}(\omega + \pi)} = 0.$$

9.10 Cumulative Energy and Entropy

We now introduce two ideas from physics that we shall use to measure the effectiveness of discrete wavelet transformations in applications. The first of these, *cumulative energy*, is a vector-valued function that returns information about how energy is stored in the input vector. The second idea, *entropy*, is used to measure the performance of image-compression algorithms. The source of this example is [RUV].

The concept of entropy is particularly interesting because it uses the logarithm function in a nice fashion.

Definition 9.10.1 (Cumulative Energy) Let $\mathbf{v} \in \mathbb{R}^N$ with $\mathbf{v} \neq 0$. Assume that $\mathbf{y}$ is the vector formed by taking the absolute value of each component of $\mathbf{v}$ and ordering the resulting nonnegative numbers from largest to smallest. Then we define the *cumulative energy vector* $C(\mathbf{v})$ to be a vector in $\mathbb{R}^N$ whose components are given by

$$C(\mathbf{v})_j = \sum_{\ell=1}^{j} \frac{y_\ell^2}{\|\mathbf{y}\|^2} \quad \text{for} \quad j = 1, 2, \ldots, N.$$

Certainly $\|\mathbf{v}\| = \|\mathbf{y}\|$—just by definition. We also know that

$$\|\mathbf{y}\|^2 = y_1^2 + y_2^2 + \cdots + y_N^2.$$

Hence we may write

$$C(\mathbf{v})_1 = \frac{y_1^2}{y_1^2 + y_2^2 + \cdots + y_N^2},$$

$$C(\mathbf{v})_2 = \frac{y_1^2 + y_2^2}{y_1^2 + y_2^2 + \cdots + y_N^2},$$

$$\vdots$$

$$C(\mathbf{v})_{N-1} = \frac{y_1^2 + y_2^2 + \cdots + y_{N-1}^2}{y_1^2 + y_2^2 + \cdots + y_N^2},$$

$$C(\mathbf{v})_N = \frac{y_1^2 + y_2^2 + \cdots + y_n^2}{y_1^2 + y_2^2 + \cdots + y_N^2} = 1.$$

Thus we see that $C(\mathbf{v})_j$ is simply the percentage of the jth largest component (in absolute value) of $\mathbf{v}$ in $\|\mathbf{v}\|$. Finally observe that $0 \leq C(\mathbf{v})_j \leq 1$ for each j.

EXAMPLE 9.10.2 Let us find the cumulative energy of each of the following vectors.

(a) $\mathbf{u} = \langle 1, 2, 3, 4, 5, 6, 7, 8 \rangle$

(b) $\mathbf{v} = \langle 1, 1, 1, 1, 1, 1, 1, 1 \rangle$

(c) $\mathbf{w} = \langle \sqrt{2}, \sqrt{2}, \sqrt{2}, \sqrt{2}, 0.0, 0, 0 \rangle$

For part (a), we calculate that

$$\|\mathbf{u}\|^2 = 1^2 + 2^2 + \cdot 8^2 = 204.$$

The components of $C(\mathbf{u})$ are then

$$C(\mathbf{u})_j = \sum_{\ell=0}^{j-1} \frac{(8-\ell)^2}{204} \quad \text{for} \quad j = 1, 2, \ldots, 8.$$

Hence

$$C(\mathbf{u}) = \left\langle \frac{64}{204}, \frac{113}{204}, \frac{149}{204}, \frac{174}{204}, \frac{190}{204}, \frac{199}{204}, \frac{203}{204}, 1 \right\rangle.$$

For part (b), observe that $\|\mathbf{v}\|^2 = 8$ and

$$C(\mathbf{v})_j = \sum_{\ell=1}^{j} \frac{1}{8} = \frac{j}{8} \quad \text{for} \quad j = 1, 2, \ldots, 8.$$

Therefore

$$C(\mathbf{v}) = \left\langle \frac{1}{8}, \frac{2}{8}, \frac{3}{8}, \frac{4}{8}, \frac{5}{8}, \frac{6}{8}, \frac{7}{8}, 1 \right\rangle.$$

For part (c), notice that

$$\|\mathbf{w}\|^2 = 2 + 2 + 2 + 2 + 0 + 0 + 0 + 0 = 8.$$

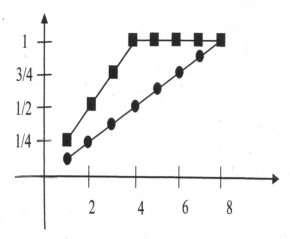

FIGURE 9.14
Cumulative energy for **v** and for **w**.

The components of $C(\mathbf{w})$ are then

$$C(\mathbf{w})_j = \sum_{\ell=1}^{j} \frac{(\sqrt{2})^2}{8} = \frac{2j}{8} = \frac{j}{4} \qquad \text{if} \quad j = 1, 2, 3, 4$$

and

$$C(\mathbf{w})_j = \sum_{\ell=1}^{4} \frac{(\sqrt{2})^2}{8} + (j-4) \cdot 0 = \frac{8}{8} = 1 \qquad \text{if} \quad j = 5, 6, 7, 8.$$

As a result,

$$C(\mathbf{w}) = \left\langle \frac{1}{4}, \frac{2}{4}, \frac{3}{4}, 1, 1, 1, 1, 1 \right\rangle.$$

Examine Figure 9.14. It shows the cumulative energy for **v** (denoted by black discs) and for **w** (denoted by black squares). We see that the energy is uniformly distributed among the elements of **v**. But, for **w**, most of the energy is stored in the four largest components.   ■

So the cumulative energy vector can be viewed as telling us which components might be important contributors to a signal. In applications such as image compression, those components not deemed important (i.e., with small cumulative energy) can be assigned a zero value. This simplifies calculations with very small loss of information.

Now, as we have seen, cumulative energy gives us a vector that tells us how the energy of a vector is distributed; by contrast, entropy tells us the average amount of information contained in each unit of measure. For instance, in the

case of digital images made up of pixels, the unit of measure is typically a *bit.*
In conclusion, the entropy of a digital image (stored as a vector) tells us, on
average, how many bits are needed to encode each pixel.[4]

Definition 9.10.3 (Entropy) Let $\mathbf{v} = \langle v_1, v_2, \ldots, v_N \rangle$ be an N-vector. Suppose that the v_j assume k distinct values, $1 \leq k \leq N$. Denote these distinct
values by $a_1, a_2, \ldots, a_k$ and let $p(a_j)$ be the relative frequency of a_j in $\mathbf{v}$. That
is to say $0 \leq p(a_j) \leq 1$ is the number of times a_j occurs in $\mathbf{v}$ divided by N.
Then the *entropy* of $\mathbf{v}$ is defined to be the quantity

$$\text{Ent}(\mathbf{v}) = \sum_{\ell=1}^{k} p(a_\ell) \cdot \log_2(1/p(a_\ell)).$$

We think of $\mathbf{v}$ now as the first row of a digital image. Then the entries of
$\mathbf{v}$ are nonnegative integers ranging from 0 (black) to 255 (white). The term
$\log_2(1/p(a_\ell))$ is an exponent that is measuring the power of 2 that is needed
to represent $1/p(a_\ell)$. This exponent can also be viewed as a bit length. We
multiply this exponent by the relative frequency of a_ℓ in $\mathbf{v}$ and sum over all
distinct values a_ℓ. As a result, entropy gives an average of the number of bits
that are needed to encode the elements of $\mathbf{v}$.

Claude Shannon (1918–2001) was one of the real pioneers of information
theory. In 1948 he showed that the best compression rate (in bits per pixel,
for instance) that we can hope for when performing lossless compression on
$\mathbf{v}$, as the length of $\mathbf{v}$ tends to infinity, is $\text{Ent}(\mathbf{v})$.

In effect, wavelet analysis has caused harmonic analysis to re-invent itself.
Wavelets and their generalizations are powerful new tools that allow localization in both the space and phase variables. They are useful in producing
unconditional bases for classical Banach spaces. They also provide flexible
methods for analyzing integral operators. The subject of wavelets promises to
be a fruitful area of investigation for many years to come.

Exercises

1. Let $c \in \mathbb{R}$, $c \neq 0$. Let N be an even positive integer. Specify the vector
   $\mathbf{v} \in \mathbb{R}^N$ componentwise by $v_j = c$, $j = 1, 2, \ldots, N$. Then

   (a) Compute one iteration $\mathbf{y}$ of the discrete Haar wavelet transform of
   $\mathbf{v}$.

   (b) Find $\text{Ent}(\mathbf{v})$ and $\text{Ent}(\mathbf{y})$.

---

[4]Remember that each pixel will have a color assigned to it and an intensity assigned to
it. This is information carried by a certain number of bits.

(c) Compute the cumulative energy vectors $\mathbf{C}(\mathbf{v})$ and $\mathbf{C}(\mathbf{y})$ and plot the components of each vector on the same set of axes.

2. Redo Exercise 1 for the vector $\mathbf{v} \in \mathbb{R}^N$, where $v_j = mj + b$. Here $m, b \in \mathbb{R}$ with $m \neq 0$.

3. Let $\mathbf{v} \in \mathbb{R}^N$ and $c \in \mathbb{R}$ with $c \neq 0$. Define $\mathbf{w} = c\mathbf{v}$. Show that $\mathbf{C}(\mathbf{v}) = \mathbf{C}(\mathbf{w})$ and $\mathrm{Ent}(\mathbf{v}) = \mathrm{Ent}(\mathbf{w})$.

4. Prove that $\mathrm{Ent}(\mathbf{v}) \geq 0$ with equality if and only if $\mathbf{v}$ is a constant vector with $v_j = c$ for $j = 1, 2, \ldots, N$ and $c \in \mathbb{R}$.

5. Let $\mathbf{v} \in \mathbb{R}^{16}$. Here we shall investigate the effect of quantizing the averages portion of the iterated Haar wavelet transform.

   (a) We can write the jth iterate of the Haar wavelet transform as

$$
\mathbf{y}^j = \begin{bmatrix} \mathbf{y}^{j,a} \\ \hline \mathbf{y}^{j,d} \\ \hline \mathbf{y}^{j-1,d} \\ \hline \mathbf{y}^{j-2,d} \\ \hline \vdots \\ \hline \mathbf{y}^{2,d} \\ \hline \mathbf{y}^{1,d} \end{bmatrix}.
$$

Compute one iteration of the Haar wavelet transform to yield

$$
\mathbf{y}^1 = \begin{bmatrix} \mathbf{y}^{1,a} \\ \hline \mathbf{y}^{1,d} \end{bmatrix}.
$$

We replace the averages portion $\mathbf{y}^{1,a}$ with the zero vector $\mathbf{0} \in \mathbb{R}^8$ to obtain

$$
\widetilde{\mathbf{y}}^1 = \begin{bmatrix} \mathbf{0} \\ \hline \mathbf{y}^{1,d} \end{bmatrix}.
$$

Next calculate the inverse transform of $\widetilde{\mathbf{y}}^1$ to obtain $\widetilde{\mathbf{v}}$. How many components of $\widetilde{\mathbf{v}}$ are the same as the corresponding components of $\mathbf{v}$? Can you give an explicit description of those components that are different?

   (b) Redo part (a), but now perform two iterations of the Haar wavelet transform and replace $\mathbf{y}^{2,a}$ with $\mathbf{0} \in \mathbb{R}^4$.

   (c) Redo part (a), but now perform three iterations of the Haar wavelet transform and replace $\mathbf{y}^{3,a}$ with $\mathbf{0} \in \mathbb{R}^2$.

   (d) Redo part (a), but now perform four iterations of the Haar wavelet transform and replace $\mathbf{y}^{4,a}$ with 0.

Problems for Review and Discovery

A. Drill Exercises

1. Calculate the Fourier transform of the function

$$f(x) = \begin{cases} N & \text{if} & 0 \le x \le 1/N \\ 0 & \text{if} & x < 0 \text{ or } 1/N < x. \end{cases}$$

Apply Fourier inversion to obtain functions that converge back to f. Notice that each of the approximants has a long tail.

2. Let φ be a C_c^∞ function which is nonnegative and such that $\int \varphi(x)\,dx = 1$. Consider the function $\varphi_\epsilon(x) = \epsilon\varphi(\epsilon x)$. Calculate the Fourier transform of φ_ϵ. How does it behave as $\epsilon \to 0^+$?

3. What is a typical element of V_5? What is a typical element of V_{-4}?

4. What is a typical element of W_7? What is a typical element of W_{-4}?

5. Calculate the Fourier series expansion of $f(x) = \chi_{[0,1]}(x) \cdot \cos 2x$. Also calculate the Haar basis expansion of f. Which is a more accurate approximation? Why?

6. Refer to Exercise 1. Calculate the Fourier *series* of f. How good an approximation to f does this series give?

7. Give an explicit example of a nonzero function that, on the interval $[0,1]$, integrates to 0 against 1, x, x^2, x^3, x^4. [**Hint:** Think of a polynomial of degree at least 5. Is there a polynomial of degree 4 that will do the job?]

8. Let $g(t) = t^3$. Calculate $P_2[g](t)$ and $P_{-1}[g](t)$.

9. Let

$$f_1(t) = 2\varphi_{1,0}(t) - 3\varphi_{1,1}(t) + 5\varphi_{1,2}(t) - 2\varphi_{1,3}(t) + 4\varphi_{1,4}(t) + \varphi_{1,5}(t).$$

Calculate that the projection f_0 of f_1 into V_0

10. Plot each of these functions. Give a verbal description of the compact support of each function.

    (a) $\psi_{1,-3}(t)$

    (b) $\psi_{-2,4}(t)$

    (c) $\psi_{3,6}(t)$

11. Plot each of these functions.

    (a) $f(t) = 3\psi(t/8 - 2) - 2\psi(t/8 + 1) + \psi(t/8 + 1)$

    (b) $g(t) = \sum_{j=1}^{4} j \cdot \psi(8t + j)$

    (c) $h(t) = \sum_{j=-3}^{3} \cos(\pi j)\psi(t/4 + j)$

    (d) $k(t) = \sum_{j=0}^{\infty} \langle e^{-t}, \psi(2t - j) \rangle \cdot \psi(2t - j)$

B. Challenge Problems

1. What can you say about the discrete Haar wavelet transform of the function $f(x) = e^{-x^2}$?

2. Let $f \in C_c^\infty(\mathbb{R})$. Then it is true that

$$|\widehat{f}(\xi)| \le C_N \cdot (1 + |\xi|)^{-N}$$

for every positive integer N. Explain why. [**Hint:** Think in terms of integration by parts.]

3. Refer to Exercise 2. Is something similar true for the discrete Haar wavelet transform?

4. Let f be a continuous function on $[0, 2\pi]$. Let $0 < \alpha < 1$. Suppose that, for each positive integer N, there is a trigonometric polynomial p_N such that
$$\sup_{x \in [0, 2\pi]} |f(x) - p_N(x)| \leq C \cdot N^{-\alpha}.$$
Prove that f is Lipschitz of order α.

5. Let $0 < \alpha < 1$. Discuss the behavior of the Hilbert transform on the Lipschitz space of order α.

6. The Hilbert transform is not bounded on the space of integrable functions on the real line. Explain why.

7. The Hilbert transform is not bounded on the space of bounded functions on the real line. Explain why.

8. Find f_0 and g_0 for each of the following functions from V_1:

   (a) $f_1 = 3\varphi_{1,0} - 5\varphi_{2,2} + 2\varphi_{2,4}$
   (b) $f_1 = 8\varphi_{1,0} - 5\varphi_{1,1} - 2\varphi_{1,3} + 3\varphi_{1,5}$
   (c) $f_1 = 2\varphi_{1,0} + 2\varphi_{1,1} + 2\varphi_{1,2} - 3\varphi_{1,5}$
   (d) $f_1 = \sum_{j=1}^{\infty} \frac{1}{j^2} \varphi_{1,j}$

9. Compute the discrete Haar wavelet transformations for each of these vectors:

   (a) $\mathbf{u} = {}^t\langle 1, 2, 1, 2, 1, 2, 1, 2 \rangle$
   (b) $\mathbf{v} = {}^t\langle 2, 1, 4, 3, 6, 5, 8, 7 \rangle$
   (c) $\mathbf{w} = {}^t\langle 1, 8, 27, 64, 125, 216, 343, 512 \rangle$

10. Compute three iterated discrete Haar wavelet transforms for each of the vectors in Exercise 9.

C. Problems for Discussion and Exploration

1. Produce a tertiary theory of wavelets modeled on the binary Haar wavelets that we discussed in the text.

2. Discuss pointwise convergence for the discrete Haar wavelet transform.

3. Give a sufficient condition on a function f for its discrete Haar wavelet transform to converge uniformly.

4. Show that every integrable function is the limit, in the topology of distributions, of functions in C_c^∞.

10

Partial Differential Equations and Boundary Value Problems

- Boundary value problems

- Ideas from physics

- The wave equation

- The heat equation

- The Laplacian

- The Dirichlet problem

- The Poisson integral

- Sturm–Liouville problems

10.1 Introduction and Historical Remarks

In the middle of the eighteenth century much attention was given to the problem of determining the mathematical laws governing the motion of a vibrating string with fixed endpoints at 0 and π (Figure 10.1). An elementary analysis of tension shows that, if $y(x,t)$ denotes the ordinate of the string at time t above the point x, then $y(x,t)$ satisfies the *wave equation*

$$\frac{\partial^2 y}{\partial t^2} = a^2 \frac{\partial^2 y}{\partial x^2}$$

(see Sections 2.5, 2.8). Here a is a parameter that depends on the tension of the string (in fact a is also the velocity of the traveling solutions of the wave equation). A change of scale will allow us to assume that $a = 1$. (A bit later we shall actually provide a formal derivation of the wave equation. See also [KRA3] for a more thorough consideration of these matters.)

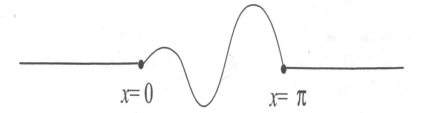

FIGURE 10.1
The wave equation.

In 1747 d'Alembert showed that solutions of this equation have the form

$$y(x,t) = \frac{1}{2}\Big(f(t+x) + g(t-x)\Big), \qquad (10.1.1)$$

where f and g are "any" functions of one variable. (The following technicality must be noted: the functions f and g are initially specified on the interval $[0, \pi]$. We extend f and g to $[-\pi, 0]$ and to $[\pi, 2\pi]$ by odd reflection. Continue f and g to the rest of the real line so that they are 2π-periodic.)

In fact the wave equation, when placed in a "well-posed" setting, comes equipped with two initial conditions:

$$\begin{aligned}\textbf{(i)} \qquad y(x,0) &= \varphi(x) \\ \textbf{(ii)} \qquad \partial_t y(x,0) &= \psi(x).\end{aligned}$$

These conditions mean **(i)** that the wave has an initial configuration that is the graph of the function φ and **(ii)** that the string is released with initial velocity ψ.

If (10.1.1) is to be a solution of this initial value problem then f and g must satisfy

$$\frac{1}{2}\left(f(x) + g(-x)\right) = \varphi(x) \qquad (10.1.2)$$

and

$$\frac{1}{2}\left(f'(x) + g'(-x)\right) = \psi(x). \qquad (10.1.3)$$

Integration of (10.1.3) gives a formula for $f(x) - g(-x)$. That and (10.1.2) give a system that may be solved for f and g with elementary algebra.

The converse statement holds as well: for any functions f and g, a function y of the form (10.1.1) satisfies the wave equation (Exercise). The work of d'Alembert brought to the fore a controversy which had been implicit in the work of Daniel Bernoulli, Leonhard Euler, and others: what is a "function"? (We recommend the article [LUZ] for an authoritative discussion of the controversies that grew out of classical studies of the wave equation. See also [LAN].)

It is clear, for instance, in Euler's writings that he did not perceive a function to be an arbitrary "rule" that assigns points of the range to points of the domain; in particular, Euler did not think that a function could be specified in a fairly arbitrary fashion at different points of the domain. Once a function was specified on some small interval, Euler thought that it could only be extended in one way to a larger interval. Therefore, on physical grounds, Euler objected to d'Alembert's work. He claimed that the initial position of the vibrating string could be specified by several different functions pieced together continuously, so that a single f could not generate the motion of the string.

Daniel Bernoulli solved the wave equation by a different method (separation of variables, which we treat below) and was able to show that there are infinitely many solutions of the wave equation having the form

$$\varphi_j(x,t) = \sin jx \cos jt , \quad j \geq 1 \text{ an integer}.$$

Proceeding formally, he posited that all solutions of the wave equation satisfying $y(0,t) = y(\pi,t) = 0$ and $\partial_t y(x,0) = 0$ will have the form

$$y = \sum_{j=1}^{\infty} a_j \sin jx \cos jt.$$

Setting $t = 0$ indicates that the initial form of the string is $f(x) \equiv \sum_{j=1}^{\infty} a_j \sin jx$. In d'Alembert's language, the initial form of the string is $\frac{1}{2}\big(f(x) - f(-x)\big)$, for we know that

$$0 \equiv y(0,t) = f(t) + g(t)$$

(because the endpoints of the string are held stationary), hence $g(t) = -f(t)$. If we suppose that d'Alembert's function is odd (as is $\sin jx$, each j), then the initial position is given by $f(x)$. Thus the problem of reconciling Bernoulli's solution to d'Alembert's reduces to the question of whether an "arbitrary" function f on $[0,\pi]$ may be written in the form $\sum_{j=1}^{\infty} a_j \sin jx$.

Since most mathematicians contemporary with Bernoulli believed that properties such as continuity, differentiability, and periodicity were preserved under (even infinite) addition, the consensus was that arbitrary f could *not* be represented as a (even infinite) trigonometric sum. The controversy extended over some years and was fueled by further discoveries (such as Lagrange's technique for interpolation by trigonometric polynomials) and more speculations.

In the 1820s, the problem of representation of an "arbitrary" function by trigonometric series was given a satisfactory answer as a result of two events. First, there is the sequence of papers by Joseph Fourier culminating with the tract [FOU]. Fourier gave a formal method of expanding an "arbitrary" function f into a trigonometric series. He computed some partial sums for some sample fs and verified that they gave very good approximations to f. Second, Dirichlet proved the first theorem giving sufficient (and very general) conditions for the Fourier series of a function f to converge pointwise to f. *Dirichlet was one of the first, in 1828, to formalize the notions of partial sum and convergence of a series*; his ideas certainly had antecedents in the work of Gauss and Cauchy.

For all practical purposes, these events mark the beginning of the mathematical theory of Fourier series (see [LAN]). Refer to our Chapter 6 as you read along.

Math Nugget

The Bernoulli family was one of the foremost in all of the history of science. In three generations this remarkable Swiss family produced eight mathematicians, three of them outstanding. These in turn produced a swarm of descendants who distinguished themselves in many fields.

James Bernoulli (1654–1705) studied theology at the insistence of his father, but soon threw it over in favor of his love for science. He quickly learned the new "calculus" of Newton and Leibniz, became Professor of Mathematics at the University of Basel, and held that position until his death. James Bernoulli studied infinite series, special curves, and many other topics. He invented polar coordinates and introduced the Bernoulli numbers

---

that appear in so many contexts in differential equations and special functions. In his book *Ars Conjectandi* he formulated what is now known as the law of large numbers (or Bernoulli's theorem). This is both an important philosophical and an important mathematical fact; it is still a source of study.

James's younger brother John (Johann) Bernoulli (1667–1748) also made a false start by first studying medicine and earning a doctor's degree at Basel in 1694 with a thesis on muscle contraction. He also became fascinated by calculus, mastered it quickly, and applied it to many problems in geometry, differential equations, and mechanics. In 1695 he was appointed Professor of Mathematics at Groningen in Holland. On James Bernoulli's death, John succeeded him in the chair at Basel. The Bernoulli brothers sometimes worked on the same problems; this was unfortunate in view of the family trait of touchiness and jealousy. On occasion their inherent friction flared up into nasty public feuds, more resembling barroom brawls than scientific debates.

In particular, both James and John were solvers of the celebrated brachistochrone problem (along with Newton and Leibniz). They quarreled for years over the relative merits of their different solutions (John's was the more elegant, James's the more general). John Bernoulli was particularly cantankerous in his personal affairs. He once threw his own son (Daniel) out of the house for winning a prize from the French Academy that he himself coveted.

Daniel Bernoulli (1700–1782) studied medicine like his father, and took a degree with a thesis on the action of the lungs. He soon yielded to his inborn talent and became Professor of Mathematics at St. Petersburg. In 1733 he returned to Basel and was, successively, professor of botany, anatomy, and physics. He won ten prizes from the French Academy (including the one that infuriated his father), and over the years published many works on physics, probability, calculus, and differential equations. His famous book *Hydrodynamica* discusses fluid mechanics and gives the earliest treatment of the kinetic theory of gases. Daniel Bernoulli was arguably the first mathematical physicist.

10.2   Eigenvalues, Eigenfunctions, and the Vibrating String

10.2.1   Boundary Value Problems

We wish to motivate the physics of the vibrating string. We begin this discussion by seeking a nontrivial solution y of the differential equation

$$y'' + \lambda y = 0 \tag{10.2.1}$$

subject to the conditions

$$y(0) = 0 \quad \text{and} \quad y(\pi) = 0 \tag{10.2.2}$$

on the interval $[0, \pi]$. Here λ is a fixed constant.

Notice that this is a different situation from the one we have studied in earlier parts of the book. In Chapter 2 on second-order linear equations, we usually had *initial conditions* $y(x_0) = y_0$ and $y'(x_0) = y_1$. Now we have what are called *boundary conditions*: we specify a condition (in this instance the *value*) for the function at two different points. For instance, in the discussion of the vibrating string in the last section, we wanted our string to be pinned down at the two endpoints. These are typical boundary conditions coming from a physical problem.

The situation with boundary conditions is quite different from that for initial conditions. The latter is a sophisticated variation of the fundamental theorem of calculus. The former is rather more subtle. So let us begin to analyze.

First, if $\lambda < 0$ then any solution of (10.2.1) has at most one zero. So it certainly cannot satisfy the boundary conditions (10.2.2). Alternatively, we could just solve the equation explicitly when $\lambda < 0$ and see that the independent solutions are a pair of exponentials, no linear combination of which can satisfy (10.2.2).

If $\lambda = 0$ then the general solution of (10.2.1) is the linear function $y = Ax + B$. Such a function cannot vanish at two points unless it is identically zero.

So the only interesting case is $\lambda > 0$. In this situation, the general solution of (10.2.1) is

$$y = A \sin \sqrt{\lambda} x + B \cos \sqrt{\lambda} x.$$

Since $y(0) = 0$, this in fact reduces to

$$y = A \sin \sqrt{\lambda} x.$$

In order for $y(\pi) = 0$, we must have $\sqrt{\lambda} \pi = n\pi$ for some positive integer n,

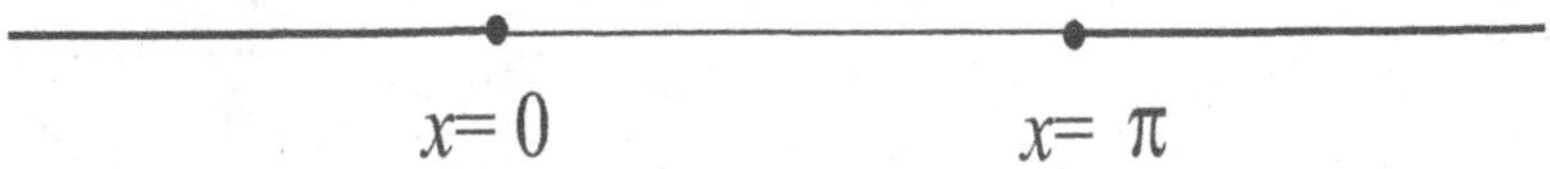

$x = 0 \qquad x = \pi$

FIGURE 10.2
The string in relaxed position.

thus $\lambda = n^2$. These values of λ are termed the *eigenvalues* of the problem (refer to Section 4.1), and the corresponding solutions

$$\sin x , \quad \sin 2x , \quad \sin 3x \ldots$$

are called the *eigenfunctions* of the problem (10.2.1), (10.2.2).

We note these immediate properties of the eigenvalues and eigenfunctions for our problem:

(i) If φ is an eigenfunction for eigenvalue λ, then so is $c \cdot \varphi$ for any constant c.

(ii) The eigenvalues $1, 4, 9, \ldots$ form an increasing sequence that approaches $+\infty$.

(iii) The nth eigenfunction $\sin nx$ vanishes at the endpoints $0, \pi$ (as we originally mandated) and has exactly $n - 1$ zeros in the interval $(0, \pi)$.

10.2.2 Derivation of the Wave Equation

Now let us re-examine the vibrating string from the last section and see how eigenfunctions and eigenvalues arise naturally in a physical problem. We consider a flexible string with negligible weight that is fixed at its ends at the points $(0, 0)$ and $(\pi, 0)$. The string is deformed into an initial position $y = f(x)$ in the x-y plane and then released. See Figure 10.1.

Our analysis will ignore damping effects, such as air resistance. We assume that, in its relaxed position, the string is as in Figure 10.2. The string is plucked in the vertical direction, and is thus set in motion in a vertical plane. We will be supposing that the oscillation has small amplitude.

We focus attention on an "element" Δx of the string (Figure 10.3) that lies between x and $x + \Delta x$. We adopt the usual physical conceit of assuming that the displacement (motion) of this string element is *small*, so that there is only a slight error in supposing that the motion of each point of the string element is strictly vertical. We let the tension of the string, at the point x at time t, be denoted by $T(x, t)$. Note that T acts only in the tangential direction (i.e., along the string). We denote the mass density (mass per unit length) of the string by ρ.

Since *there is no horizontal component of acceleration*, we see that

$$T(x + \Delta x, t) \cdot \cos(\theta + \Delta\theta) - T(x, t) \cdot \cos(\theta) = 0. \qquad (10.2.3)$$

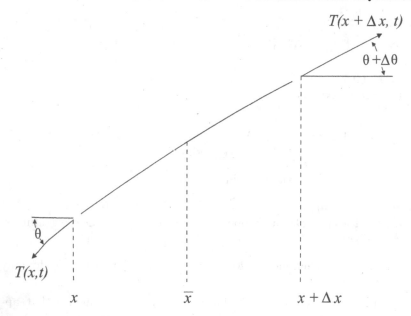

FIGURE 10.3
An element of the plucked string.

(Refer to Figure 10.4: The expression $T(\star)\cdot\cos(\star)$ denotes $H(\star)$, the horizontal component of the tension.) Thus equation (10.2.3) says that H is independent of x.

Now we look at the vertical component of force:

$$T(x + \Delta x, t) \cdot \sin(\theta + \Delta\theta) - T(x,t) \cdot \sin(\theta) = \rho \cdot \Delta x \cdot u_{tt}(\bar{x}, t). \qquad (10.2.4)$$

Here $\bar{x}$ is the mass center of the string element and we are applying Newton's second law—that the external force is the mass of the string element times the acceleration of its center of mass. We use subscripts to denote derivatives. We denote the vertical component of $T(\star)$ by $V(\star)$. Thus equation (10.2.4) can be written as

$$\frac{V(x + \Delta x, t) - V(x,t)}{\Delta x} = \rho \cdot u_{tt}(x,t).$$

Letting $\Delta x \to 0$ yields

$$V_x(x,t) = \rho \cdot u_{tt}(x,t). \qquad (10.2.5)$$

We would like to express equation (10.2.5) entirely in terms of u, so we notice that

$$V(x,t) = H(t)\tan\theta = H(t) \cdot u_x(x,t).$$

(We have used the fact that the derivative in x is the slope of the tangent line, which is $\tan\theta$.) Substituting this expression for V into (10.2.5) yields

$$(Hu_x)_x = \rho \cdot u_{tt}.$$

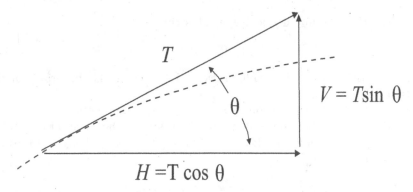

FIGURE 10.4
The horizontal component of the tension.

But H is independent of x, so this last line simplifies to

$$H \cdot u_{xx} = \rho \cdot u_{tt}.$$

For small displacements of the string, θ is nearly zero, so $H = T\cos\theta$ is nearly T. We are most interested in the case where T is constant. And of course ρ is constant. Thus we finally write our equation as

$$\frac{T}{\rho} u_{xx} = u_{tt}.$$

It is traditional to denote the constant T/ρ on the left by a^2. We finally arrive at the *wave equation*

$$a^2 u_{xx} = u_{tt}.$$

10.2.3 Solution of the Wave Equation

We consider the wave equation

$$a^2 y_{xx} = y_{tt} \qquad (10.2.6)$$

with the boundary conditions

$$y(0, t) = 0$$

and

$$y(\pi, t) = 0.$$

Physical considerations dictate that we also impose the initial conditions

$$\left. \frac{\partial y}{\partial t} \right|_{t=0} = 0 \qquad (10.2.7)$$

(indicating that the initial velocity of the string is 0) and

$$y(x,0) = f(x) \qquad (10.2.8)$$

(indicating that the initial configuration of the string is the graph of the function f).

We solve the wave equation using a classical technique known as "separation of variables." For convenience, we assume that the constant $a = 1$. We guess a solution of the form $y(x,t) = \varphi(x) \cdot \psi(t)$. Putting this guess into the differential equation

$$y_{xx} = y_{tt}$$

gives

$$\varphi''(x)\psi(t) = \varphi(x)\psi''(t).$$

We may obviously separate variables, in the sense that we may write

$$\frac{\varphi''(x)}{\varphi(x)} = \frac{\psi''(t)}{\psi(t)}.$$

The left-hand side depends only on x while the right-hand side depends only on t. The only way this can be true is if

$$\frac{\varphi''(x)}{\varphi(x)} = \lambda = \frac{\psi''(t)}{\psi(t)}$$

for some constant λ. But this gives rise to two second-order linear, ordinary differential equations that we can solve explicitly:

$$\varphi'' = \lambda \cdot \varphi \qquad (10.2.9)$$

$$\psi'' = \lambda \cdot \psi. \qquad (10.2.10)$$

Observe that this is the *same* constant λ in both of these equations. Now, as we have already discussed, we want the initial configuration of the string to pass through the points $(0,0)$ and $(\pi, 0)$. We can achieve these conditions by solving (10.2.9) with $\varphi(0) = 0$ and $\varphi(\pi) = 0$. But of course this is the eigenvalue problem that we treated at the beginning of the section. The problem has a nontrivial solution if and only if $\lambda = -n^2$ for some positive integer n, and the corresponding eigenfunction is

$$\varphi_n(x) = \sin nx.$$

For this same λ, the general solution of (10.2.10) is

$$\psi(t) = A \sin nt + B \cos nt.$$

If we impose the requirement that $\psi'(0) = 0$, so that (10.2.7) is satisfied, then $A = 0$ and we find the solution

$$\psi(t) = B \cos nt.$$

This means that the solution we have found of our differential equation with boundary and initial conditions is

$$y_n(x, t) = B \cdot \sin nx \cos nt. \tag{10.2.11}$$

And in fact any finite sum with real coefficients (or *linear combination*) of these solutions will also be a solution:

$$y = \alpha_1 \sin x \cos t + \alpha_2 \sin 2x \cos 2t + \cdots \alpha_k \sin kx \cos kt.$$

Ignoring the rather delicate issue of convergence (which was discussed a bit in Section 6.2), we may claim that any *infinite* linear combination of the solutions (10.2.11) will also be a solution:

$$y = \sum_{j=1}^{\infty} b_j \sin jx \cos jt. \tag{10.2.12}$$

Now we must examine the initial condition (10.2.8). The mandate $y(x, 0) = f(x)$ translates to

$$\sum_{j=1}^{\infty} b_j \sin jx = y(x, 0) = f(x) \tag{10.2.13}$$

or

$$\sum_{j=1}^{\infty} b_j \varphi_j(x) = y(x, 0) = f(x). \tag{10.2.14}$$

Thus we demand that f have a valid Fourier series expansion. We know from our studies in Chapter 6 that such an expansion is correct for a rather broad class of functions f. Thus the wave equation is solvable in considerable generality.

Now fix $m \neq n$. We know that our eigenfunctions φ_j satisfy

$$\varphi_m'' = -m^2 \varphi_m \quad \text{and} \quad \varphi_n'' = -n^2 \varphi_n.$$

Multiply the first equation by φ_n and the second by φ_m and subtract. The result is

$$\varphi_n \varphi_m'' - \varphi_m \varphi_n'' = (n^2 - m^2) \varphi_n \varphi_m$$

or

$$[\varphi_n \varphi_m' - \varphi_m \varphi_n']' = (n^2 - m^2) \varphi_n \varphi_m.$$

We integrate both sides of this last equation from 0 to π and use the fact that $\varphi_j(0) = \varphi_j(\pi) = 0$ for every j. The result is

$$0 = [\varphi_n \varphi_m' - \varphi_m \varphi_n'] \Big|_0^{\pi} = (n^2 - m^2) \int_0^{\pi} \varphi_m(x) \varphi_n(x) \, dx.$$

Thus

$$\int_0^\pi \sin mx \sin nx \, dx = 0 \qquad \text{for } n \neq m \qquad (10.2.15)$$

or

$$\int_0^\pi \varphi_m(x)\varphi_n(x) \, dx = 0 \qquad \text{for } n \neq m. \qquad (10.2.16)$$

Of course this is a standard fact from calculus. But now we understand it as an orthogonality condition (see Sections 4.1, 6.5), and we see how the condition arises naturally from the differential equation. As we have seen in Chapter 4, these ideas fit rather naturally into the general context of Sturm–Liouville problems.

In view of the orthogonality condition (10.2.16), it is natural to integrate both sides of (10.2.14) against $\varphi_k(x)$. The result is

$$\int_0^\pi f(x) \cdot \varphi_k(x) \, dx = \int_0^\pi \left(\sum_{j=0}^\infty b_j \varphi_j(x) \right) \cdot \varphi_k(x) \, dx$$

$$= \sum_{j=0}^\infty b_j \int_0^\pi \varphi_j(x)\varphi_k(x) \, dx$$

$$= \frac{\pi}{2} b_k.$$

Here we use the fact that the integral is 0 when $j \neq k$ and is equal to $\pi/2$ otherwise.

The b_k are the Fourier coefficients that we studied in Chapter 6. Using these coefficients, we have *Bernoulli's solution* (10.2.12) of the wave equation.

Exercises

1. Find the eigenvalues λ_n and the eigenfunctions y_n for the equation $y'' + \lambda y = 0$ in each of the following instances.

    (a) $y(0) = 0$, $y(\pi/2) = 0$
    (b) $y(0) = 0$, $y(2\pi) = 0$
    (c) $y(0) = 0$, $y(1) = 0$
    (d) $y(0) = 0$, $y(L) = 0$ for $L > 0$
    (e) $y(-L) = 0$, $y(L) = 0$ for $L > 0$
    (f) $y(a) = 0$, $y(b) = 0$ for $a < b$

    Solve the following two exercises without worrying about convergence of series or differentiability of functions.

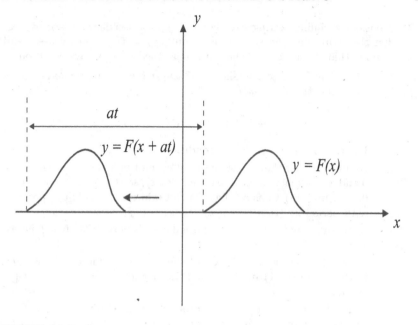

FIGURE 10.5
Wave of fixed shape moving to the left.

2. If $y = F(x)$ is an arbitrary function, then $y = F(x + at)$ represents a wave of fixed shape that moves to the left along the x-axis with velocity a (Figure 10.5).

    Similarly, if $y = G(x)$ is another arbitrary function, then $y = G(x - at)$ is a wave moving to the right, and the most general one-dimensional wave with velocity a is

    $$y(x, t) = F(x + at) + G(x - at). \qquad (*)$$

    (a) Show that $(*)$ satisfies the wave equation.

    (b) It is easy to see that the constant a in the wave equation has the dimension of velocity. Also, it is intuitively clear that if a stretched string is disturbed, then the waves will move in both directions away from the source of the disturbance. These considerations suggest introducing the new variables $\alpha = x + at$, $\beta = x - at$. Show that with these independent variables, the wave equation becomes

    $$\frac{\partial^2 y}{\partial \alpha \partial \beta} = 0.$$

    From this derive $(*)$ by integration. Formula $(*)$ is called *d'Alembert's solution* of the wave equation. It was also obtained, slightly later and independently, by Euler.

3. Consider an infinite string stretched taut on the x-axis from $-\infty$ to $+\infty$. Let the string be drawn aside into a curve $y = f(x)$ and released, and assume that its subsequent motion is described by the wave equation.

   (a) Use (∗) in Exercise 2 to show that the string's displacement is given by d'Alembert's formula

$$y(x,t) = \frac{1}{2}[f(x + at) + f(x - at)]. \qquad (∗∗)$$

   Hint: Remember the initial conditions.

   (b) Assume further that the string remains motionless at the points $x = 0$ and $x = \pi$ (such points are called *nodes*), so that $y(0, t) = y(\pi, t) = 0$, and use (∗∗) to show that f is an odd function that is periodic with period 2π (that is, $f(-x) = f(x)$ and $f(x + 2\pi) = f(x)$).

   (c) Show that since f is odd and periodic with period 2π then f necessarily vanishes at 0 and π.

   (d) Show that Bernoulli's solution of the wave equation can be written in the form (∗∗). **Hint:** Note that $2 \sin nx \cos nat = \sin[n(x + at)] + \sin[n(x - at)]$.

4. Solve the vibrating string problem in the text if the initial shape $y(x, 0) = f(x)$ is specified by the given function. In each case, sketch the initial shape of the string on a set of axes.

   (a)
$$f(x) = \begin{cases} 2cx/\pi & \text{if} \quad 0 \le x \le \pi/2 \\ 2c(\pi - x)/\pi & \text{if} \quad \pi/2 \le x \le \pi \end{cases}$$

   (b)
$$f(x) = \frac{1}{\pi}x(\pi - x)$$

   (c)
$$f(x) = \begin{cases} x & \text{if} \quad 0 \le x \le \pi/4 \\ \pi/4 & \text{if} \quad \pi/4 < x < 3\pi/4 \\ \pi - x & \text{if} \quad 3\pi/4 \le x \le \pi \end{cases}$$

5. Solve the vibrating string problem in the text if the initial shape $y(x, 0) = f(x)$ is that of a single arch of the sine curve $f(x) = c \sin x$. Show that the moving string always has the same general shape, regardless of the value of c. Do the same for functions of the form $f(x) = c \sin nx$. Show in particular that there are $n - 1$ points between $x = 0$ and $x = \pi$ at which the string remains motionless; these points are called *nodes*, and these solutions are called *standing waves*. Draw sketches to illustrate the movement of the standing waves.

6. The problem of the *struck string* is that of solving the wave equation with the boundary conditions

$$y(0, t) = 0, \quad y(\pi, t) = 0$$

and the initial conditions

$$\left.\frac{\partial y}{\partial t}\right|_{t=0} = g(x) \quad \text{and} \quad y(x, 0) = 0.$$

(These initial conditions mean that the string is initially in the equilibrium position, and has an initial velocity $g(x)$ at the point x as a result of being struck.) By separating variables and proceeding formally, obtain the solution

$$y(x,t) = \sum_{j=1}^{\infty} c_j \sin jx \sin jat \,,$$

where

$$c_j = \frac{2}{\pi j a} \int_0^{\pi} g(x) \sin jx \, dx \,.$$

10.3   The Heat Equation

Fourier's Point of View

In [FOU], Fourier considered variants of the following basic question. Let there be given an insulated, homogeneous rod of length π with initial temperature at each $x \in [0, \pi]$ given by a function $f(x)$ (Figure 10.6). Assume that the endpoints are held at temperature 0, and that the temperature of each cross-section is constant. The problem is to describe the temperature $u(x, t)$ of the point x in the rod at time t. Fourier [FOU] perceived the fundamental importance of this problem as follows:

> Primary causes are unknown to us; but are subject to simple and constant laws, which may be discovered by observation, the study of them being the object of natural philosophy.
>
> Heat, like gravity, penetrates every substance of the universe, its rays occupying all parts of space. The object of our work is to set forth the mathematical laws which this element obeys. The theory of heat will hereafter form one of the most important branches of general physics ....
>
> I have deduced these laws from prolonged study and attentive comparison of the facts known up to this time; all these facts I have observed afresh in the course of several years with the most exact instruments that have hitherto been used.

Let us now describe the manner in which Fourier solved his problem. First, it is required to write a differential equation which u satisfies. We shall derive such an equation using three physical principles:

(1) The density of heat energy is proportional to the temperature u, hence the amount of heat energy in any interval $[a, b]$ of the rod is proportional to $\int_a^b u(x, t) \, dx$.

(2) **(Newton's law of cooling)** The rate at which heat flows from a

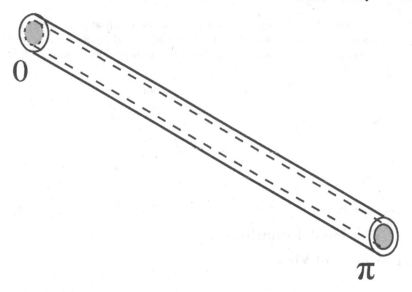

0

π

FIGURE 10.6
The insulated rod.

hot place to a cold one is proportional to the difference in temperature. The infinitesimal version of this statement is that the rate of heat flow across a point x (from left to right) is some negative constant times $\partial_x u(x, t)$.

(3) (Conservation of Energy)   Heat has no sources or sinks.

Now **(3)** tells us that the only way that heat can enter or leave any interval portion $[a, b]$ of the rod is through the endpoints. And **(2)** tells us exactly how this happens. Using **(1)**, we may therefore write

$$\frac{d}{dt} \int_a^b u(x, t) \, dx = \eta^2 [\partial_x u(b, t) - \partial_x u(a, t)].$$

We may rewrite this equation as

$$\int_a^b \partial_t u(x, t) \, dx = \eta^2 \int_a^b \partial_x^2 u(x, t) \, dx.$$

Differentiating in b, we find that

$$\partial_t u = \eta^2 \partial_x^2 u, \tag{10.3.1}$$

and that is the heat equation.

Math Nugget

The English biologist J. B. S. Haldane (1892–1964) had this
remark about the one-dimensional heat equation: "In scientific thought we adopt the simplest theory which will explain
all the facts under consideration and enable us to predict
new facts of the same kind. The catch in this criterion lies
in the word 'simplest.' It is really an aesthetic canon such as
we find implicit in our criticism of poetry or painting. The
layman finds such a law as

$$a^2 \frac{\partial^2 w}{\partial x^2} = \frac{\partial w}{\partial t}$$

much less simple than 'it oozes,' of which it is the mathematical statement. The physicist reverses this judgment,
and his statement is certainly the more fruitful of the two,
so far as prediction is concerned. It is, however, a statement
about something very unfamiliar to the plain man, namely,
the rate of change of a rate of change."

Suppose for simplicity that the constant of proportionality η^2 equals 1.
Fourier guessed that equation (10.3.1) has a solution of the form $u(x,t) = \alpha(x)\beta(t)$. Substituting this guess into the equation yields

$$\alpha(x)\beta'(t) = \alpha''(x)\beta(t)$$

or

$$\frac{\beta'(t)}{\beta(t)} = \frac{\alpha''(x)}{\alpha(x)}.$$

Since the left side is independent of x and the right side is independent of t,
it follows that there is a constant K such that

$$\frac{\beta'(t)}{\beta(t)} = K = \frac{\alpha''(x)}{\alpha(x)}$$

or

$$\beta'(t) = K\beta(t)$$
$$\alpha''(x) = K\alpha(x).$$

We conclude that $\beta(t) = Ce^{Kt}$. The nature of β, and hence of α, thus
depends on the sign of K. But physical considerations tell us that the temperature will dissipate as time goes on, so we conclude that $K \leq 0$. Therefore

$\alpha(x) = \cos\sqrt{-K}x$ and $\alpha(x) = \sin\sqrt{-K}x$ are solutions of the differential equation for α. The initial conditions $u(0, t) = u(\pi, t) = 0$ (since the ends of the rod are held at constant temperature 0) eliminate the first of these solutions and force $K = -j^2$, j an integer. Thus Fourier found the solutions

$$u_j(x, t) = e^{-j^2 t} \sin jx, \quad j \in \mathbb{N}$$

of the heat equation. By linearity, any finite linear combination

$$u(x, t) = \sum_{j=1}^{k} b_j e^{-j^2 t} \sin jx \tag{10.3.2}$$

of these solutions is also a solution. It is plausible to extend this assertion to infinite linear combinations. Using the initial condition $u(x, 0) = f(x)$ again raises the question of whether "any" function $f(x)$ on $[0, \pi]$ can be written as a (infinite) linear combination of the functions $\sin jx$.

Fourier's solution to this last problem (of the sine functions spanning essentially everything) is roughly as follows. Suppose f is a function that is so representable:

$$f(x) = \sum_{j} b_j \sin jx. \tag{10.3.3}$$

Setting $x = 0$ gives

$$f(0) = 0.$$

Differentiating both sides of (10.3.3) and setting $x = 0$ gives

$$f'(0) = \sum_{j=1}^{\infty} j b_j. \tag{10.3.4}$$

Successive differentiation of (10.3.3), and evaluation at 0, gives

$$f^{(k)}(0) = \sum_{j=1}^{\infty} j^k b_j (-1)^{\lfloor k/2 \rfloor}$$

for k odd (by oddness of f, the even derivatives must be 0 at 0). Here $\lfloor \ \rfloor$ denotes the greatest integer function. Thus Fourier devised a system of infinitely many equations in the infinitely many unknowns $\{b_j\}$. He proceeded to solve this system by truncating it to an $N \times N$ system (the first N equations restricted to the first N unknowns), solving that truncated system, and then letting N tend to ∞. Suffice it to say that Fourier's arguments contained many dubious steps (see [FOU] and [LAN]).

The upshot of Fourier's intricate and lengthy calculations was that

$$b_j = \frac{2}{\pi} \int_0^{\pi} f(x) \sin jx \, dx. \tag{10.3.5}$$

By modern standards, Fourier's reasoning was specious; for he began by assuming that f possessed an expansion in terms of sine functions. The formula (10.3.5) hinges on that supposition, together with steps in which one compensated division by zero with a later division by ∞. Nonetheless, Fourier's methods give an actual *procedure* for endeavoring to expand any given f in a series of sine functions.

Fourier's abstract arguments constitute the first part of his book. The bulk, and remainder, of the book consists of separate chapters in which the expansions for particular functions are computed.

EXAMPLE 10.3.1 Suppose that the thin rod in the setup of the heat equation is first immersed in boiling water so that its temperature is uniformly 100°C. Then imagine that it is removed from the water at time $t = 0$ with its ends immediately put into ice so that these ends are kept at temperature 0°C. Find the temperature $u = u(x,t)$ under these circumstances.

Solution:
The initial temperature distribution is given by the constant function

$$f(x) = 100, \quad 0 < x < \pi.$$

The two boundary conditions, and the other initial condition, are as usual. Thus our job is simply this: to find the sine series expansion of this function f. We calculate that

$$
\begin{aligned}
b_j &= \frac{2}{\pi} \int_0^\pi 100 \sin jx \, dx \\
&= -\frac{200}{\pi} \frac{\cos jx}{j} \Big|_0^\pi \\
&= -\frac{200}{\pi} \left[\frac{(-1)^j}{j} - \frac{1}{j} \right] \\
&= \begin{cases} 0 & \text{if} \quad j = 2\ell \text{ is even} \\[2mm] \dfrac{400}{\pi j} & \text{if} \quad j = 2\ell - 1 \text{ is odd}. \end{cases}
\end{aligned}
$$

Thus

$$f(x) = \frac{400}{\pi} \left(\sin x + \frac{\sin 3x}{3} + \frac{\sin 5x}{5} + \cdots \right).$$

Now, referring to formula (10.3.2) from our general discussion of the heat equation, we know that

$$u(x,t) = \frac{400}{\pi} \left(e^{-t} \sin x + \frac{1}{3} e^{-9t} \sin 3x + \frac{1}{5} e^{-25t} \sin 5x + \cdots \right). \quad ■$$

EXAMPLE 10.3.2 Find the steady-state temperature of the thin rod from our analysis of the heat equation if the fixed temperatures at the ends $x = 0$ and $x = \pi$ are w_1 and w_2 respectively.

Solution:
The phrase "steady-state" means that $\partial u/\partial t = 0$, so that the heat equation reduced to $\partial^2 u/\partial x^2 = 0$ or $d^2 u/dx^2 = 0$. The general solution is then $u = Ax + B$. The values of these two constants A and B are forced by the two boundary conditions.

In fact a little high school algebra tells us that

$$u = w_1 + \frac{1}{\pi}(w_2 - w_1)x.$$ ∎

The steady-state version of the 3-dimensional heat equation

$$a^2\left(\frac{\partial^2 u}{\partial x^2} + \frac{\partial^2 u}{\partial y^2} + \frac{\partial^2 u}{\partial z^2}\right) = \frac{\partial u}{\partial t}$$

is

$$\frac{\partial^2 u}{\partial x^2} + \frac{\partial^2 u}{\partial y^2} + \frac{\partial^2 u}{\partial z^2} = 0.$$

This last is called *Laplace's equation*. The study of this equation and its solutions and subsolutions and their applications is a deep and rich branch of mathematics called *potential theory*. There are applications to heat, to gravitation, to electromagnetics, and to many other parts of physics. The equation plays a central role in the theory of partial differential equations, and is also an integral part of complex variable theory.

Exercises

1. Solve the boundary value problem

$$\begin{aligned}
a^2\frac{\partial^2 w}{\partial x^2} &= \frac{\partial w}{\partial t} \\
w(x,0) &= f(x) \\
w(0,t) &= 0 \\
w(\pi,t) &= 0
\end{aligned}$$

if the last three conditions—the boundary conditions—are changed to

$$\begin{aligned}
w(x,0) &= f(x) \\
w(0,t) &= w_1 \\
w(\pi,t) &= w_2.
\end{aligned}$$

[**Hint:** Write $w(x,t) = W(x,t) + g(x)$, where $g(x)$ is the function that we produced in Example 10.3.2.]

2. Suppose that the lateral surface of the thin rod that we analyzed in the text is not insulated, but in fact radiates heat into the surrounding air. If Newton's law of cooling (that a body cools at a rate proportional to the difference of its temperature with the temperature of the surrounding air) is assumed to apply, then show that the 1-dimensional heat equation becomes

$$a^2 \frac{\partial^2 w}{\partial x^2} = \frac{\partial w}{\partial t} + c(w - w_0)$$

where c is a positive constant and w_0 is the temperature of the surrounding air.

3. In Exercise 2, find $w(x, t)$ if the ends of the rod are kept at $0°C$, $w_0 = 0°C$, and the initial temperature distribution on the rod is $f(x)$.

4. In Example 10.3.1, suppose that the ends of the rod are insulated instead of being kept fixed at $0°C$. What are the new boundary conditions? Find the temperature $w(x, t)$ in this case by using just common sense—and *not* calculating.

5. Solve the problem of finding $w(x, t)$ for the rod with insulated ends at $x = 0$ and $x = \pi$ (see the preceding exercise) if the initial temperature distribution is given by $w(x, 0) = f(x)$.

6. The 2-dimensional heat equation is

$$a^2 \left(\frac{\partial^2 w}{\partial x^2} + \frac{\partial^2 w}{\partial y^2} \right) = \frac{\partial w}{\partial t} .$$

Use the method of separation of variables to find a steady-state solution of this equation in the infinite half-strip of the x-y plane bounded by the lines $x = 0$, $x = \pi$, and $y = 0$ if the following boundary conditions are satisfied:

$$w(0, y, t) = 0 \qquad\qquad w(\pi, y, t) = 0$$
$$w(x, 0, 0) = f(x) \qquad\qquad \lim_{y \to +\infty} w(x, y, t) = 0 .$$

7. Derive the 3-dimensional heat equation

$$a^2 \left(\frac{\partial^2 w}{\partial x^2} + \frac{\partial^2 w}{\partial y^2} + \frac{\partial^2 w}{\partial z^2} \right) = \frac{\partial w}{\partial t}$$

by adapting the reasoning in the text to the case of a small box with edges Δx, Δy, Δz contained in a region R in x-y-z space where the temperature function $w(x, y, z, t)$ is sought. [**Hint:** Consider the flow of heat through two opposite faces of the box, first perpendicular to the x-axis, then perpendicular to the y-axis, and finally perpendicular to the z-axis.]

10.4   The Dirichlet Problem for a Disc

We now study the two-dimensional Laplace equation, which is

$$\triangle w = \frac{\partial^2 w}{\partial x^2} + \frac{\partial^2 w}{\partial y^2} = 0.$$

It will be useful for us to write this equation in polar coordinates. To do so, recall that

$$r^2 - x^2 + y^2 \ , \quad x = r\cos\theta \ , \quad y = r\sin\theta.$$

Thus

$$\frac{\partial}{\partial r} = \frac{\partial x}{\partial r}\frac{\partial}{\partial x} + \frac{\partial y}{\partial r}\frac{\partial}{\partial y} = \cos\theta\frac{\partial}{\partial x} + \sin\theta\frac{\partial}{\partial y}$$

$$\frac{\partial}{\partial \theta} = \frac{\partial x}{\partial \theta}\frac{\partial}{\partial x} + \frac{\partial y}{\partial \theta}\frac{\partial}{\partial y} = -r\sin\theta\frac{\partial}{\partial x} + r\cos\theta\frac{\partial}{\partial y}.$$

We may solve these two equations for the unknowns $\partial/\partial x$ and $\partial/\partial y$. The result is

$$\frac{\partial}{\partial x} = \cos\theta\frac{\partial}{\partial r} - \frac{\sin\theta}{r}\frac{\partial}{\partial \theta} \quad \text{and} \quad \frac{\partial}{\partial y} = \sin\theta\frac{\partial}{\partial r} + \frac{\cos\theta}{r}\frac{\partial}{\partial \theta}.$$

A tedious calculation now reveals that

$$\begin{aligned}
\triangle = \frac{\partial^2}{\partial x^2} + \frac{\partial^2}{\partial y^2} &= \left(\cos\theta\frac{\partial}{\partial r} - \frac{\sin\theta}{r}\frac{\partial}{\partial \theta}\right)\left(\cos\theta\frac{\partial}{\partial r} - \frac{\sin\theta}{r}\frac{\partial}{\partial \theta}\right) \\
&\quad + \left(\sin\theta\frac{\partial}{\partial r} - \frac{\cos\theta}{r}\frac{\partial}{\partial \theta}\right)\left(\sin\theta\frac{\partial}{\partial r} - \frac{\cos\theta}{r}\frac{\partial}{\partial \theta}\right) \\
&= \frac{\partial^2}{\partial r^2} + \frac{1}{r}\frac{\partial}{\partial r} + \frac{1}{r^2}\frac{\partial^2}{\partial \theta^2}.
\end{aligned} \tag{10.4.1}$$

Let us fall back once again on the separation of variables method. We shall seek a solution $w = w(r,\theta) = u(r)\cdot v(\theta)$ of the Laplace equation. Using the polar form (10.4.1) of the Laplacian, we find that this leads to the equation

$$u''(r)\cdot v(\theta) + \frac{1}{r}u'(r)\cdot v(\theta) + \frac{1}{r^2}u(r)\cdot v''(\theta) = 0.$$

Thus

$$\frac{r^2 u''(r) + ru'(r)}{u(r)} = -\frac{v''(\theta)}{v(\theta)}.$$

Since the left-hand side depends only on r, and the right-hand side only on θ, both sides must be constant. Denote the common constant value by λ.

Then we have

$$v'' + \lambda v = 0 \qquad (10.4.2)$$

and

$$r^2 u'' + r u' - \lambda u = 0. \qquad (10.4.3)$$

If we demand that v be continuous and periodic, then we must have (because of equation (10.4.2)) that $\lambda > 0$ and in fact that $\lambda = n^2$ for some nonnegative integer n (so that we end up with solutions $v = \sin n\theta$ and $v = \cos n\theta$). We have studied this situation in detail in Section 10.2. For $n = 0$ the only suitable solution is $v \equiv$ constant and for $n > 0$ the general solution (with $\lambda = n^2$) is

$$y = A \cos n\theta + B \sin n\theta.$$

We set $\lambda = n^2$ in equation (10.4.3) and obtain[1]

$$r^2 u'' + r u' - n^2 u = 0.$$

We may solve this equation by guessing $u(r) = r^m$. Plugging that guess into the differential equation, we obtain

$$r^m(m^2 - n^2) = 0.$$

Now there are two cases:

(i)   If $n = 0$ then we obtain the repeated root $m = 0, 0$. Now we proceed by analogy with our study of second-order linear equations with constant coefficients, and hypothesize a second solution of the form $u(r) = \ln r$. This works, so we obtain the general solution

$$u(r) = A + B \ln r.$$

(ii) If $n > 0$ then $m = \pm n$ and the general solution is

$$u(r) = A r^n + B r^{-n}.$$

We are most interested in solutions u that are continuous at the origin, so we take $B = 0$ in all cases. The resulting solutions are

$$
\begin{aligned}
n = 0, & \qquad w = a_0/2 \ \text{ a constant} \\
n = 1, & \qquad w = r(a_1 \cos \theta + b_1 \sin \theta) \\
n = 2, & \qquad w = r^2(a_2 \cos 2\theta + b_2 \sin 2\theta) \\
n = 3, & \qquad w = r^3(a_3 \cos 3\theta + b_3 \sin 3\theta)
\end{aligned}
$$

$\cdots$

---

[1] This is Euler's equidimensional equation. The change of variables $r = e^z$ transforms this equation to a linear equation with constant coefficients, and that can in turn be solved with our standard techniques.

Of course any finite sum of solutions of Laplace's equation is also a solution. The same is true for infinite sums. Thus we are led to consider

$$w = w(r, \theta) = \frac{1}{2}a_0 + \sum_{j=0}^{\infty} r^j(a_j \cos j\theta + b_j \sin j\theta).$$

On a formal level, letting $r \to 1^-$ in this last expression gives

$$\frac{1}{2}a_0 + \sum_{j=1}^{\infty}(a_j \cos j\theta + b_j \sin j\theta).$$

We draw all these ideas together with the following physical rubric. Consider a thin aluminum disc of radius 1, and imagine applying a heat distribution to the boundary of that disc. In polar coordinates, this distribution is specified by a function $f(\theta)$. We seek to understand the steady-state heat distribution on the entire disc. So we seek a function $w(r, \theta)$, continuous on the closure of the disc, which agrees with f on the boundary and which represents the steady-state distribution of heat inside. Some physical analysis shows that such a function w is the solution of the boundary value problem

$$\Delta w = 0$$
$$w\big|_{\partial D} = f.$$

Here we use the notation ∂D to denote the boundary of D.

According to the calculations we performed prior to this last paragraph, a natural approach to this problem is to expand the given function f in its Fourier series:

$$f(\theta) = \frac{1}{2}a_0 + \sum_{j=1}^{\infty}(a_j \cos j\theta + b_j \sin j\theta)$$

and then posit that the w we seek is

$$w(r, \theta) = \frac{1}{2}a_0 + \sum_{j=1}^{\infty} r^j(a_j \cos j\theta + b_j \sin j\theta).$$

This process is known as solving *the Dirichlet problem on the disc with boundary data f*.

EXAMPLE 10.4.1 Follow the paradigm just sketched to solve the Dirichlet problem on the disc with $f(\theta) = 1$ on the top half of the boundary and $f(\theta) = -1$ on the bottom half of the boundary.

Solution:
The data function f is odd on the interval $[-\pi, \pi]$. It is straightforward to calculate that the Fourier series (sine series) expansion for this f is

$$f(\theta) = \frac{4}{\pi}\left(\sin\theta + \frac{\sin 3\theta}{3} + +\frac{\sin 5\theta}{5} + \cdots\right).$$

The solution of the Dirichlet problem is therefore

$$w(r, \theta) = \frac{4}{\pi}\left(r\sin\theta + \frac{r^3 \sin 3\theta}{3} + + \frac{r^5 \sin 5\theta}{5} + \cdots \right).$$

■

10.4.1   The Poisson Integral

We have presented a formal procedure with series for solving the Dirichlet problem. But in fact it is possible to produce a closed formula (i.e., an integral formula) for this solution. We now make the construction explicit.

Referring back to our Fourier series expansion for f, and the resulting expansion for the solution of the Dirichlet problem, we recall that

$$a_j = \frac{1}{\pi}\int_{-\pi}^{\pi} f(\varphi)\cos j\varphi\, d\varphi \quad \text{and} \quad b_j = \frac{1}{\pi}\int_{-\pi}^{\pi} f(\varphi)\sin j\varphi\, d\varphi.$$

Thus

$$
\begin{aligned}
w(r,\theta) \;=\;& \frac{1}{2}a_0 + \sum_{j=1}^{\infty} r^j \left(\frac{1}{\pi}\int_{-\pi}^{\pi} f(\varphi)\cos j\varphi\, d\varphi \cos j\theta \right.\\
& \left. + \frac{1}{\pi}\int_{-\pi}^{\pi} f(\varphi)\sin j\varphi\, d\varphi \sin j\theta \right).
\end{aligned}
$$

This, in turn, equals

$$\frac{1}{2}a_0 + \frac{1}{\pi}\sum_{j=1}^{\infty}\int_{-\pi}^{\pi} f(\varphi)r^j\left(\cos j\varphi\cos j\theta + \sin j\varphi\sin j\theta \right)d\varphi$$

$$= \frac{1}{2}a_0 + \frac{1}{\pi}\sum_{j=1}^{\infty}\int_{-\pi}^{\pi} f(\varphi)r^j\left(\cos j(\theta - \varphi)d\varphi \right).$$

We finally simplify our expression to

$$w(r,\theta) = \frac{1}{\pi}\int_{-\pi}^{\pi} f(\varphi)\left(\frac{1}{2} + \sum_{j=1}^{\infty} r^j \cos j(\theta - \varphi) \right) d\varphi.$$

It behooves us, therefore, to calculate the expression inside the large parentheses. For simplicity, we let $\alpha = \theta - \varphi$ and then we let

$$z = re^{i\alpha} = r(\cos\alpha + i\sin\alpha).$$

Likewise

$$z^n = r^n e^{in\alpha} = r^n(\cos n\alpha + i\sin n\alpha).$$

In what follows, if $z = x + iy$, then we let $\mathrm{Re}\, z = x$ denote the *real part* of z and $\mathrm{Im}\, z = y$ denote the *imaginary part* of z. Also $\bar{z} = x - iy$ is the *conjugate* of z.

Then

$$
\begin{aligned}
\frac{1}{2} + \sum_{j=1}^{\infty} r^j \cos j\alpha \;&=\; \operatorname{Re}\left(\frac{1}{2} + \sum_{j=1}^{\infty} z^j\right) \\
&=\; \operatorname{Re}\left(-\frac{1}{2} + \sum_{j=0}^{\infty} z^j\right) \\
&=\; \operatorname{Re}\left(-\frac{1}{2} + \frac{1}{1-z}\right) \\
&=\; \operatorname{Re}\left(\frac{1+z}{2(1-z)}\right) \\
&=\; \operatorname{Re}\left(\frac{(1+z)(1-\bar{z})}{2|1-z|^2}\right) \\
&=\; \frac{1-|z|^2}{2|1-z|^2} \\
&=\; \frac{1-r^2}{2(1-2r\cos\alpha+r^2)} \, .
\end{aligned}
$$

Putting the result of this calculation into our original formula for w we finally obtain the Poisson integral formula:

$$
w(r,\theta) = \frac{1}{2\pi} \int_{-\pi}^{\pi} \frac{1-r^2}{1-2r\cos(\theta-\varphi)+r^2} f(\varphi)\, d\varphi . \tag{10.4.4}
$$

Observe what this formula does for us: It expresses the solution of the Dirichlet problem with boundary data f as an explicit integral of a universal expression (called a *kernel*) against that data function f. To be very plain about this, the kernel here is

$$
P(r,\theta) = \frac{1}{2\pi} \cdot \frac{1-r^2}{1-2r\cos(\theta-\varphi)+r^2} .
$$

There is a great deal of information about w and its relation to f contained in formula (10.4.4). As just one simple instance, we note that when r is set equal to 0 then we obtain

$$
w(0,\theta) = \frac{1}{2\pi} \int_{-\pi}^{\pi} f(\varphi)\, d\varphi .
$$

This says that the value of the steady-state heat distribution at the origin is just the average value of f around the circular boundary.

Math Nugget

Siméon Denis Poisson (1781–1840) was an eminent French mathematician and physicist. He succeeded Fourier in 1806 as Professor at the École Polytechnique. In physics, Poisson's equation describes the variation of the potential inside a continuous distribution of mass or in an electric charge. Poisson made important theoretical contributions to the study of elasticity, magnetism, heat, and capillary action. In pure mathematics, the Poisson summation formula is a major tool in analytic number theory, and the Poisson integral pointed the way to many significant developments in Fourier analysis. In addition, Poisson worked extensively in probability theory. It was he who identified and named the "law of large numbers;" and the Poisson distribution—or "law of small numbers"—has fundamental applications in all parts of statistics and probability.

According to Abel, Poisson was a short, plump man. His family tried to encourage him in many directions, from being a doctor to being a lawyer, this last on the theory that perhaps he was fit for nothing better. But at last he found his place as a scientist and produced over 300 works in a relatively short lifetime. "La vie, c'est le travail (Life is work)," said Poisson—and he had good reason to know.

EXAMPLE 10.4.2 Consider an initial heat distribution on the boundary of the unit disc which is given by a "point mass." That is to say, there is a "charge of heat" at the point $(1, 0)$ of total mass 1 and with value 0 elsewhere on ∂D. What will be the steady-state heat distribution on the entire disc?

Solution:
Think of the point mass as the limit of functions that take the value N on a tiny interval of length $1/N$. Convince yourself that the Poisson integral of such a function tends to the Poisson kernel itself. So the steady-state heat distribution in this case is given by the Poisson kernel. This shows, in particular, that the Poisson kernel is a harmonic function. ∎

Exercises

1. Solve the Dirichlet problem for the unit disc when the boundary function $f(\theta)$ is defined by

   (a) $f(\theta) = \cos\theta/2, \quad -\pi \le \theta \le \pi$

   (b) $f(\theta) = \theta, \quad -\pi < \theta \le \pi$

   (c) $f(\theta) = \begin{cases} 0 & \text{if} & -\pi \le \theta < 0 \\ \sin\theta & \text{if} & 0 \le \theta \le \pi \end{cases}$

   (d) $f(\theta) = \begin{cases} 0 & \text{if} & -\pi \le \theta < 0 \\ 1 & \text{if} & 0 \le \theta \le \pi \end{cases}$

   (e) $f(\theta) = \theta^2/4, \quad -\pi \le \theta \le \pi$

2. Show that the Dirichlet problem for the disc $\{(x,y) : x^2 + y^2 \le R^2\}$, where $f(\theta)$ is the boundary function, has the solution

   $$w(r,\theta) = \frac{1}{2}a_0 + \sum_{j=1}^{\infty} \left(\frac{r}{R}\right)^j (a_j \cos j\theta + b_j \sin j\theta)$$

   where a_j and b_j are the Fourier coefficients of f. Show also that the Poisson integral formula for this more general disc setting is

   $$w(r,\theta) = \frac{1}{2\pi} \int_{-\pi}^{\pi} \frac{R^2 - r^2}{R^2 - 2Rr\cos(\theta - \varphi) + r^2} f(\varphi)\, d\varphi .$$

   [**Hint:** Do not solve this problem from first principles. Rather, do a change of variables to reduce this new problem to the already-understood situation on the unit disc.]

3. Let w be a harmonic function in a planar region, and let C be any circle entirely contained (along with its interior) in this region. Prove that the value of w at the center of C is the average of its values on the circumference.

4. If $w = F(x,y) = \mathcal{F}(r,\theta)$, with $x = r\cos\theta$ and $y = r\sin\theta$, then show that

   $$\frac{\partial^2 w}{\partial x^2} + \frac{\partial^2 w}{\partial y^2} = \frac{1}{r}\left\{ \frac{\partial}{\partial r}\left(r\frac{\partial w}{\partial r}\right) + \frac{1}{r}\frac{\partial^2 w}{\partial \theta^2} \right\}$$

   $$= \frac{\partial^2 w}{\partial r^2} + \frac{1}{r}\frac{\partial w}{\partial r} + \frac{1}{r^2}\frac{\partial^2 w}{\partial \theta^2} .$$

   [**Hint:** We can calculate that

   $$\frac{\partial w}{\partial r} = \frac{\partial w}{\partial x}\cos\theta + \frac{\partial w}{\partial y}\sin\theta \quad \text{and} \quad \frac{\partial w}{\partial \theta} = \frac{\partial w}{\partial x}(-r\sin\theta) + \frac{\partial w}{\partial y}(r\cos\theta).$$

   Similarly, compute $\dfrac{\partial}{\partial r}\left(r\dfrac{\partial w}{\partial r}\right)$ and $\dfrac{\partial^2 w}{\partial \theta^2}$.]

5. Use your symbol manipulation software, such as Maple or Mathematica, to calculate the Poisson integral of the given function on $[-\pi, \pi]$.

   (a) $f(\theta) = \ln^2\theta$

(b) $f(\theta) = \theta^3 \cdot \cos\theta$

(c) $f(\theta) = e^\theta \cdot \sin\theta$

(d) $f(\theta) = e^\theta \cdot \ln\theta$

Historical Note

Fourier

Jean Baptiste Joseph Fourier (1768–1830) was a mathematical physicist of some note. He was an acolyte of Napoleon Bonaparte and accompanied the fiery leader to Egypt in 1798. On his return, Fourier became the prefect of the district of Isère in southeastern France; in that post he built the first real road from Grenoble to Turin. He also became the friend and mentor of the boy Champollion, who later was to be the first to decipher the Rosetta Stone.

During these years he worked on the theory of the conduction of heat. Euler, Bernoulli, d'Alembert, and many others had studied the heat equation and made conjectures on the nature of its solutions. The most central issues hinged on the problem of whether an "arbitrary function" could be represented as a sum of sines and cosines. In those days, nobody was very sure what a function was and the notion of convergence of a series had not yet been defined, so the debate was largely metaphysical.

Fourier actually came up with a formula for producing the coefficients of a cosine or sine series of any given function. He presented it in the first chapter of his book *The Analytic Theory of Heat*. Fourier's ideas were controversial, and he had a difficult time getting the treatise published. In fact he only managed to do so when he became the Secretary of the French National Academy of Sciences and published the book himself.

The series that Fourier studied, and in effect put on the map, are now named after him. The subject area has had a profound influence on mathematics as a whole. Riemann's theory of the integral—the one that is used in most every calculus book—was developed specifically in order to study certain questions of the convergence of Fourier series. Cantor's theory of sets was cooked up primarily to address issues of sets of convergence for Fourier series. Many of the modern ideas in functional analysis—the uniform boundedness principle, for example—grew out of questions of the convergence of Fourier series. Dirichlet invented the modern rigorous notion of "function" as part of his study of Fourier series. As we have indicated in this chapter, Fourier analysis is a powerful tool in the study of partial differential equations (and ordinary differential equations as well).

Fourier's name has become universally known in modern analytical science. His ideas have been profound and influential. Harmonic analysis is the modern generalization of Fourier analysis, and wavelets are the latest implementation of these ideas.

Historical Note
Dirichlet

Peter Gustav Lejeune Dirichlet (1805–1859) was a German mathematician who was deeply influenced by the works of the Parisians—Cauchy, Fourier, Legendre, and many others. He was strongly influenced by Gauss's *Disquisitiones Arithmeticae*. This was quite a profound but impenetrable work, and Dirichlet was not satisfied until he had worked through the ideas himself in detail. He was not only the first to understand Gauss's famous book, but also the first to explain it to others.

In later life Dirichlet became a friend and disciple of Gauss, and also a friend and advisor to Riemann. In 1855, after lecturing in Berlin for many years, he succeeded Gauss in the professorship at Göttingen.

In 1829 Dirichlet achieved two milestones. One is that he gave a rigorous definition of the convergence of series. The other is that he gave the definition, that we use today, of a function. In particular, he freed the idea of function from any dependence on formulas or laws or mathematical operations. He applied both these ideas to the study of the convergence of Fourier series, and gave the first rigorously proved convergence criterion.

Between 1837 and 1839, Dirichlet developed some very remarkable applications of mathematical analysis to number theory. In particular, he proved that there are infinitely many primes in any arithmetical progression of the form $a + bn$ with a and b relatively prime. His studies of absolutely convergent series also appeared in 1837. Dirichlet's important convergence test for series was not published until after his death.

Dirichlet also engaged in studies of mathematical physics. These led, in part, to the important *Dirichlet principle* in potential theory. This idea establishes the existence of certain extremal harmonic functions. It was important historically, because it was the key to finally obtaining a rigorous proof of the Riemann mapping theorem. It is still used today in partial differential equations, the calculus of variations, differential geometry, and mathematical physics.

Dirichlet is remembered today for the Dirichlet problem, for his results in number theory (the useful "pigeonhole principle" was originally called the

"Dirichletscher Schubfachschluss" or "Dirichlet's drawer-shutting principle").
He is one of the important mathematicians of the nineteenth century.

Problems for Review and Discovery

A. Drill Exercises

1. Find the eigenvalues λ_n and eigenfunctions y_n for the equation $y'' + \lambda y = 0$ in each of the following cases.

   (a) $y(-2) = 0$, $y(2) = 0$
   (b) $y(0) = 0$, $y(3) = 0$
   (c) $y(1) = 0$, $y(4) = 0$
   (d) $y(-3) = 0$, $y(0) = 0$

2. Solve the vibrating string problem in Section 10.2 if the initial shape $y(x,0) = f(x)$ is specified by the function $f(x) = x + |x|$. Sketch the initial shape of the string on a set of axes.

3. Solve the Dirichlet problem for the unit disc when the boundary function $f(\theta)$ is defined by

   (a) $f(\theta) = \sin \theta/2$, $-\pi \le \theta \le \pi$
   (b) $f(\theta) = \theta + |\theta|$, $-\pi \le \theta \le \pi$
   (c) $f(\theta) = \theta^2$, $-\pi \le \theta \le \pi$

4. Find the solution to the Dirichlet problem on the unit disc with boundary data

   (a) $f(\theta) = |\theta|$
   (b) $g(\theta) = \sin^2 \theta$.
   (c) $h(\theta) = \cos \theta/2$
   (d) $f(\theta) = \theta/2$

5. Find a solution to this Dirichlet problem for a half-annulus:

$$\frac{\partial^2 u}{\partial r^2} + \frac{1}{r}\frac{\partial u}{\partial r} + \frac{1}{r^2}\frac{\partial^2 u}{\partial \theta^2} = 0, \; 1 < r < 2, \; 0 < \theta < \pi$$
$$u(r,0) = \sin \pi r, \; 1 \le r \le 2$$
$$u(r,\pi) = 0, \; 1 \le r \le 2$$
$$u(1,\theta) = u(2,\theta) = 0, \; 0 \le \theta \le \pi.$$

B. Challenge Problems

1. Is it possible for a harmonic function to vanish on an entire line segment? Give an example.

2. Use methods introduced in this chapter to find a solution of the boundary value problem

$$\frac{\partial u}{\partial t} = 4\frac{\partial^2 u}{\partial x^2} ,\ 0 < x < \pi ,\ t > 0$$

$$\frac{\partial u}{\partial x}(0,t) = \frac{\partial u}{\partial x}(\pi,t) = 0 ,\ t > 0$$

$$u(x,0) = 3x ,\ 0 < x < \pi .$$

3. Use methods introduced in this chapter to find a solution of the boundary value problem

$$\frac{\partial u}{\partial t} = \frac{\partial^2 u}{\partial x^2} ,\ 0 < x < 1 ,\ t > 0$$

$$u(0,t) = 0 ,\ u(1,t) + \frac{\partial u}{\partial x}(1,t) = 0 ,\ t > 0 .$$

$$u(x,0) = f(x) ,\ 0 < x < 1$$

4. Use methods introduced in this chapter to find a solution of the boundary value problem

$$\frac{\partial u}{\partial t} = 2\frac{\partial^2 u}{\partial x^2} + 4 ,\ 0 < x < 1 ,\ t > 0$$

$$u(0,t) = u(1,t) = 1 ,\ t > 0$$

$$u(x,0) = 1 ,\ 0 < x < 1 .$$

5. Use methods introduced in this chapter to find a solution of the boundary value problem

$$\frac{\partial^2 u}{\partial t^2} = 9\frac{\partial^2 u}{\partial x^2} ,\ 0 < x < 1 ,\ t > 0$$

$$u(0,t) = u(1,t) = 0 ,\ t > 0$$

$$u(x,0) = 1 - \cos^2 \pi x ,\ 0 < x < 1$$

$$\frac{\partial u}{\partial t}(x,0) = 1 - \sin x ,\ 0 < x < 1 .$$

6. Use methods introduced in this chapter to find a solution of the boundary value problem

$$\frac{\partial^2 u}{\partial t^2} = \frac{\partial^2 u}{\partial x^2} ,\ 0 < x < 2 ,\ t > 0$$

$$u(0,t) = u(2,t) = 0 ,\ t > 0$$

$$u(x,0) = x(2 - x) ,\ 0 < x < 2$$

$$\frac{\partial u}{\partial t}(x,0) = \cos 4\pi x ,\ 0 < x < 2 .$$

C. Problems for Discussion and Exploration

1. Let w be a harmonic function in a planar region, and let D be any disc entirely contained in this region. Prove that the value of w at the center of D is the average of the values of w on D.

2. Let w be a real-valued, twice continuously differentiable function on planar region U. Suppose that both w and w^2 are harmonic. Prove that w must be constant.

3. It is a fact (not obvious) that if u is harmonic on a connected region U in the plane and if $(x_0, y_0) \in U$ then u has a convergent power series about (x_0, y_0) of the form

    $$u(x, y) = \sum_{j,k} a_{j,k}(x - x_0)^j (y - y_0)^k$$

    $$\text{for } |(x - x_0)^2 + (y - y_0)^2| < \epsilon, \text{ some small } \epsilon > 0.$$

    Use this information to show that if u vanishes on some disc in U then u is identically 0 on all of U.

4. Use methods introduced in this chapter to find a solution of the boundary value problem

    $$\frac{\partial u}{\partial t} = \frac{\partial^2 u}{\partial x^2} + \frac{\partial^2 u}{\partial y^2}, \quad 0 < x < 1, 0 < y < 1, t > 0$$

    $$\frac{\partial u}{\partial x}(0, y, t) = \frac{\partial u}{\partial x}(1, y, t) = 0, \quad 0 < y < 1, t > 0$$

    $$u(x, 0, t) = u(x, 1, t) = 0, \quad 0 < x < 1, t > 0$$

    $$u(x, y, 0) = f(x, y), \quad 0 < x < 1, 0 < y < 1.$$

5. A vibrating circular membrane, or drum, of radius 1 with edges held fixed in the plane and with displacement $u(r, t)$ (r is radius and t is time) is given. That is to say, the displacement of any point of the drum depends only on the distance from the center of the drum and the elapsed time. This situation is described by the boundary value problem

    $$\frac{\partial^2 u}{\partial t^2} = \alpha^2 \left(\frac{\partial^2 u}{\partial r^2} + \frac{1}{r}\frac{\partial u}{\partial r} \right), \quad 0 < r < 1, t > 0$$

    $$u(1, t) = 0, \quad t > 0$$

    $$u(r, t) \quad \text{remains bounded as } r \to 0^+$$

    $$u(r, 0) = f(r), \quad 0 < r < 1$$

    $$\frac{\partial u}{\partial t}(r, 0) = g(r), \quad 0 < r < 1.$$

    Here f is the initial displacement and g is the initial velocity. Use the method of separation of variables, as introduced in this chapter, to find a solution of this boundary value problem. [**Hint:** Bessel functions will be involved.]

Table of Notation

Notation	Section	Meaning
m	1.1	mass
$\mathbf{F}$	1.1	force
g	1.1	gravitational constant
k	1.1	constant of proportionality
$F(x, y, dy/dx, d^2y/dx^2, \ldots, d^n f/dx^n)$	1.3	differential equation (ODE)
dy/dx	1.4	Leibniz notation
$y' + a(x)y = b(x)$	1.5	first-order, linear equation
$e^{\int a(x)\,dx}$	1.5	integrating factor
$M(x, y)dx + N(x, y)dy = 0$	1.6	exact equation
$\partial f/\partial x = M,\ \partial f/\partial y = N$	1.6	exactness condition
$\partial M/\partial y = \partial N/\partial x$	1.6	exactness criterion
$g(tx, ty) = t^\alpha g(x, y)$	1.8	homogeneity condition
$\mu(x) = e^{\int g(x)\,dx}$	1.9	integrating factor
$\triangle s$	1.11	increment of arc length
E	1.12	electromotive force
I	1.12	current
R	1.12	resistance
L	1.12	inductance
C	1.12	capacitance
Q	1.12	charge
$ay'' + by' + cy = d$	2.1	second-order, linear ODE
$v_1'y_1 + v_2'y_2 = 0$ $v_1'y_1' + v_2'y_2' = r$	2.3	variation of parameters
$y_2 = \left[\int 1/y_1^2 \cdot e^{-\int p(x)\,dx}\,dx\right] \cdot y_1$	2.4	second solution in terms of first
F_s	2.5	spring force
M	2.5	mass
k	2.5	Hooke's constant

Notation	Section	Meaning
x_0	2.5	amplitude
f	2.5	frequency
T	2.5	period
F_d	2.5	damping force
F_e	2.5	external force
$\mathcal{R}(t)$	2.6	position of planet
$\mathbf{F}$	2.6	force
m	2.6	mass
a	2.6	acceleration
G	2.6	universal gravitational constant
$y^{(n)} + a_{n-1}y^{(n-1)} + \cdots$ $+a_1 y^{(1)} + a_0 y = f$	2.7	higher-order linear equation
y_p	2.7	a particular solution
y_g	2.7	general solution to the homogeneous equation
$y_g = A_1 y_1 + A_2 y_2 + \cdots$ $+A_{n-1}y_{n-1} + A_n y_n$	2.7	the general solution
J_n	2.8	nth Bessel function
$p(x) = a_0 + a_1 x + a_2 x^2 + \cdots + a_n x^n$	3.1	a polynomial
$\sum_{j=0}^{\infty} a_j x^j$	3.1	a power series
$\sum_{j=0}^{\infty} a_j (x-a)^j$	3.1	a power series centered at a
$a_j = f^{(j)}(0)/j!$	3.1	formula for power series coefficients
$n!$	3.1	n factorial
$f(x) = \sum_{j=0}^{n} f^{(j)}(0)/j! \cdot x^j + R_n(x)$	3.1	Taylor expansion
$R_n(x) = f^{(n+1)}(\xi)/(n+1)! \cdot x^{n+1}$	3.1	remainder term for Taylor expansion
$c_m = \sum_{j=0}^{m} a_j b_{m-j}$	3.1	Cauchy product
$y = x^m \sum_{j=0}^{\infty} a_j x^j$	3.5	Frobenius solution at a regular singular point
$f(m) = m(m-1) + mp_0 + q_0 = 0$	3.5	equation for m in Frobenius solution
T	3.6	steady-state temperature
$\int_a^b y_m(x) y_n(x)\, dx = 0$	4.1	orthogonal functions

Notation	Section	Meaning
$d/dx(p(x)dy/dx) + [\lambda q(x) + r(x)]y = 0$	4.1	a Sturm-Liouville equation
λ_m	4.1	eigenvalues
y_m	4.1	eigenfunctions
$d/dx(xdy/dx) + (k^2 x - n^2/x)y = 0$	4.2	Bessel's equation
$y'' + (\lambda + 16d\cos 2x)y = 0$	4.2	Mathieu's equation
$u_{tt} = a^2 u_{xx}$	4.3	the wave equation
$(1 - x^2)y'' - 2xy' + \ell(\ell + 1)y = 0$	4.4	Legendre's equation
$(1 - x^2)y'' - xy' + n^2 y = 0$	4.4	Chebyshev's equation
$y'' - 2xy' + 2\alpha y = 0$	4.4	Hermite's equation
$xy'' + (1 - x)y' + \alpha y = 0$	4.4	Lagrange's equation
$\psi(\mathbf{r}, t)$	4.5	state function in quantum mechanics
J	4.5	Joule
H	4.5	linear operator from quantum mechanics
$\overline{E}_n$	5.2	total relative error
$f(x) = \frac{1}{2}a_0 + \sum_{j=1}^{\infty} (a_j \cos jx + b_j \sin jx)$	6.1	a Fourier series
a_j	6.1	a Fourier coefficient
b_j	6.1	a Fourier coefficient
$a_0 = \frac{1}{\pi} \int_{-\pi}^{\pi} f(x)\, dx$	6.1	formula for a_0
$a_j = \frac{1}{\pi} \int_{-\pi}^{\pi} f(x) \cos jx\, dx$	6.1	formula for a_j
$b_j = \frac{1}{\pi} \int_{-\pi}^{\pi} f(x) \sin jx\, dx$	6.1	formula for b_j
$S_N(f)(x)$	6.2	partial sum of Fourier series
$\sigma_N(f)(x)$	6.2	Cesàro means of a Fourier series
$\widetilde{g}(x)$	6.3	odd extension of the function g
$\widetilde{\widetilde{g}}$	6.3	even extension of the function g
L	6.4	the length of an interval
$\|\mathbf{u} \cdot \mathbf{v}\| \leq \|\mathbf{u}\| \cdot \|\mathbf{v}\|$	6.5	Cauchy-Schwarz -Buniakovski inequality
$\|\mathbf{u} + \mathbf{v}\| \leq \|\mathbf{u}\| + \|\mathbf{v}\|$	6.5	triangle inequality
$C[0, 1]$	6.5	the space of continuous functions on $[0, 1]$
$\widehat{f}(\xi)$	6.6	the Fourier transform of f
ρ	6.6	a rotation
$\overline{f}$	6.6	the conjugate of f
$\widetilde{f}$	6.6	the reflection of f

Notation	Section	Meaning
$L[f](s)$	7.1	the Laplace transform of f
$L[y'](s) = pL[y](s) - y(0)$	7.1	Laplace transform of the derivative
$L[y''](s) = s^2 L[y](s) - py(0) - y'(0)$	7.1	Laplace transform of the second derivative
$L[e^{ax}f(x)](s) = F(p-a)$	7.1	the Laplace transform and translations
$\frac{d}{ds}F(s) = L[-xf(x)](s)$	7.2	derivative of the Laplace transform
$\frac{d^2}{ds^2}F(s) = L[x^2 f(x)](s)$	7.2	second derivative of the Laplace transform
$\frac{d^j}{ds^j}F(s) = L[(-1)^j x^j f(x)](s)$	7.2	jth derivative of the Laplace transform
$L[xy] = -dY/ds$	7.2	the Laplace transform and the derivative
$L[xy'] = -(d/ds)[pY]$	7.2	the Laplace transform and the derivative
$L[xy''] = -(d/ds)[s^2 Y] + y(0)$	7.2	the Laplace transform and the second derivative
$f * g(x) = \int f(x-t)g(t)\,dt$	7.3	convolution
$L[f*g](s) = L[f](s) \cdot L[g](s)$	7.3	Laplace transform of the convolution
φ_ϵ	7.3	approximation to the impulse function
$\mathrm{erfc}(x) = (2/\sqrt{\pi})\int_0^x e^{-t^2}\,dt$	7.5	the erf function
$\mathcal{U}$	7.5	state constant
$\mathcal{S}$	8.1	the space of Schwartz distributions
$\rho_{\alpha,\beta}$	8.1	seminorms on the Schwartz space
α	8.1	a Schwartz distribution
T_f	8.1	the distribution induced by the function f
$\eta_{K,\alpha}$	8.1	norm on a space of distributions
$\mathcal{D}$	8.1	the space C_c^∞
$\mathcal{E}$	8.1	the space of C^∞ functions
$\mathcal{D}'$	8.1	a space of distributions
$\mathcal{E}'$	8.1	a space of distributions
$\varphi = \chi_{[0,1)}$	9.3	the scaling function
$\psi(x) = \varphi(2x) - \varphi(2x-1)$	9.3	the wavelet

Notation	Section	Meaning
$\{V_j\}$	9.3	scale of function spaces
$\tau_a f(x) = f(x - a)$	9.3	the translation operator
$\alpha_\delta f(x) = f(\delta x)$	9.3	the dilation operator
W_j	9.3	the orthogonal complement of V_j in V_{j+1}
$\{2^{j/2} \alpha_{2^j} \tau_m \psi\}$	9.3	an explicit orthonormal basis for W_j
$L^2 = \bigoplus_{j \in \mathbb{Z}} W_j$	9.3	orthonormal decomposition of L^2
$\{2^{j/2} \alpha_{2^j} \tau_m \psi\}$	9.3	an orthonormal basis for L^2
$L^2 = V_0 \oplus \bigoplus_{j=0}^{\infty} W_j$	9.4	orthonormal decomposition for L^2
$a_n(m), b_n(m), c_k$	9.4	coefficients of the wavelet expansion
$m(\xi) = \frac{1}{\sqrt{2}} \sum_j c_j e^{ijt}$	9.5	a low-pass filter
$\{e_j\}$	9.5	an unconditional basis
$\varphi_{j,k}(t) = 2^{j/2} \varphi(2^j t - k)$	9.6	an orthonormal basis for V_j
$I_{j,k} = \left[\frac{k}{2^j}, \frac{k+1}{2^j}\right)$	9.6	the support of $\psi_{j,k}$
W_0	9.6	the Haar wavelet space
$\varphi(t) = \varphi(2t) + \varphi(2t - 1)$ $\psi(t) = \varphi(2t) - \varphi(2t - 1)$	9.6	dilation equations
$V \oplus W$	9.6	direct sum of V and W
$\psi_{j,k}(t) = 2^{j/2} \psi(2^j t - k)$	9.6	a wavelet function
W_j	9.6	the Haar wavelet space
$\{\psi_{j,k}\}$	9.6	an orthonormal basis for W_j
$\mathbf{h} = {}^t\left[\frac{\sqrt{2}}{2}, \frac{\sqrt{2}}{2}\right], \mathbf{g} = {}^t\left[\frac{\sqrt{2}}{2}, -\frac{\sqrt{2}}{2}\right]$	9.6	basis vectors for dilation equation
$\varphi_{j,k}(t) = \frac{\sqrt{2}}{2} \varphi_{j+1,2k}(t) + \frac{\sqrt{2}}{2} \varphi_{j+1,2k+1}(t)$	9.7	dilation equation

Notation	Section	Meaning
$\psi_{j,k}(t) = \frac{\sqrt{2}}{2}\varphi_{j+1,2k}(t) - \frac{\sqrt{2}}{2}\varphi_{j+1,2k+1}(t)$	9.7	dilation equation
$g_j = f_{j+1} - f_j$	9. 7	error term
$V_{j+1} = V_j \oplus W_j$	9.7	fundamental orthogonal decomposition
f_j	9.8	an element of V_j
g_j	9.8	an element of W_j
Q_j, G_j, H_j	9.9	the wavelet transform in matrix form
MAD	9.9	median absolute deviation
$C(\mathbf{v})$	9.10	cumulative energy vector
$\text{Ent}(\mathbf{v})$	9.10	the entropy of $\mathbf{v}$
λ_m	10.2	eigenvalues
φ_m	10.2	eigenfunctions
$\partial_t u = \eta^2 \partial_x^2 u$	10.3	the heat equation
$\triangle = \partial^2/\partial x^2 + \partial^2/\partial y^2$	10.4	the Laplacian
$\partial^2/\partial r^2 + (1/r)\partial/\partial r + (1/r^2)\partial^2/\partial\theta^2$	10.4	the Laplacian in polar coordinates
$P_r(\theta) = \frac{1}{2\pi} \cdot \frac{1-r^2}{1-2r\cos\theta+r^2}$	10.4	the Poisson kernel

Glossary

Abel's mechanics problem Given a nonnegative function $T(y)$ that describes how long it takes a bead to slide down a wire from height y, determine the shape of the curved wire that gives rise to the function T.

additive identity An element $\mathbf{0}$ of the vector space V such that $\mathbf{v} + \mathbf{0} = \mathbf{v}$ for every $\mathbf{v} \in V$.

additive inverse Given an element $\mathbf{v}$ of the vector space V, this is an element $-\mathbf{v} \in V$ such that $\mathbf{v} + (-\mathbf{v}) = \mathbf{0}$.

associated homogeneous equation Given the differential equation

$$ay'' + by' + cy = f,$$

the associated homogeneous equation is

$$ay'' + by' + cy = 0.$$

balls In a normed linear space, we have open balls

$$B(\mathbf{x}, r) = \{\mathbf{t} \in X : \|\mathbf{t} - \mathbf{x}\| < r\}$$

and closed balls

$$\overline{B}(\mathbf{x}, r) = \{\mathbf{t} \in X : \|\mathbf{t} - \mathbf{x}\| \leq r\}.$$

Bessel function A special function that arises in the solution of the Bessel differential equation

$$xy'' + xy' + (x^2 - p^2)y = 0.$$

Bessel's equation The differential equation

$$xy'' + xy' + (x^2 - p^2)y = 0.$$

Bessel's inequality

Theorem: If $\{u_\alpha : \alpha \in A\}$ is any orthonormal set in the Hilbert space H, and if $\widehat{x}(\alpha) \equiv \langle x, u_\alpha \rangle$ for each α, then

$$\sum_{\alpha \in A} |\widehat{x}(\alpha)|^2 \le \|x\|^2 \,.$$

boundary conditions   If we are given a differential equation on an interval $[a, b]$, then boundary conditions specify values for the solution at the points a and b.

capacitance   The force which stores electrical energy.

catenary   The curve that describes the shape of a hanging chain.

Cauchy product   A device for calculating the product of two power series.

Cauchy-Schwarz-Buniakovski inequality   This is the inequality

$$|\langle \mathbf{v}, \mathbf{w} \rangle| \le \|\mathbf{v}\| \cdot \|\mathbf{w}\| \,.$$

Chebyshev's equation   This is the differential equation

$$(1 - x^2)y'' - xy' + n^y = 0 \,.$$

C^∞ Urysohn lemma

Lemma: Let K and L be disjoint closed sets in $\mathbb{R}^N$. Then there is a C^∞ function φ on $\mathbb{R}^N$ such that $\varphi \equiv 0$ on K and $\varphi \equiv 1$ on L.

complex power series   A power series with complex coefficients and a complex variable.

constant perturbation method for linear, second-order equations   A numerical method for solving differential equations in which one approximates the given equation by a constant-coefficient equation.

convergence of a power series   A power series converges at a point x if

$$\lim_{k \to +\infty} \sum_{j=0}^{k} a_j x^j$$

converges.

convex set A set E in a linear space X is convex if, whenever $x, y \in E$ and $0 \le t \le 1$, then

$$(1 - t)x + ty \in E.$$

convolution If f and g are integrable functions then their convolution is defined to be

$$f * g(x) = \int f(x - t)g(t)\, dt = \int f(t)g(x - t)\, dt.$$

critical point A point at which both dx/dt and dy/dt equal 0.

cumulative energy Let $\mathbf{v} \in \mathbb{R}^N$ with $\mathbf{v} \ne 0$. Assume that $\mathbf{y}$ is the vector formed by taking the absolute value of each component of $\mathbf{v}$ and ordering the resulting nonnegative numbers from largest to smallest. Then we define the cumulative energy vector $C(\mathbf{v})$ to be a vector in $\mathbb{R}^N$ whose components are given by

$$C(\mathbf{v})_j = \sum_{\ell=1}^{j} \frac{y_\ell^2}{\|\mathbf{y}\|^2} \qquad \text{for} \quad j = 1, 2, \ldots, N.$$

d'Alembert's solution to the wave equation This solution has the form

$$y(x, t) = \frac{1}{2}(f(t + x) + g(t - x)).$$

damped vibrations A simple harmonic motion with a damping force present.

Daubechies wavelet A wavelet that is smooth and compactly supported.

dialysis machine A machine designed by biomedical engineers to emulate the function of a kidney.

differential equation An equation involving a function and some of its derivatives.

dilation An action on Euclidean space induced by multiplying each variable by a fixed constant. Dilation by δ is often denoted by α_δ.

dilation operators of Euclidean analysis   These are the operators, for f a function on $\mathbb{R}^N$, given by

$$\alpha_\delta f(x) = f(\delta x) \qquad \text{and} \qquad \alpha^\delta f(x) = \delta^{-N} f(x/\delta).$$

Dirac delta mass   This is the distribution that evaluates each test function at the origin. Intuitively this is a function that takes the value 0 at $x \neq 0$ and takes the value $+\infty$ at the origin.

Dirichlet kernel   A kernel for summing Fourier series:

$$D_N(t) \equiv \frac{\sin\left(N + \frac{1}{2}\right)t}{\sin\frac{1}{2}t}.$$

Dirichlet problem   The basic boundary value problem for the Laplacian. On the disc, for instance, we specify boundary values for a function u on the unit circle and ask that u be harmonic on the interior.

Dirichlet's theorem

Theorem: Let f be a function on $[-\pi, \pi]$ which is piecewise continuous. Assume that each piece of f is monotone. Then the Fourier series of f converges at each point of continuity c of f in $[-\pi, \pi]$ to $f(c)$. At other points x it converges to $[f(x^-) + f(x^+)]/2$.

discretization error   This is

$$\epsilon_n = y(x_n) - y_n.$$

divergence of a power series   A power series diverges at a point x if it does not converge at x.

eigenfunction   If $T : X \to X$ is a linear operator, then an eigenfunction or eigenvector for T is a vector $\mathbf{v}$ such that $T\mathbf{v} = \lambda\mathbf{v}$ for some scalar λ.

eigenvalue   If $T : X \to X$ is a linear operator, then an eigenvalue for T is a scalar λ such that $T\mathbf{v} = \lambda\mathbf{v}$ for some vector $\mathbf{v}$.

electromotive force   The force that produces an electrical current.

elementary transcendental function   The familiar calculus functions sine, cosine, exponential, logarithm, their inverses and combinations.

entropy   Let $\mathbf{v} = \langle v_1, v_2, \ldots, v_N \rangle$ be an N-vector. Suppose that the v_j assume k distinct values, $1 \leq k \leq N$. Denote these distinct values by $a_1, a_2, \ldots, a_k$ and let $p(a_j)$ be the relative frequency of a_j in $\mathbf{v}$. That is to say $0 \leq p(a_j) \leq 1$ is the number of times a_j occurs in $\mathbf{v}$ divided by N. Then the entropy of $\mathbf{v}$ is defined to be the quantity

$$\text{Ent}(\mathbf{v}) = \sum_{\ell=1}^{k} p(a_\ell) \cdot \log_2(1/p(a_\ell)).$$

equation of biological growth   The differential equation that describes a petri dish of bacteria or other life form that reproduces regularly.

equation of exponential decay   The differential equation that describes radioactive decay.

essentially diagonal   A linear operator T is essentially diagonal with respect to a wavelet basis if the matrix entries die off rapidly away from the diagonal.

Euclidean 3-space   The collection of all triples (x, y, z), where $x, y, z \in \mathbb{R}$.

Euler method   A numerical method for solving differential equations that is encapsulated in the formula

$$y_{k+1} = y_k + h \cdot f(x_k, y_k).$$

Euler's formula   This is the formula

$$e^{iy} = \cos y + i \sin y.$$

even function   We say that f is even if $f(-x) = f(x)$ for every x.

exact equation   A differential equation that can be written in the form $M(x, y)dx + N(x, y)dy = 0$, with $M = \partial f / \partial x$ and $N = \partial f / \partial y$ for some function f.

exactness condition   This is the condition

$$\frac{\partial M}{\partial y} = \frac{\partial N}{\partial x}.$$

Fejér's theorem

Theorem: Let f be a piecewise continuous function on $[-\pi, \pi]$—meaning that the graph of f consists of finitely many continuous curves. Let s be the endpoint of one of those curves, and assume that $\lim_{x \to p^-} f(x) \equiv f(s^-)$ and $\lim_{x \to s^+} f(x) \equiv f(s^+)$ exist. Then the Cesàro means of the Fourier series of f at s converges to $[f(s^-) + f(s^+)]/2$.

first-order, linear equation   A differential equation of the form $y' + a(x)y = b(x)$.

forced vibrations   A simple harmonic motion with an external force acting on it.

Fourier coefficients   These are

$$a_j = \frac{1}{\pi} \int_{-\pi}^{\pi} f(x) \cos jx \, dx \qquad \text{and} \qquad b_j = \frac{1}{\pi} \int_{-\pi}^{\pi} f(x) \sin jx \, dx \,.$$

Fourier coefficients with respect to an orthonormal system   Let H be a Hilbert space and $\{u_\alpha\}$ an orthonormal system in H. If $x \in H$, then the Fourier coefficients of x with respect to the orthonormal system are

$$\widehat{x}(\alpha) = \langle x, u_\alpha \rangle \,.$$

Fourier inversion   The Fourier transform is univalent on the space of integrable functions. Its inverse, in Euclidean N-space, is given by

$$f(y) = (2\pi)^{-N} \int \widehat{f}(\xi) e^{-iy \cdot \xi} \, d\xi \,.$$

Fourier series of a function   The expansion of a given function f into sines and cosines.

Fourier-Stieltjes series   The Fourier series of a measure.

Fourier transform   Let f be an integrable function on Euclidean space. Then the Fourier transform of f is

$$\widehat{f}(\xi) = \int_{\mathbb{R}^N} f(t) e^{it \cdot \xi} \, dt \,.$$

Fourier transform of a Schwartz distribution   Let λ be a Schwartz distribution and φ a Schwartz function. Then

$$\widehat{\lambda}(\varphi) \equiv \lambda(\widehat{\varphi}) \,.$$

Frobenius solution of a differential equation   A differential equation
solution of the form

$$y = y(x) = x^m \cdot \left(a_0 + a_1 x + a_2 x^2 + \cdots\right),$$

where m may not be an integer.

function defined by a power series   On the interval of convergence, the
power series

$$\sum_{j=0}^{\infty} a_j x^j$$

defines a function f.

Gaussian white noise   A signal composed of independent samples from a
normal distribution with mean 0 and variance σ^2.

Gauss-Weierstrass summation   A technique for summing the inverse
Fourier transform.

general solution of a differential equation   Usually a family of functions
that describes all possible solutions to the differential equation.

geometric interpretation of the solution   This is usually the interpretation of the solution of a differential equation in terms of vector fields.

Green's function   Let $\mathcal{L}$ be a linear differential operator. A Green's function
is any solution of the equation

$$\mathcal{L}_x G(x, y) = \delta(y - x).$$

Haar basis   The wavelet basis generated by step functions.

hanging chain   An interesting physical system is a chain, fixed at both
ends, and hanging under its own weight. We can analyze such a system using
a differential equation.

heat equation   This is the partial differential equation from physics that
says

$$\frac{\partial y}{\partial t} = \eta^2 \frac{\partial^2 y}{\partial x^2}.$$

Hermite equation   This is the differential equation

$$y'' - 2xy' + 2\alpha y = 0.$$

Hermite polynomial   A function defined by

$$H_j(x) = (-1)^j e^{x^2} \frac{d^j}{dx^j} \left(e^{-x^2}\right).$$

Heun's method   See *improved Euler method.*

higher-order coupled harmonic oscillators   A motion described by the equations

$$m_1 \frac{d^2 x_1}{dt^2} = -k_1 x_1 + k_3(x_2 - x_1).$$
$$m_2 \frac{d^2 x_2}{dt^2} = -k_2 x_2 - k_3(x_2 - x_1).$$

These can be combined to yield a fourth-order differential equation.

higher transcendental functions   Transcendental functions defined by power series and that are not elementary.

Hilbert transform   The important linear operator given by

$$f \longmapsto \int \frac{f(t)}{x - t} \, dt.$$

homogeneous equation   This could mean a differential equation whose right-hand side, or forcing term, is 0. It could also mean a differential equation whose coefficients satisfy a certain homogeneity condition.

homogeneous of degree α   A function g is homogeneous of degree α if $g(bx, by) = b^\alpha g(x, y)$.

Hooke's law   The force exerted by a spring is proportional to its displacement.

imaginary number Any multiple of i.

implicitly defined solution The solution of a differential equation that is expressed as an equation in x and y, but *not* in the form $y = f(x)$.

improved Euler method The numerical technique encapsulated by

$$y_{j+1} = y_j + \frac{h}{2} \cdot [f(x_j, y_j) + f(x_{j+1}, z_{j+1})],$$

where

$$z_{j+1} = y_j + h \cdot f(x_j, y_j)$$

and $j = 0, 1, 2, \ldots$.

impulse function Another name for the Dirac delta mass. Intuitively this is a function that equals 0 for $x \neq 0$ and takes the value $+\infty$ at the origin.

impulsive response The solution of a differential equation when the input is the Dirac delta mass.

independent solutions These are usually solutions that are linearly independent.

indicial equation The equation

$$2m(m-1) + m - 1 = 0$$

for the index m in the Frobenius method. More generally, for the differential equation

$$y'' + py' + qy = 0,$$

the indicial equation is

$$m(m-1) + mp_0 + q_0.$$

inductance The force which opposes any change in current.

initial conditions Restrictions, usually imposed through values of the function or its derivatives at a particular point, that restricts the choice of solution to a differential equation. Often the initial condition(s) are physically motivated.

inner product   If V is a vector space, then an inner product on V is a mapping
$$\langle \bullet, \bullet \rangle : V \times V \to \mathbb{R}$$
with these properties:

  (a) $\langle \mathbf{v}, \mathbf{v} \rangle \geq 0$;

  (b) $\langle \mathbf{v}, \mathbf{v} \rangle = 0$ if and only if $\mathbf{v} = 0$;

  (c) $\langle \alpha \mathbf{v} + \beta \mathbf{w}, \mathbf{u} \rangle = \alpha \langle \mathbf{v}, \mathbf{u} \rangle + \beta \langle \mathbf{v}, \mathbf{u} \rangle$ for any vectors $\mathbf{u}, \mathbf{v}, \mathbf{w} \in V$ and scalars α, β.

inner product space   A linear space equipped with an inner product.

input function   The stimulus of a system.

integrating factor   A factor which we can multiply through a differential equation to make it exact.

interval of convergence   The interval on which a power series converges.

Kepler's laws of planetary motion

   1. The orbit of each planet is an ellipse with the sun at one focus.

   2. The segment from the center of the sun to the center of an orbiting planet sweeps out area at a constant rate.

   3. The square of the period of revolution of a planet is proportional to the cube of the length of the major axis of its elliptical orbit, with the same constant of proportionality for any planet.

Lagrange's equation   This is the differential equation
$$xy'' + (1 - x)y' + \alpha y = 0.$$

Laplace's equation   This is the differential equation
$$\triangle u = f \qquad \text{or} \qquad \left(\frac{\partial^2}{\partial x^2} + \frac{\partial^2}{\partial y^2} \right) u = f.$$

Laplace transform   This is the transform given by
$$L[f](s) = F(s) = \int_0^\infty e^{-sx} f(x)\, dx.$$

Liénard's equation This is the differential equation

$$\frac{dx^2}{dt^2} + f(x)\frac{dx}{dt} + g(x) = 0.$$

linear damping A spring for which the damping force is a linear function of dx/dt.

linear operator A function $T : V \to W$ between vector spaces such that

$$T(c_1\mathbf{x}_1 + c_2\mathbf{x}_2) = c_1 T(\mathbf{x}_1) + c_2 T(\mathbf{x}_2)$$

for any $\mathbf{x}_1, \mathbf{x}_2 \in V$ and any scalars c_1, c_2.

linearization Approximation of a nonlinear system by a linear one.

linear spring A spring for which the restoring force is a linear function of x.

linear transformation A mapping of vector spaces $T : V \to W$ such that

(a) $T(\mathbf{v} + \mathbf{w}) = T(\mathbf{v}) + T(\mathbf{w})$;
(b) $T(c\mathbf{v}) = cT(\mathbf{v})$ for any scalar c.

Maxwell's field equations The law which forms the foundations of classical electromagnetism, classical optics, and electric circuits.

mean absolute deviation The mean absolute deviation or MAD is defined to be

$$\mathrm{MAD}(\mathbf{a}) = \mathrm{median}(|a_1 - a_{\mathrm{med}}|, |a_2 - a_{\mathrm{med}}|, \ldots, |a_N - a_{\mathrm{med}}|).$$

Minkowski functional If E is a convex set in a linear space X, then the Minkowski functional μ_E of E is defined to be

$$\mu_E(x) = \inf\{t > 0 : t^{-1}x \in E\}.$$

multi-resolution analysis A collection of subspaces $\{V_j\}_{j\in\mathbb{Z}}$ of $L^2(\mathbb{R})$ is called a Multi-Resolution Analysis or MRA if the following are true:

MRA_1 **(Scaling)** For each j, the function $f \in V_j$ if and only if $\alpha_2 f \in V_{j+1}$.

MRA_2 **(Inclusion)** For each j, $V_j \subseteq V_{j+1}$.

MRA_3 **(Density)** The union of the V_j's is dense in L^2:

$$\text{closure} \left\{ \bigcup_{j \in \mathbb{Z}} V_j \right\} = L^2(\mathbb{R}) \, .$$

MRA_4 **(Maximality)** The spaces V_j have no non-trivial common intersection:

$$\bigcap_{j \in \mathbb{Z}} V_j = \{0\} \, .$$

MRA_5 **(Basis)** There is a function φ such that $\{\tau_j \varphi\}_{j \in \mathbb{Z}}$ is an orthonormal basis for V_0.

Newton's binomial formula   A formula for $(1 + x)^p$:

$$(1 + x)^p = 1 + px + \frac{p(p-1)}{2!}x + \frac{p(p-1)(p-2)}{3!} + \cdots$$
$$+ \frac{p(p-1)(p-2) \cdots (p-j+1)}{j!} x^j + \cdots \, .$$

Newton's law of universal gravitation   Newton's inverse square law governing the gravitational attraction between two planets.

odd function   We say that f is odd if $f(-x) = -f(x)$ for every f.

Ohm's law   The law that governs electrical resistance.

order of a differential equation   The maximal order derivative that appears in the equation.

ordinary differential equations   A differential equation that involves functions of one variable and ordinary derivatives.

ordinary point   A point at which the coefficients of a differential equation have convergent power series expansions.

orthogonal complement   If x is an element of the Hilbert space H, then its orthogonal complement is

$$x^{\perp} = \{y \in H : \langle x, y \rangle\} = 0 \, .$$

If $E \subseteq H$ is a subspace, then its orthogonal complement is

$$E^\perp = \{y \in H : \langle y, e \rangle = 0 \quad \text{for all} \ e \in E\}.$$

orthogonal functions Two functions f, g are orthogonal if

$$\int f(x)g(x)\, dx = 0.$$

orthogonality condition Two functions f, g are orthogonal if

$$\int f(x)g(x)\, dx = 0.$$

orthogonality condition with weight q Two functions f, g are orthogonal with weight q if

$$\int f(x)g(x)q(x)\, dx = 0.$$

orthogonal trajectory A curve that is pointwise orthogonal to a given family of curves.

orthonormal system Let H be a Hilbert space. Then an orthonormal system in H is a collection of elements $\{u_\alpha\}$ such that $\|u_\alpha\| = 1$ for each α and also $\langle u_\alpha, u_\beta \rangle = 0$ whenever $\alpha \neq \beta$.

output function The response in a physical system.

parallelogram law In an inner product space,

$$\|x + y\|^2 + \|x - y\|^2 = 2\|x\|^2 + 2\|y\|^2.$$

partial differential equation A differential equation that involves functions of several variables and partial derivatives.

particular solution This is a solution of the inhomogeneous equation

$$ay'' + by' + cy = f.$$

path of the system A path of the autonomous system is a function $x = x(t)$,

$y = y(t)$.

Picard's existence and uniqueness theorem　A theorem that expresses the fact that most ordinary differential equations possess solutions. Also, under certain initial conditions, the solution is unique.

Plancherel's theorem　If f is a square integrable function on $\mathbb{R}^N$, then

$$(2\pi)^{-N} \int |\widehat{f}(\xi)|^2 \, d\xi = \int |f(x)|^2 \, dx.$$

Poisson kernel　The integration kernel that solves the Dirichlet problem. On the disc in the plane it is given by

$$P_r(\theta) = \frac{1}{2\pi} \frac{1 - r^2}{1 - 2r \cos\theta + r^2}.$$

potential theory　The study of harmonic functions, subharmonic functions, and related ideas.

power series　A series of the form

$$\sum_{j=0}^{\infty} a_j x^j \quad \text{or} \quad \sum_{j=0}^{\infty} a_j (x - c)^j.$$

pursuit curve　If a dog is chasing a rabbit then its path describes a pursuit curve.

quantum mechanics　This is a fundamental theory in physics which describes nature at the smallest scales of energy levels of atoms and subatomic particles.

radius of convergence　The radius of the interval on which a power series converges.

Rayleigh's problem　The problem of studying the two-dimensional flow of a semi-infinite extent of viscous fluid, supported on a flat plate, caused by the sudden motion of the flat plate in its own plane.

real analytic function　A function that is locally representable by a convergent power series.

real power series   A power series with real coefficients.

recursion   One or more equations that relate later indexed a_js to earlier indexed a_js.

reduction of order   There are various algebraic and notational tricks for reducing the order of a differential equation from 2 to 1, or more generally from order k to order $k-1$. This usually facilitates the solving of the equation.

regular singular point   We say that a singular point x_0 for the differential equation

$$y'' + p \cdot y' + q \cdot y = 0$$

is a regular singular point if

$$(x - x_0) \cdot p(x) \qquad \text{and} \qquad (x - x_0)^2 q(x)$$

are analytic at x_0.

resistance   The force which opposes an electrical current.

Riemann-Lebesgue lemma   If f is an integrable function, then

$$\lim_{\xi \to \pm\infty} |\widehat{f}(\xi)| = 0.$$

Riesz–Fisher theorem

Theorem: Let $\{u_\alpha\}_{\alpha \in A}$ be a complete orthonormal system in H. Let $\varphi \in \ell^2(A)$. Then $\varphi = \widehat{x}$ for some $x \in H$.

Riesz representation theorem

Theorem: If λ is a bounded linear functional on the Hilbert space H, then there is a unique element $y \in H$ such that

$$\lambda x = \langle x, y \rangle$$

for all $x \in H$.

rotation   An action on Euclidean space induced by a special orthogonal matrix. Rotations are often denoted by ρ.

round-off error   The error that results from rounding off decimals.

Runge–Kutta method A numerical method for solving differential equations that can be described as follows: Let

$$m_1 = h \cdot f(x_k, y_k)$$

$$m_2 = h \cdot f\left(x_k + \frac{h}{2}, y_k + \frac{m_1}{2}\right)$$

$$m_3 = h \cdot f\left(x_k + \frac{h}{2}, y_k + \frac{m_2}{2}\right)$$

$$m_4 = h \cdot f(x_k + h, y_k + m_3).$$

Then y_{k+1} is given by

$$y_{k+1} = y_k + \frac{1}{6}(m_1 + 2m_2 + 2m_3 + m_4).$$

scaling function The function φ in the development of the Haar basis.

Schwartz distribution A continuous linear functional on the Schwartz space.

Schwartz space The space of functions

$$\mathcal{S} = \left\{\varphi \in C^\infty(\mathbb{R}^N) : \rho_{\alpha,\beta}(\varphi) \equiv \sup_x \left|x^\alpha \cdot \frac{\partial^\beta}{\partial x^\beta}\varphi(x)\right| < \infty, \right.$$

$$\left. \alpha = (\alpha_1, \ldots, \alpha_N), \beta = (\beta_1, \ldots, \beta_N)\right\}.$$

Schwarz inequality This is the inequality

$$|\langle \mathbf{v}, \mathbf{w} \rangle| \leq \|\mathbf{v}\| \cdot \|\mathbf{w}\|.$$

second-order, linear equation A differential equation of the form

$$ay'' + by' + cy = d.$$

separable equation An ordinary differential equation which can be written so that all the independent variables are on one side of the equation and all the dependent variables are on the other side of the equation.

simple critical point An isolated critical point at which linearization is a good approximation.

singular Sturm-Liouville problem   A Sturm-Liouville problem in which we allow p and q to vanish at the endsoints of the interval of study.

solution of a differential equation   A function, or family of functions, that satisfy the differential equation.

state function   In quantum mechanics, the function which represent the state of a point in space at time t.

stationary fluid motion   The solution of an autonomous system (which is independent of t).

steady-state solution   This is a solution u that does not change with time, so that $\partial u / \partial t = 0$.

step function   Also known as the heaviside function, this is the function

$$u(t) = \begin{cases} 0 & \text{if} \quad t < 0 \\ 1 & \text{if} \quad t \geq 0. \end{cases}$$

Sturm-Liouville problem   A differential equation of the form

$$\frac{d}{dx}\left(p(x)\frac{dy}{dx}\right) + [\lambda q(x) + r(x)]y = 0.$$

We typically assume that p and q are non-vanishing.

support of a distribution   The complement of the union of all open sets U such that $\mu(\varphi) = 0$ for all elements of C_c^∞ that are supported in U.

time-independent Schrödinger equation   This is the equation

$$-\frac{\overline{h}^2}{2m}\nabla^2\alpha + (V(\mathbf{r}) - \mu)\alpha = 0.$$

total relative error   The quantity

$$\overline{E}_n = \frac{|y(x_n) - y_n|}{|y(x_n)|}.$$

tractrix   An example of a pursuit curve.

transcendental function   A function that is not polynomial, rational, or a root.

translation   An action on Euclidean space induced by adding a fixed constant to each variable. Translation by a is often denoted τ_a.

translation-invariant operator   A linear operator T on functions such that
$$T(\tau_a f)(x) = (\tau_a T f)(x).$$

triangle inequality   This is the inequality
$$\|\mathbf{v} + \mathbf{w}\| \leq \|\mathbf{v}\| + \|\mathbf{w}\|.$$

trigonometric series   A series of the form
$$f(x) = \frac{1}{2}a_0 + \sum_{n=1}^{\infty} \left(a_n \cos nx + b_n \sin nx \right).$$

unconditional basis   A set of vectors $\{e_0, e_1, \dots\}$ in a Banach space X is called an *unconditional basis* if it has the following properties:

For each $x \in X$ there is a unique sequence of scalars $\alpha_0, \alpha_1, \dots$ such that
$$x = \sum_{j=0}^{\infty} \alpha_j e_j,$$

in the sense that the partial sums $S_N \equiv \sum_{j=0}^{N} \alpha_j e_j$ converge to x in the topology of the Banach space.

There exists a constant C such that, for each integer m, for each sequence $\alpha_0, \alpha_1, \dots$ of coefficients as above, and for any sequence $\beta_0, \beta_1, \dots$ satisfying $|\beta_k| \leq |\alpha_k|$ for all $0 \leq k \leq m$, we have
$$\left\| \sum_{k=0}^{m} \beta_k e_k \right\| \leq C \cdot \left\| \sum_{k=0}^{m} \alpha_k e_k \right\|.$$

undamped simple harmonic motion   The motion of a mass attached to a spring with no interference or damping force.

undetermined coefficients A method of organized guessing used to solve differential equations.

van der Pol equation This is the equation

$$\frac{d^2x}{dt^2} + \mu(x^2 - 1)\frac{dx}{dt} + x = 0.$$

variation of parameters A solution method for differential equations that consists of applying algebraic techniques to the solutions of the associated homogeneous equation.

wave equation This is the partial differential equation from physics that says

$$\frac{\partial^2 y}{\partial t^2} = a^2 \frac{\partial^2 y}{\partial x^2}.$$

wavelet The function ψ in the development of the Haar basis.

wavelet basis The collection

$$\mathcal{H} \equiv \left\{ 2^{j/2}\alpha_{2^j}\tau_m\psi : m, j \in \mathbb{Z} \right\}$$

is an orthonormal basis for L^2, and will be called a *wavelet basis* for L^2.

wavelet transform The decomposition of a function f in terms of the spaces V_j, W_j.

Wronskian determinant If y_1, y_2 are solutions of a differential equation, then their Wronskian is

$$W = \det \begin{pmatrix} y_1 & y_2 \\ y_1' & y_2' \end{pmatrix}.$$

Bibliography

[BLK]   B. E. Blank and S. G. Krantz, *Calculus with Analytic Geometry*, Key Curriculum Press, Emeryville, CA, 2005.

[DAU]   I. Daubechies, *Ten Lectures on Wavelets*, SIAM, Philadelphia, PA, 1992.

[DES]   M. H. DeGroot and M. J. Schervish, *Probability and Statistics*, 3rd ed., Addison-Wesley, Reading, MA, 2002.

[DER]   J. Derbyshire, *Prime Obsession: Bernhard Riemann and the Greatest Unsolved Problem in Mathematics*, Joseph Henry Press, Washington, DC, 2003.

[DOJ]   D. Donoho and I. Johnstone, Ideal spatial adaptation via wavelet shrinkage, *Biometrika* 81(1994), 425–455.

[FEF]   C. Fefferman, The uncertainty principle, *Bull. A.M.S.* 9(1983), 129–206.

[FOU]   J. Fourier, *The Analytical Theory of Heat*, G. E. Stechert & Co., New York, 1878.

[GER]   C. F. Gerald, *Applied Numerical Analysis*, Addison-Wesley, Reading, MA, 1970.

[GRK]   R. E. Greene and S. G. Krantz, *Function Theory of One Complex Variable*, 2nd ed., American Mathematical Society, Providence, RI, 2002.

[HAM]   F. Hampel, The influence curve and its role in robust estimation, *J. Am. Stat. Association* 69(1974), 383–393.

[HERG]   E. Hernandez and G. Weiss, *A First Course in Wavelets*, CRC Press, Boca Raton, FL, 1996.

[HIL]   F. B. Hildebrand, *Introduction to Numerical Analysis*, Dover, New York, 1987.

[HIR]   M. Hirsch, *Differential Topology*, Springer-Verlag, 1976.

[HOR]   L. Hörmander, *Linear Partial Differential Operators*, Springer-Verlag, Heidelberg, 1963.

[ISK]   E. Isaacson and H. Keller, *Analysis of Numerical Methods*, J. Wiley and Sons, New York, 1966.

[KAT]   Y. Katznelson, *An Introduction to Harmonic Analysis*, John Wiley & Sons, New York, 1968.

[KBO]   A. C. King, J. Billingham, and S. R. Otto, *Differential Equations: Linear, Nonlinear, Ordinary, Partial*, Cambridge University Press, Cambridge, 2003.

[KNO]   K. Knopp, *Elements of the Theory of Functions*, Dover, New York, 1952.

[KRA1] S. G. Krantz, *Complex Analysis: The Geometric Viewpoint*, 2nd ed., Mathematical Association of America, Washington, DC, 2003.

[KRA2] S. G. Krantz, *Real Analysis and Foundations*, CRC Press, Boca Raton, FL, 1992.

[KRA3] S. G. Krantz, *A Panorama of Harmonic Analysis*, Mathematical Association of America, Washington, DC, 1999.

[KRA4] S. G. Krantz, *Partial Differential Equations and Complex Analysis*, CRC Press, Boca Raton, FL, 1992.

[KRP1] S. G. Krantz and H. R. Parks, *A Primer of Real Analytic Functions*, 2nd ed., Birkhäuser, Boston, 2002.

[LAN]   R. E. Langer, *Fourier Series: The Genesis and Evolution of a Theory*, Herbert Ellsworth Slaught Memorial Paper I, *Am. Math. Monthly* 54(1947).

[LUZ]   N. Luzin, The evolution of "Function", Part I, Abe Shenitzer, ed., *Am. Math. Monthly* 105(1998), 59–67.

[MEY1] Y. Meyer, *Wavelets and Operators*, Cambridge University Press, Cambridge, 1992.

[MEY2] Y. Meyer, *Wavelets, Vibrations and Scalings*, American Mathematical Society, Providence, RI, 1998.

[RUV]   D. K. Ruch and P. J. Van Fleet, *Wavelet Theory: An Elementary Approach with Applications*, John Wiley and Sons, New York, 2009.y

[RUD1] W. Rudin, *Real and Complex Analysis*, McGraw-Hill, New York, 1966.

[RUD2] W. Rudin, *Functional Analysis*, 2nd ed., McGraw-Hill, New York, 1991.

[SAB]   K. Sabbagh, *The Riemann Hypothesis: The Greatest Unsolved Problem in Mathematics*, Farrar, Straus and Giroux, New York, 2003.

[SCH]   L. Schwarz, *Théorie des distributions*, Hermann, Paris 1966.

[STA]   P. Stark, *Introduction to Numerical Methods*, Macmillan, New York, 1970.

[STE] J. Stewart, *Calculus. Concepts and Contexts*, Brooks/Cole Publishing, Pacific Grove, CA, 2001.

[STR] R. Strichartz, How to make wavelets, *Am. Math. Monthly* 100(1993), 539–556.

[TIT] E. C. Titchmarsh, *Introduction to the Theory of Fourier Integrals*, The Clarendon Press, Oxford, 1948.

[TOD] J. Todd, *Basic Numerical Mathematics*, Academic Press, New York, 1978.

[WAL] J. Walker, Fourier analysis and wavelet analysis, *Notices of the AMS* 44(1997), 658–670.

[WAT] G. N. Watson, *A Treatise on the Theory of Bessel Functions*, 2nd ed., Cambridge University Press, Cambridge, 1958.

Index